W9-DDN-511

Quaternary Stereocenters

Edited by Jens Christoffers, Angelika Baro

Further Titles of Interest:

A. Berkessel, H. Gröger

Asymmetric Organocatalysis

From Biomimetic Concepts to Applications in Asymmetric Synthesis

2005, ISBN 3-527-30517-3

R. Mahrwald (Ed.)

Modern Aldol Reactions

2 Volumes

2004, ISBN 3-527-30714-1

A. de Meijere, F. Diederich (Eds.)

Metal-Catalyzed Cross-Coupling Reactions

2 Volumes, 2nd Edition

2004, ISBN 3-527-30518-1

K.C. Nicolaou, S.A. Snyder

Classics in Total Synthesis II

More Targets, Strategies, Methods

2003, ISBN 3-527-30685-4

Quaternary Stereocenters

Challenges and Solutions for Organic Synthesis

Edited by Jens Christoffers, Angelika Baro

WILEY-VCH

WILEY-VCH Verlag GmbH & Co. KGaA

The Editors

Prof. Dr. Jens Christoffers
Institut für Organische Chemie
Universität Stuttgart
Pfaffenwaldring 55
70569 Stuttgart
Germany

Dr. Angelika Baro
Institut für Organische Chemie
Universität Stuttgart
Pfaffenwaldring 55
70569 Stuttgart
Germany

■ All books published by Wiley-VCH are carefully produced. Nevertheless, authors, editors and publisher do not warrant the information contained in these books, including this book, to be free of errors. Readers are advised to keep in mind that statements, data, illustrations, procedural details or other items may inadvertently be inaccurate.

Library of Congress Card No.:
applied for.

British Library Cataloguing-in-Publication Data
A catalogue record for this book is available from the British Library.

**Bibliographic information published by
Die Deutsche Bibliothek**
Die Deutsche Bibliothek lists this publication in the Deutsche Nationalbibliografie; detailed bibliographic data is available in the Internet at
<http://dnb.ddb.de>.

© 2005 WILEY-VCH Verlag GmbH & Co. KGaA, Weinheim

All rights reserved (including those of translation into other languages). No part of this book may be reproduced in any form – by photoprinting, microfilm, or any other means – nor transmitted or translated into a machine language without written permission from the publishers. Registered names, trademarks, etc. used in this book, even when not specifically marked as such, are not to be considered unprotected by law.

Typesetting   SMI Tech. Pvt. Ltd., Mohali, India
Printing   betz-druck GmbH, Darmstadt
Bookbinding   J. Schäffer GmbH, Grünstadt
Cover Design   Matthes + Traut, Darmstadt

Printed in the Federal Republic of Germany
Printed on acid-free paper

ISBN-13:   978-3-527-31107-1
ISBN-10:   3-527-31107-6

Contents

Quaternary Stereocenters: Challenges and Solutions for Organic Synthesis. Edited by Jens Christoffers, Angelika Baro
Copyright © 2005 WILEY-VCH Verlag GmbH & Co. KGaA, Weinheim
ISBN: 3-527-31107-6

Foreword

This book, for the first time, addresses a key issue at the heart of complex molecule synthesis namely the stereoselective formation of quaternary centers. This often overlooked topic is very timely and in desperate need of an effective review since the area is commonly very fragmented and tends to be covered as a part of wider synthetic issues. Nevertheless good synthesis planning must always address the problems associated with the installation of quaternary centers. Failure to do so early on in the process can lead to later disaster. Consequently I welcome this text as a useful addition to our bookshelf. It will undoubtedly be widely read and consulted by synthesis chemists looking either for specific solutions to a problem or more generally to realise the gaps in our current knowledge and ability to effectively synthesise these important architectural elements. The editors have brought together a refreshing team of contributors representing a wide user community. The chapters cover quaternary bond forming processes in terms of traditional reaction classes such as aldol, conjugate addition, cycloaddition, rearrangements etc. but have not excluded the importance of enzymes, phase transfer processes or radical reactions.

While there is a tendency these days for books to focus on trends and perceived hot areas it is pleasing to have new texts that bring together "bread and butter" topics such as quaternary stereogenic center formation to the way we assemble molecules and is a vital component of our synthesis toolbox today.

Cambridge, June 2005   *Steven V. Ley*

Quaternary Stereocenters: Challenges and Solutions for Organic Synthesis. Edited by Jens Christoffers, Angelika Baro
Copyright © 2005 WILEY-VCH Verlag GmbH & Co. KGaA, Weinheim
ISBN: 3-527-31107-6

Preface

This work on stereoselective synthesis is not about a particular type of reaction or class of catalysts, but shines light on a specific structural issue from different points of view. Quaternary stereocenters, being fully substituted asymmetric carbon centers, are a challenging task for synthetic organic chemists, as is already indicated in the subtitle of this book. In contrast to tertiary stereocenters, where a wide variety of chiral auxiliaries, reagents and catalysts nowadays form the basis for modern asymmetric synthesis and are a guarantee for high selectivity, the construction of a quaternary stereocenter remains the touchstone of every enantioselective procedure.

This book collects review articles from authors with different scientific backgrounds and thus a different focus on quaternary stereocenters in synthetic targets. Most of the chapters concentrate on a specific type of reaction or methodology, with the relevant authors chosen for their expertise in this area. The chapters do not aim to cover the topic comprehensively, but rather the authors have compiled highlighting examples that reflect their personal choice. Some of the chapters are based on a liberal definition of a quaternary stereocenter, since they also include tertiary alcohols, ethers, or amines. All of the chapters include intentionally diastereoselective reactions, in some cases even kinetic resolutions, alongside enantioselective procedures.

We hope that this monograph will be considered a reader for organic chemists with different backgrounds. The focus on quaternary stereocenters will guide the reader through different areas of stereoselective synthesis, starting with an overview on natural compounds and important industrial products, and ending with modern concepts of biotechnology.

Stuttgart, February 2005

Jens Christoffers
Angelika Baro

Quaternary Stereocenters: Challenges and Solutions for Organic Synthesis. Edited by Jens Christoffers, Angelika Baro
Copyright © 2005 WILEY-VCH Verlag GmbH & Co. KGaA, Weinheim
ISBN: 3-527-31107-6

List of Contributors

Hirokazu Arimoto
Graduate School of Life Sciences
Tohoku University
Tsutsumidori-Amamiyamachi,
Aoba-Ku
Senda: 981-8555
Japan

Angelika Baro
Institut für Organische Chemie
Universität Stuttgart
Pfaffenwaldring 55
70569 Stuttgart
Germany

Louis Barriault
Department of Chemistry
University of Ottawa
D'Iorio Hall
10, Marie Curie
Ottawa (Ontario) K1N 6N5
Canada

Uwe T. Bornscheuer
Institut für Technische Chemie
und Biotechnologie
Ernst-Moritz-Arndt-Universität
Greifswald
Soldmannstraße 16
17487 Greifswald
Germany

Manfred Braun
Institut für Organische und
Makromolekulare Chemie I
Heinrich-Heine-Universität Düsseldorf
Universitätsstraße 1
40225 Düsseldorf
Germany

Jens Christoffers
Institut für Organische Chemie
Universität Stuttgart
Pfaffenwaldring 55
70569 Stuttgart
Germany

Giovanni Desimoni
Dipartimento di Chimica Organica
Universita di Pavia
Viale Taramelli 10
27100 Pavia
Italy

Johannes G. deVries
DSM Pharma Chemicals
Advanced Synthesis, Catalysis &
Development
P.O. Box 18
6160 MD Geleen
The Netherlands

Quaternary Stereocenters: Challenges and Solutions for Organic Synthesis. Edited by Jens Christoffers, Angelika Baro
Copyright © 2005 WILEY-VCH Verlag GmbH & Co. KGaA, Weinheim
ISBN: 3-527-31107-6

Giuseppe Faita
Dipartimento di Chimica Organica
Universita di Pavia
Viale Taramelli 10
27100 Pavia
Italy

Erik Henke
Institut für Chemie und
Biochemie
Ernst-Moritz-Arndt-Universität
Greifswald
Soldmannstraße 16
17487 Greifswald
Germany

Martin Hiersemann
Institut für Organische Chemie
Technische Universität Dresden
Bergstraße 66
01069 Dresden
Germany

Rainer Mahrwald
Institut für Chemie der
Humboldt-Universität zu Berlin
Brook-Taylor-Straße 2
12489 Berlin
Germany

Keiji Maruoka
Department of Chemistry
Graduate School of Science
Kyoto University, Sakyo
Kyoto 606-8502
Japan

Takashi Ooi
Department of Chemistry
Graduate School of Science
Kyoto University, Sakyo
Kyoto 606-8502
Japan

Kalyani Patil
Department of Chemistry
North Dakota State University
Fargo, North Dakota 58105
USA

Jürgen Pleiss
Institut für Technische Biochemie
Universität Stuttgart
Allmandring 31
D-70569 Stuttgart
Germany

Annett Pollex
Institut für Organische Chemie
Technische Universität Dresden
Bergstraße 66
01069 Dresden
Germany

Diego G. Ramón
Instituto de Síntesis Orgánica and
Departamento de Química Orgánica
Facultad de Ciencias
Universidad de Alicante, Apdo.
Correos 99
E-03080-Alicante
Spain

Effiette L. O. Sauer
Department of Chemistry
University of Ottawa
D'Iorio Hall to Marie Curie
Ottawa (Ontario)
K1N 6N5
Canada

Bernd Schetter
Institut für Chemie der
Humboldt-Universität zu Berlin
Brook-Taylor-Straße 2
12489 Berlin
Germany

Mukund B. Sibi
Department of Chemistry
North Dakota State University
Fargo, North Dakota 58105
USA

Daisuke Uemura
Department of Chemistry
Graduate School of Science
Nagoya University
Furo-cho, Chikusa
Nagoya 464-8602
Japan

Miguel Yus
Instituto de Síntesis Orgánica and
Departamento de Química Orgánica
Facultad de Ciencias
Universidad de Alicante, Apdo.
Correos 99
E-03080-Alicante
Spain

Symbols and Abbreviations

acac	2,4-pentanedionato
acam	actetamidato
Acc	acceptor (group)
ACCN	azobis(cyclohexanecarbonitrile)
AChE	acetylcholinesterase
AD	asymmetric dihydroxylation
AH	asymmetric Heck (reaction)
AIBN	azobis(isobutyronitrile)
AIDS	acquired immune deficiency syndrome
ALB	aluminum lithium BINOLate
AZT	azidothymidine
Ala	alanine
All	allyl
bAChE	banded krait acetylcholinesterase
BHT	2,6-di-*tert*-butyl-4-methylphenyl
BINAP	2,2'-bis(diphenylphosphano)-1,1'-binaphthyl
BINOL	2,2'-dihydroxy-1,1'-binaphthyl
BIPI	Boehringer Ingelheim phosphanoimidazolines
Bn	benzyl
Boc	*tert*-butyloxycarbonyl
BOM	benzyloxymethyl
box	2,2'-methylenebis(2-oxazoline)
Bs	benzenesulfonyl (Chapter 7)
Bs	4-bromobenzenesulfonyl (Chapter 1)
BSA	*N,O*-bis(trimethylsilyl)acetamide
BsubpNBE	*Bacillus subtilis p*-nitrobenzylesterase
Bz	benzoyl
c	conversion
CAL	*Candida antarctica* lipase
CAN	ammoniumcerium(IV)nitrate
Cbz	benzyloxycarbonyl
CDI	carbonyldiimidazole

Quaternary Stereocenters: Challenges and Solutions for Organic Synthesis. Edited by Jens Christoffers, Angelika Baro
Copyright © 2005 WILEY-VCH Verlag GmbH & Co. KGaA, Weinheim
ISBN: 3-527-31107-6

cod	1,5-cyclooctadiene
coe	cyclooctene
Cp	cyclopentadienyl
Cp*	pentamethylcyclopentadienyl
CRL	*Candida rugosa* lipase
Cy	cyclohexyl
DA	Diels–Alder (reaction)
dba	dibenzylideneacetone
DBB	di-*tert*-butylbiphenyl
DBFOX	dibenzofuranebis(2-oxazoline)
DBU	1,8-diazabicyclo[5.4.0]undec-7-ene
DCM	dichloromethane
DDQ	dichlorodicyanobenzoquinone
(DHQ)$_2$PHAL	bis(dihydroquinine)phthalazine
DIOP	2,3-*O*-isopropylidene-2,3-dihydroxy-1,4-bis (diphenylphosphano)butane
DIPEA	ethyldiisopropylamine
DIPT	diisopropyl tartrate
DKR	dynamic kinetic resolution
DMA	*N*,*N*-dimethylacetamide
DMAc	*N*,*N*-dimethylacetamide
DMAP	4-dimethylaminopyridine
DMB	2,3-dimethyl-1,3-butadiene
DME	1,2-dimethoxyethane
DMF	dimethylformamide
DMPU	1,3-dimethyl-3,4,5,6-tetrahydro-2(1*H*)-pyrimidone
DMS	dimethylsulfane
DMSO	dimethylsulfoxide
Don	donor (group)
dppb	1,4-bis(diphenylphosphano)butane
dppe	1,2-bis(diphenylphosphano)ethane
DTBMP	2,6-di-*tert*-butyl-4-methylpyridine
duphos	1,2-bis(2,5-dialkylphospholano)benzene
E	*tert*-butyloxycarbonyl (Chapter 3)
E	enantioselectivity (Chapter 12)
E.C.	international enzyme classification
EDCI	1-(3-dimethylaminopropyl)-3-ethylcarbodiimide
eeAChE	electric eel acetylcholinesterase
ee_P	enantiomeric excess of the product
ee_S	enantiomeric excess of the substrate
EPHP	1-ethylpiperidine hypophosphite
EWG	electron-withdrawing group
FMO	frontier molecular orbital
GCL	*Geotrichum candidum* lipase
GGA	double glycine alanine

GGGX	triple glycine and one hydrophobic amino acid
GX	one glycine and one hydrophobic amino acid
hAChE	human acetylcholinesterase
HDA	hetero-Diels–Alder (reaction)
hfc	3-heptafluorobutyryl-*d*-camphorato
His	histidine
HIV	human immunodeficiency virus
HMPA	hexamethylphosphoric acid triamide
HOCSAC	1,2-*trans*-bis(isoborneolsulfonamido)cyclohexane
HOMO	highest occupied molecular orbital
HSAB	hard and soft acid and base
ICD	β-isocupreidine
IMDA	intramolecular Diels–Alder (reaction)
ISC	in solid state
k_{cat}	catalytic constant (reaction rate constant)
KDA	potassium diisopropylamide
KHMDS	potassium hexamethylsiloxane
K_M	Michaelis–Menten constant
LA	Lewis acid
LDA	lithium diisopropylamide
LiDBB	lithium di-*tert*-butylbiphenylide
LiHMDS	lithium hexamethyldisilazide
LiTMP	lithium 2,2,6,6-tetramethylpiperidide
lk	like, see *ul*
LSB	lanthanum sodium BINOLate
LUMO	lowest occupied molecular orbital
MAD	methyl aluminum bis(2,6-di-*tert*-butyl-4-methylphenoxide)
MAPh	methyl aluminum bis(2,6-diphenylphenoxide)
MD	molecular dynamics
MEM	2-methoxyethoxymethyl
MMPP	magnesium monoperoxyphthalate
MNDO	modiguid neglect of diatomic overlap
MO	molecular orbital
MOM	methoxymethyl
Ms	methanesulfonyl
MS	molecular sieve
MVK	methyl vinyl ketone
NaHMDS	sodium hexamethyldisilazide
Naph	naphthyl (Chapter 6)
naphth	naphthyl (Chapter 7)
NBS	*N*-bromosuccinimide
NeuA	Neuraminidase A
NMO	*N*-methyl-morpholine-*N*-oxide
NMP	*N*-methyl-2-pyrrolidone

NOBIN	2-amino-2'-hydroxy-1,1'-binaphthyl
NOE	nuclear Overhauser effect
Ox	oxidant
PCL	*Pseudomonas cepacia* lipase
PCR	polymerase chain reaction
PHAL	phthalazine
Phe	phenylalanine
Phebox	2,6-bis(2-oxazolinyl)phenyl
PHOX	diarylphosphanophenyloxazoline
Pht	phthalimido
PhTRAP	2,2''-bis[(diphenylphosphanyl)ethyl]-1,1''-biferrocene
PG	protecting group
Piv	pivaloyl
PKR	Pauson–Khand reaction
PLE	pig liver esterase
PMB	4-methoxybenzyl
PMP	1,2,2,6,6-pentamethylpiperidine
PNB	4-nitrobenzyl
Pro	proline
PS	polystyrene
PS-DVB	styrene-divinylbenzene-copolymer
PTC	phase-transfer catalysis
Pht	phthaldioyl
Py	pyridyl
py	pyridine
pybox	2,6-bis(2-oxazolinyl)pyridine
RAMP	(*R*)-aminomethoxymethylpyrrolidine
Red-Al	sodium dihydridobis(2-methoxyethoxy)aluminate
$Rh_2(acam)_4$	dirhodium (II) tetrakisacetamide
rhaP	rhamnose inducible promoter
ROL	*Rhizopus oryzae* lipase
salen	*N*,*N*'-bis(salicylidene)-1,2-ethylenediamine
SAMP	(*S*)-aminomethoxymethylpyrrolidine
SEM	2-(trimethylsilyl)ethoxymethyl
SMB	simulated moving bed
TADA	transannular Diels–Alder (reaction)
TADDOL	tetraaryl-1,3-dioxolane-4,5-dimethanol (Chapter 6)
TADDOL	(4*R*,5*R*) or (4*S*,5*S*)-2,2-dimethyl- α,α,α',α'-tetraphenylne-1,3-dioxolane-4,5-dimethanol (Chapter 10)
TASF	tris(dimethylamino)sulfur(trimethylsilyl)difluoride
TBAF	tetrabutylammonium fluoride
TBDMS	*tert*-butyldimethylsilyl
TBDPS	*tert*-butyldiphenylsilyl
TBHP	*tert*-butylhydroperoxide
TBS	*tert*-butyldimethylsilyl

TCC	tandem cascade cycloaddition
TDS	dimethylthexylsilyl
TE	transesterification
TEA	triethylamine
TES	triethylsilyl
Tf	trifluoromethanesulfonyl
tfa	trifluoroacetate
tfc	3-trifluoroacetyl-*d*-camphorato
THF	tetrahydrofuran
TIPS	triisopropylsilyl
TIPSO	(triisopropylsilyl)oxy
TLC	thin-layer chromatography
TMS	trimethylsilyl
tol	tolyl (Chapter 7)
Tol	tolyl (Chapter 2, Chapter 8)
Tol-BINAP	2,2′-bis(di-2-tolylphosphano)-1,1′-binaphthyl
TPAP	tetrapropylammonium perruthenate
TPS	triphenylsilyl (Chapter 1)
TPS	*tert*-BuPh$_2$Si (Chapter 5)
Tr	triphenylmethyl
Ts	4-toluenesulfonyl
ul	unlike, for a definition see: D. Seebach, V. Prelog, *Angew. Chem.* **1982**, *94*, 696; *Angew. Chem. Int. Ed. Engl.* **1982**, *21*, 654.
ν	reaction rate
WT	wild type

1
Important Natural Products

Hirokazu Arimoto and Daisuke Uemura

1.1
Introduction

The development of synthetic methodologies in the last quarter of the twentieth century has been truly impressive, but the stereoselective construction of quaternary centers remains a significant challenge in the total synthesis of natural products. It is quite difficult to invert undesired configurations of quaternary centers to the desired ones, so the stereoselectivities of reactions on quaternary carbons often govern the total efficiency of the syntheses. In this chapter, we highlight recent natural product syntheses, with emphasis on the stereoselective preparation of the quaternary carbons [1, 2].

Unfortunately, the term "quaternary center" is a cause of confusion in terminology, because it is also used to mean "quaternary-substituted carbon", which includes tri-carbon-substituted carbon such as occurs in tertiary alcohols. The term "all-carbon quaternary centers" is also found in the literature. Quaternary-substituted carbons and quaternary carbons in the full sense must be distinguished carefully. Thus, in this chapter we use the term "quaternary carbon" to designate those carbon centers that are substituted with four carbon substituents.

Quaternary-substituted carbons can be prepared routinely by the face selective addition of carbon nucleophiles to carbon–heteroatom double bonds such as asymmetrical ketones or imines [3]. Stereochemical induction may be achieved based either on neighboring functional groups in the substrate or on chiral catalysts.

On the contrary, the stereoselective synthesis of quaternary carbon (all-carbon quaternary centers) is still challenging and only limited options are available. The most popular and powerful concept at present is the chirality transfer of configurations at neighboring heteroatom-substituted asymmetric centers to the quaternary carbons. As described above, enantioselective preparation of heteroatom-substituted centers has become much easier. Asymmetric oxidations of tri- or tetrasubstituted alkenes to diols, epoxides, or amino alcohols also give quaternary-substituted centers, which are now regarded as conventional

Quaternary Stereocenters: Challenges and Solutions for Organic Synthesis. Edited by Jens Christoffers, Angelika Baro
Copyright © 2005 WILEY-VCH Verlag GmbH & Co. KGaA, Weinheim
ISBN: 3-527-31107-6

approaches. A variety of asymmetric reagents and catalysts is now available for these purposes [4].

Total syntheses of natural products are usually achieved through multi-step transformations from simple starting materials. Needless to say, the efficiency of a total synthesis depends not only on the yield or selectivity of each synthetic operation (strategy) but also on the overall synthetic plan. In the case of target compounds with a small number of asymmetric centers, each quaternary carbon center could be built directly by asymmetric catalysis (such as an asymmetric Heck reaction or an asymmetric Diels–Alder reaction, *vide infra*). Although wide varieties of chiral starting materials are available either from natural sources (e.g. sugars, amino acids) or from asymmetric reactions [5], chiral building blocks with a quaternary center are rare.

When there are two or more asymmetric centers in the target molecule, the timing of the construction of quaternary centers should be considered in the whole scheme of the synthetic route. Thus, catalytic asymmetric methods are not always the choice in the synthesis of quaternary centers in multi-stereogenic compounds. The introduction of a small number of non-quaternary stereocenters in the early stage, followed by a series of diastereoselective transformations to control quaternary center(s), is a popular approach in the synthesis of complex natural products owing to the overall total efficiency. The central issue in the context of quaternary centers seems to be how to induce quaternary carbons efficiently and stereospecifically, based on the preexisting secondary or tertiary stereocenters. Another challenge exists in the construction of quaternary carbons. This often suffers from hindered chemical environments around reaction centers, and the use of intramolecular reactions is a popular approach to enhance reactivity.

1.2
Alkylation of Tertiary Carbon Centers

Alkylations, or acylations, of enolate equivalents are straightforward approaches to the construction of quaternary carbons. Stereoinductions at newly formed quaternary carbons are often achieved diastereoselectively based on preexisting chiral centers.

Omura et al. have reported an elegant asymmetric total synthesis of madindolines [6]. An impressive remote diastereomeric induction was observed in the acylation of the lithium enolate of ester **1**. When hexamethylphosphoramide (HMPA) was employed as a co-solvent in the acylation, or potassium diisopropylamide (KDA) was used as base, the selectivity of the acylation dropped considerably. Thus, the authors suggested that oxygen in both enolate and tetrahydrofuran coordinated to a lithium cation, and this chelation directed the electrophile attack on the less hindered face of enolate. The acylation product was converted in one step to the natural product.

Incorporation of a chiral auxiliary into carbonyl compounds is also a popular and reliable method, although it requires additional steps to introduce and

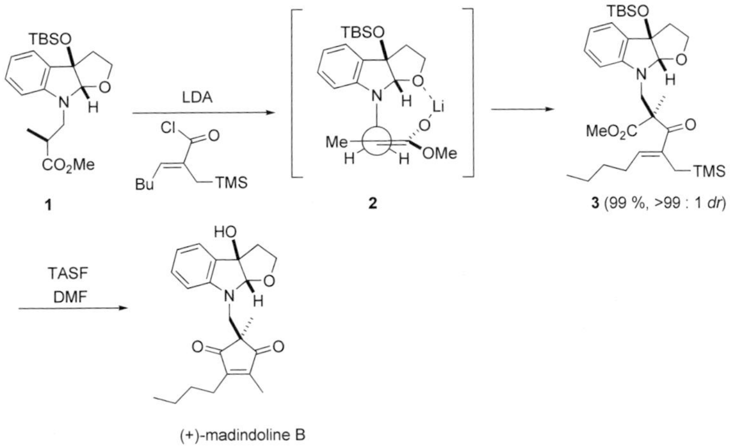

Scheme 1.1 Omura's remote diastereomeric alkylation in the synthesis of (+)-madindoline.

remove the auxiliary. The pioneering contributions of Meyers, for instance, have been employed extensively in natural product synthesis, where amino acid-derived amino alcohols were the chiral sources [7].

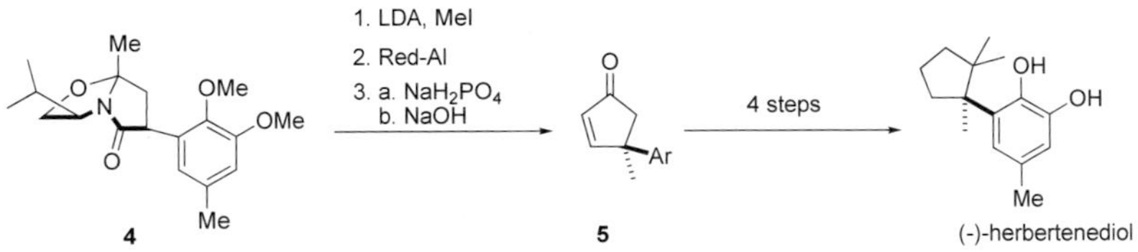

Scheme 1.2 Meyers' asymmetric synthesis of (–)-herbertenediol.

In the synthesis of (–)-herbertenediol [8], a chiral bicyclic lactam was methylated from the *endo* face to construct a quaternary stereogenic center with complete stereocontrol. Reduction and hydrolytic cleavage afforded the cyclopentenone, which was subsequently converted to herbertenediol, containing vicinal quaternary centers.

Catalytic methods of generating a chiral enolate have also been explored in order to construct quaternary centers. Sodeoka et al. have reported asymmetric Michael addition .catalyzed by a palladium aqua complex (see Chapter 4, Scheme 4.19) [9]. Cyclic and acyclic achiral β-ketoesters added to methyl vinyl ketones via chiral palladium enolate with high yields and *ees* (in most cases *ee* > 90%). Metal-catalyzed asymmetric Michael-type reactions have been investigated, but the enantioselectivity and reactivity were, in general, not sufficient. Sodeoka's landmark results will be applied in natural products synthesis shortly.

Although less frequently employed, asymmetric alkylation of non-chiral enolate equivalents with chiral electrophiles has also been investigated as a complementary approach. MacMillan employed his chiral organocatalyst for asymmetric Michael-type addition in the total synthesis of (–)-flustramine B [10, 11].

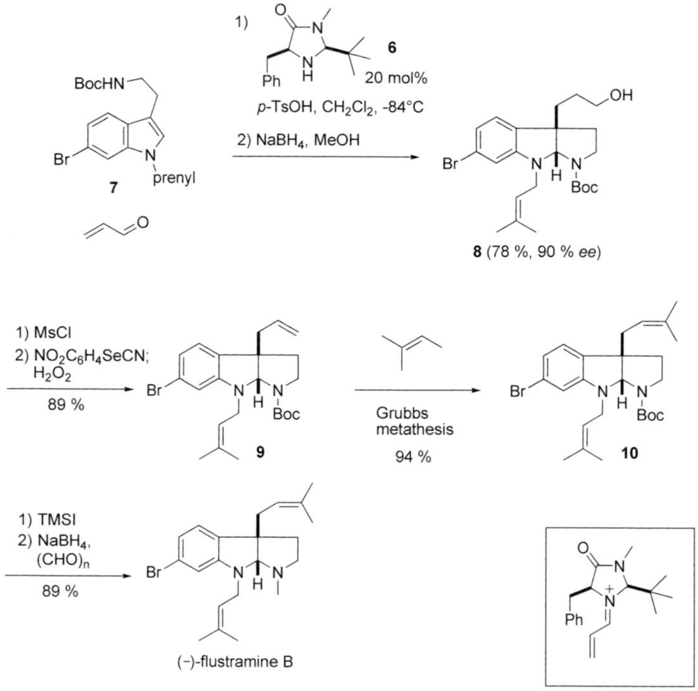

Scheme 1.3 MacMillan's organocatalyzed asymmetric Michael-type addition in the total synthesis of (–)-flustramine B.

The chiral organocatalyst **6** generated *in situ* an iminium salt (in box) with acrolein, and the alkylation occurred to give a quaternary center enantioselectively. The iminium ion was expected to be formed with (*E*)-isomer selectively to avoid the repulsion of the bulky *tert*-butyl group. The benzyl and *tert*-butyl groups shield the upper face (*Si*-face) of the substrate, leaving the down face (*Re*-face) exposed for enantioselective bond formation. However, the cause of enantiofacial selectivity on indole **7** was not clear. The choice of solvent has been shown to be the major controlling factor. The indolium ion thus formed was attacked intramolecularly by a Boc-protected amine to afford pyrroloindoline **8**.

In addition to carbanion-mediated reactions, radical alkylations can also be employed. The high reactivity of radical species is an advantage for the generation

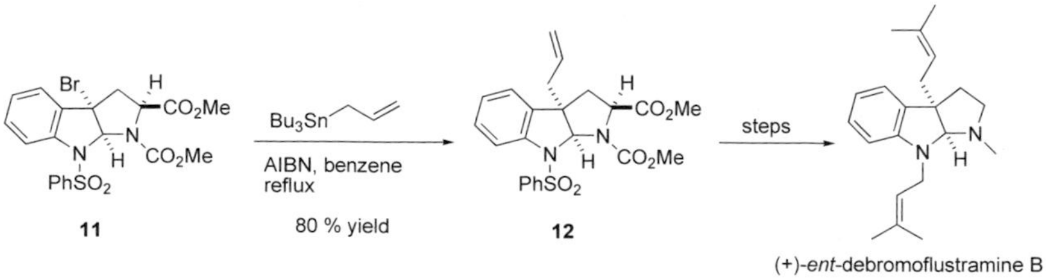

Scheme 1.4 Radical allylation by Crich et al.

of congested quaternary carbons. Thus Crich's synthesis of a marine alkaloid, (+)-*ent*-debromoflustramine B, involved radical allylation of a tertiary bromide with allyltributyltin to **12** [12].

Clive achieved total synthesis of (+)-puraquinoic acid using a route based on radical cyclization of the Stork bromoacetal **13** [13]. The chirality of the quaternary carbon center was controlled by a temporary adjacent tertiary asymmetric center.

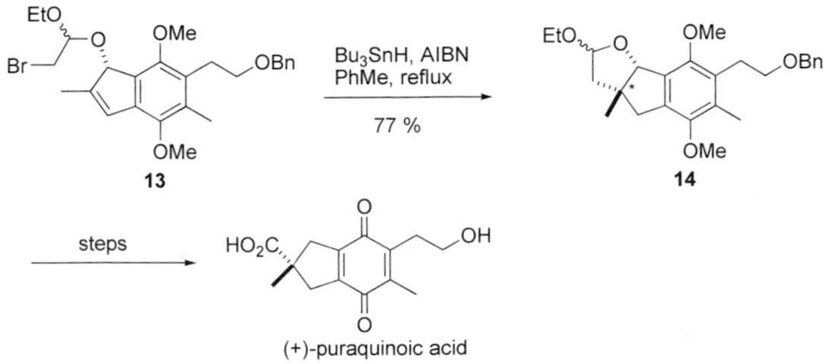

Scheme 1.5 Generation of an asymmetric quaternary center based on a temporary adjacent chiral center.

1.3
Cycloaddition to Alkenes

1.3.1
Diels–Alder Reaction

There have been significant numbers of applications of the Diels–Alder reaction for the synthesis of polycyclic natural products, inspired by their biogenesis. The Diels–Alder reaction provides one of the most powerful strategies for forming a quaternary carbon within a cyclohexane system.

Nicolaou et al. reported the total synthesis of colombiasin A, a marine diterpenoid, using an intramolecular Diels–Alder reaction (IMDA) [14]. Six stereogenic centers, two of which are adjacent quaternary carbons, exist in the compact tetracyclic framework of colombiasin. The construction of contiguous stereogenic quaternary carbons posed a synthetic challenge arising from the steric congestion imposed by the four attached substituents. Stereoselective generation of both quaternary centers was accomplished via an *endo*-specific Diels–Alder cycloaddition with quinone as dienophile (Scheme 1.6).

It has been proposed since the early 1960s that the Diels–Alder reaction is involved in the biosynthetic pathways of natural products. Oikawa proved for the first time that the decaline skeleton of solanapyrone was biosynthesized by a "Diels–Alderase" [15].

Following Oikawa's report, two additional Diels–Alderases, lovastatin nonaketide synthase [16] and macrophomate synthase [17] were purified.

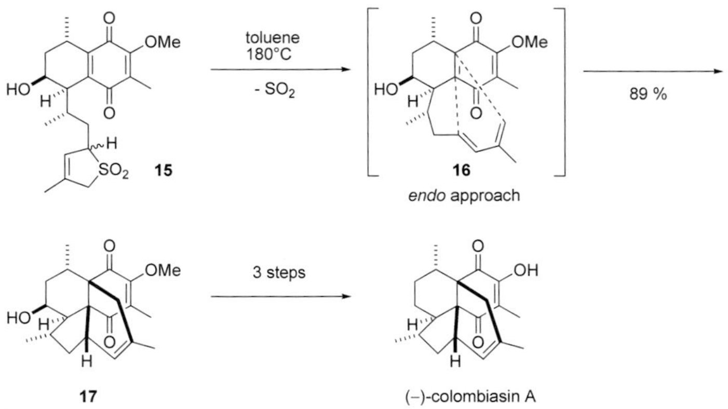

Scheme 1.6 Construction of contiguous quaternary centers in
(–)-colombiasin A by an intramolecular Diels–Alder reaction.

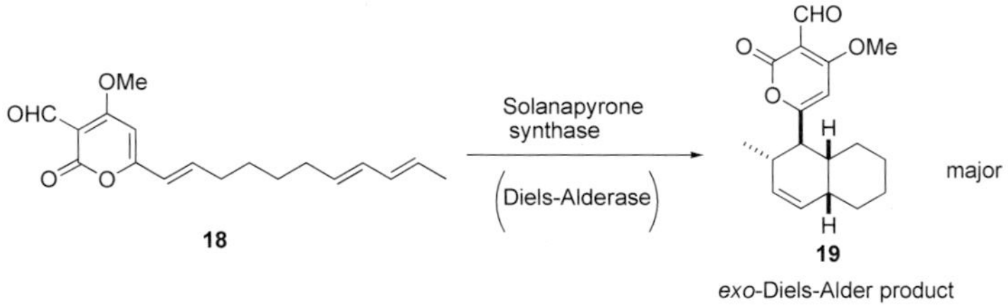

Scheme 1.7 The first identified Diels–Alderase by Oikawa et al.

X-ray crystallographic analysis of a Diels–Alderase complex with its substrate
analogue has been reported, which allowed the reaction mechanism to be
analyzed at the molecular level [18]. Owing to the limited number of
Diels–Alderases identified so far, there has been no confirmed example of an
enzyme-catalyzed Diels–Alder reaction forming quaternary centers. However, the
advance of investigations into biosynthesis did inspire synthetic chemists to consider biomimetic approaches to quaternary carbon centers in polycyclic natural
products.

It is also noteworthy that the Diels–Alderases described above catalyze not
only the cycloaddition but also the oxidation of allylic alcohols to enals, the
dienophile. It is not certain at this stage if all Diels–Alderases must have oxidase
activity. The fact that all Diels–Alderases found have oxidase activity seems
suggestive in designing the Diels–Alder precursors for biomimetic synthesis.

Norzoanthamine, a marine polyketide that was isolated by Uemura in 1995
[19], exhibits promising antiosteoporotic activity in ovariectomized mice.
Miyashita achieved the first total synthesis of norzoanthamine in 2004 [20].
Among the three quaternary carbons, the C-12 and C-22 stereocenters were constructed through an IMDA reaction, which was designed by taking into account
the proposed biogenetic pathway [19b].

As shown in Scheme 1.8, this IMDA reaction of **20** proceeded at 240°C and
gave rise to a 72:28 ratio of the *exo* and *endo* adducts. The remaining quaternary

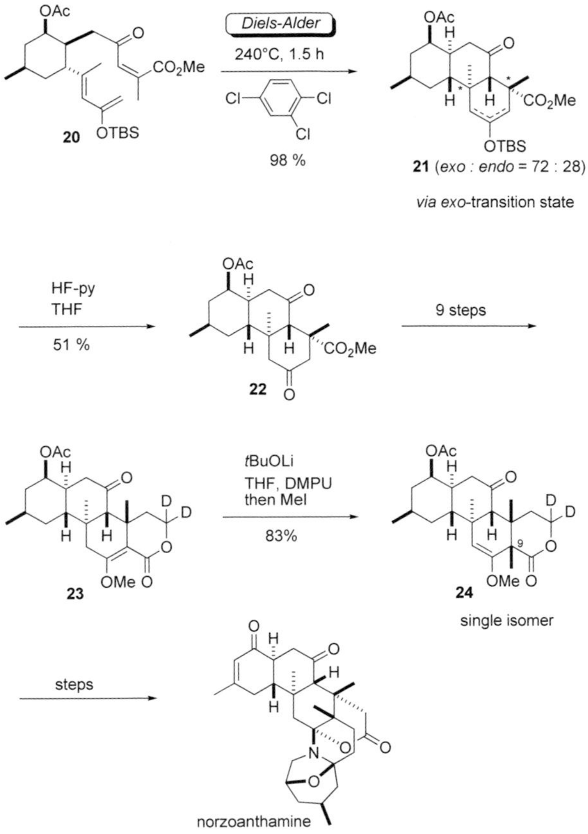

Scheme 1.8 Miyashita's synthesis of norzoanthamine via an IMDA reaction.

center at C-9 was then constructed by a diastereoselective methylation of enolate from the β side. The methylated products were obtained as a single isomer in 83% yield.

The discovery of Diels–Alderases allowed these biomimetic total syntheses to give an insight into whether the natural product is biosynthesized by the enzyme.

In the biomimetic total synthesis of longithrone by Shair, combinations of inter- and intramolecular Diels–Alder reactions were employed to construct two quaternary centers [21].

The first Diels–Alder reaction of Shair's synthesis (from **25** and **26** to **27**) required activation of the dienophile by Lewis acids, and the desired stereoisomer was obtained as the minor isomer. Heating at 80°C without Lewis acid did not give the intermolecular Diels–Alder adducts, but the second intramolecular (transannular) Diels–Alder (TADA) reaction of **28** proceeded at room temperature with higher yield. Low reactivity and lack of substrate-induced stereoselectivity might suggest the involvement of a Diels–Alderase in the former Diels–Alder

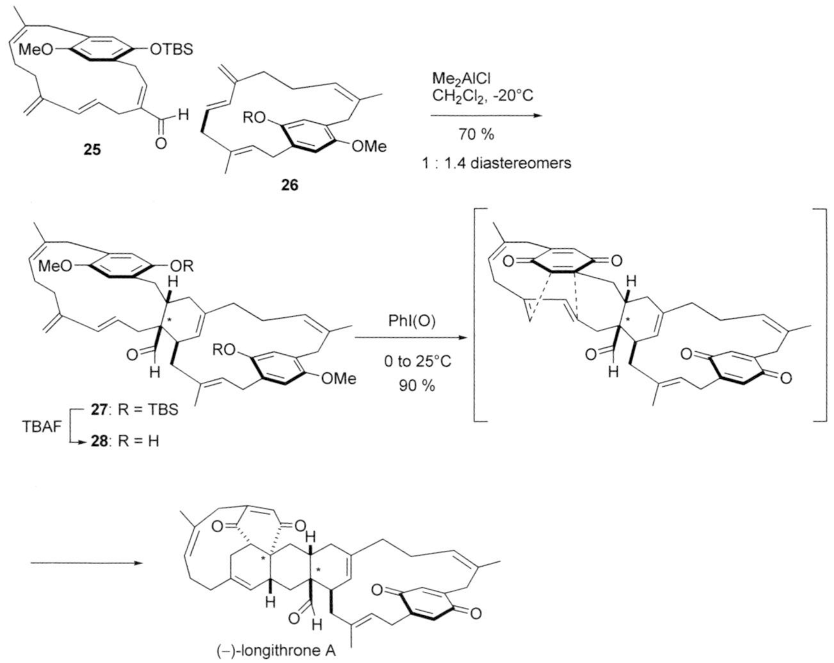

Scheme 1.9 Biomimetic synthesis of (–)-longithrone A by Shair et al.

reaction. Similar observations in the total synthesis of keramaphidine B might also suggest that the biosynthesis is Diels–Alderase catalyzed [22].

As demonstrated in the total synthesis of longithrone described above, a well-designed TADA approach can reduce the entropic requirement $\Delta S^{\ddagger}$ in the transition state; thus it has a potential advantage in reactivity over the IMDA reaction [23]. Another application of a TADA reaction in the stereoselective construction of a quaternary spirocenter was reported in a synthetic study on a marine sesterterpene mangicol [24, 25]. In this case, the design of the synthetic scheme was not based on biogenesis of the natural product.

In Uemura's synthetic study of mangicol, the configuration of the C-3 secondary alcohol in the triene precursor **32** had an unexpectedly significant effect on the stereocontrol of the Diels–Alder reaction. A calculation (B3LYP level, basis set 6-31G*) of the transition states revealed that intramolecular hydrogen bonding of the alcohol with the cyclopentenone carbonyl oxygen stabilizes the transition state to the desired product **33** from the 3*S*-triene precursor **32** [26], whereas with 3*R*-triene the hydrogen bond destabilized the desired transition state. In general, detailed analysis of the relative energies among the possible modes of cycloaddition is the key to the success of the TADA approach.

The use of Diels–Alder reactions for stereoselective construction of quaternary carbons will continue to be popular in the synthesis of polycyclic natural products.

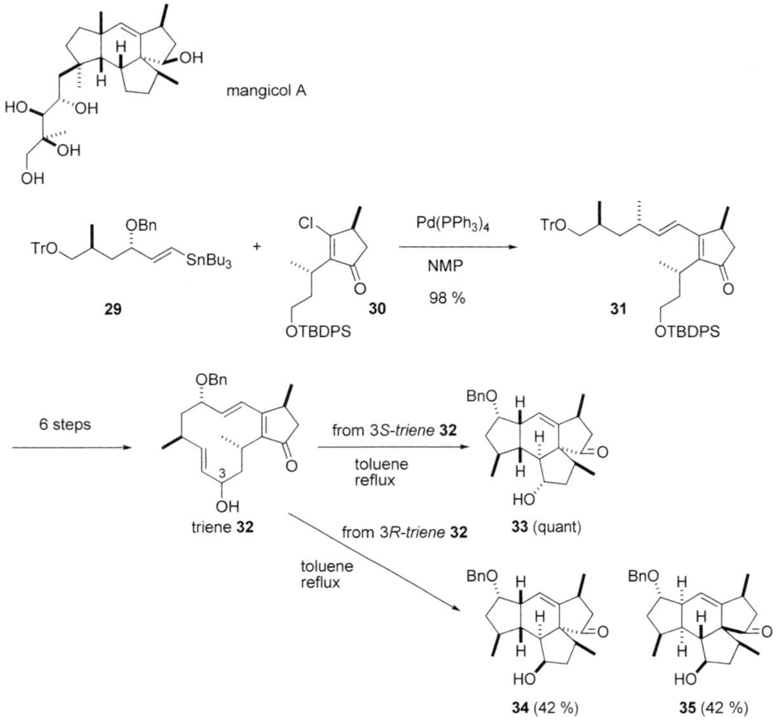

Scheme 1.10 Uemura's TADA approach to mangicol A.

1.3.2
Other Types of Cycloaddition

The Pauson–Khand reaction (PKR) is a powerful tool for assembling polycyclic natural products that involve cyclopentane rings [27]. An alkyne–cobalt complex, derived from $Co_2(CO)_8$ and alkyne, reacts with an alkene to generate cyclopentenone. Additives, such as amine N-oxide, are used to lower the reaction temperature to room temperature.

The PKR is known to be sensitive to steric factors in the transition states. For instance, PKR of the ene-yne **36** afforded the desired product **39** with two adjacent quaternary centers in excellent yield [27b]. However, neither a similar substrate with a longer alkyl tether **37** [27b] nor the tetrasubstituted *endo*-cyclic alkene **38** [27c] gave the cyclized product.

A formal total synthesis of magellanine, a lipodium alkaloid, is shown in Scheme 1.12 [28].

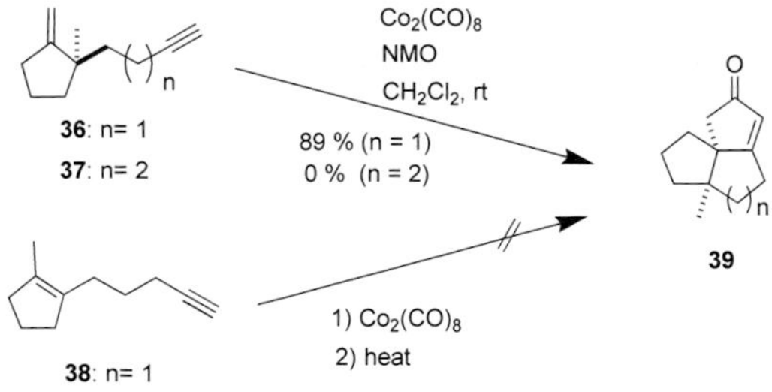

Scheme 1.11 Construction of adjacent quaternary carbon centers by a PKR.

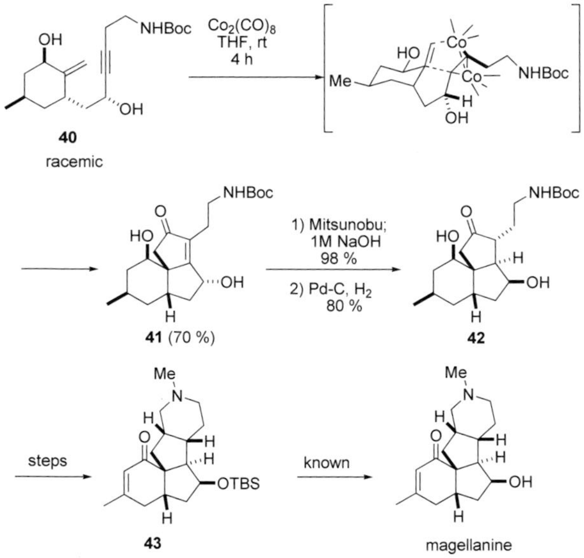

Scheme 1.12 Hoshino's formal total synthesis of magellanine with a PKR as a key step.

1.4
Rearrangement Reactions

The sterically hindered nature of quaternary centers makes the rearrangement reaction an attractive strategy for their synthesis. Danishefsky employed two [3,3]-sigmatropic rearrangements (a Johnson–Claisen rearrangement, **44** to **45**; an Eschenmoser amide acetal Claisen rearrangement, **46** to **47**) for stereocontrol of the quaternary carbon centers in gelsemine [29].

The quaternary carbon center at C-7 of gelsemine was constructed also by a [3,3]-sigmatropic rearrangement in a more straightforward manner (Fukuyama, Scheme 1.14) [30]. The stereochemical information about a quaternary center in a cyclopropane ring in **49** was transferred to the spiro-quaternary carbon center of **51**.

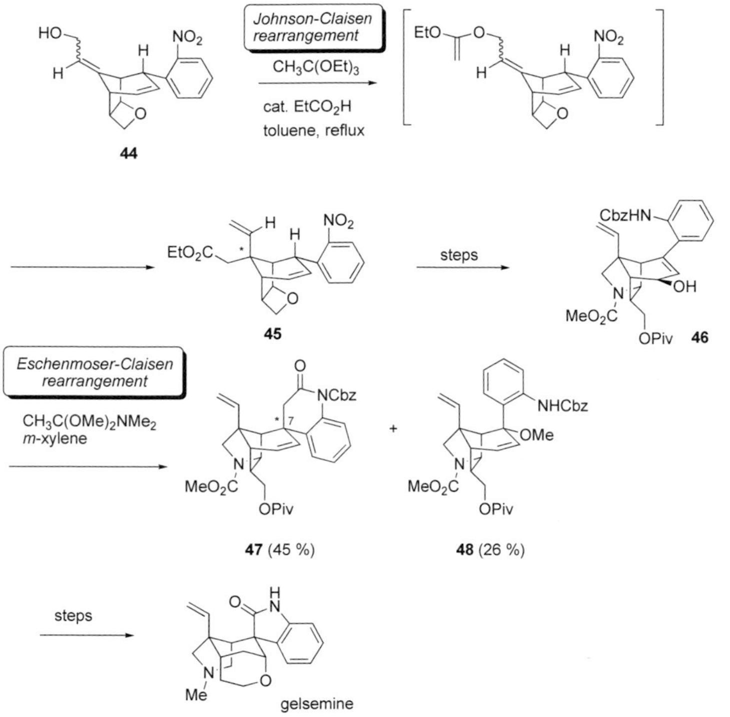

Scheme 1.13 Danishefsky's total synthesis of gelsemine.

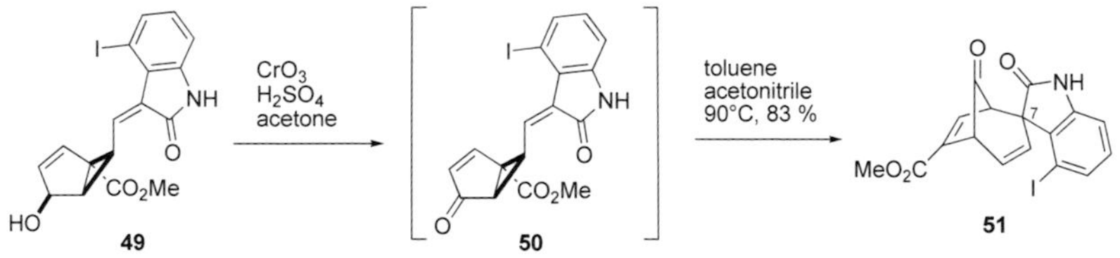

Scheme 1.14 Construction of the C-7 quaternary center of gelsemine in Fukuyama's total synthesis.

Both Fleming [31] and Fukuyama [30] employed intramolecular alkylation in the construction of the C-20 stereocenter in their gelsemine synthesis (Scheme 1.15).

Besides the [3,3]-sigmatropic rearrangements, 1,2-rearrangements are also very important approaches to quaternary carbon centers. It is noteworthy that 1,2-diols or 2,3-epoxy alcohols, which are frequently employed as substrates, can be prepared easily by methods such as those of Sharpless, Katsuki, and Jacobsen [4].

In their asymmetric total synthesis of (+)-asteltoxin Cha et al. used the Suzuki–Tsuchihashi 1,2-rearrangement [32]. Chiral epoxy alcohol **59** was prepared by Sharpless asymmetric epoxidation and was rearranged in the stereospecific manner under Suzuki–Tuchihashi conditions to afford the quaternary carbon center.

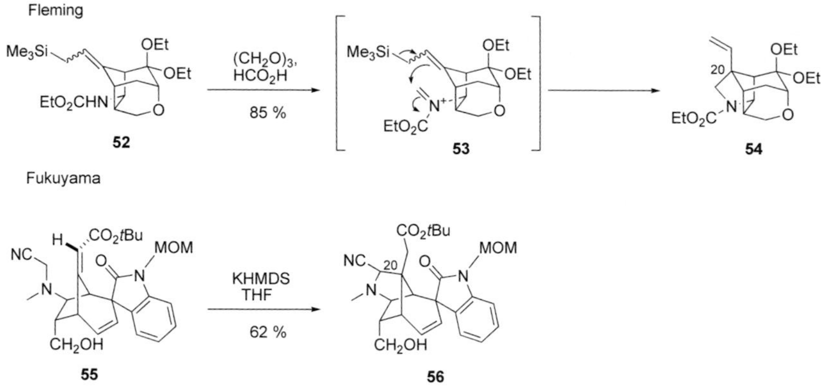

Fleming

52 → **53** → **54**

Fukuyama

55 → **56**

Scheme 1.15 Construction of the C-20 stereocenter of gelsemine by Fleming et al. and Fukuyama et al.

The product aldehyde **60** was further converted to (+)-asteltoxin via the Sharpless asymmetric dihydroxylation of **63**. Cha's total synthesis was inspired by the proposed biogenesis of asteltoxin by Vleggar [34].

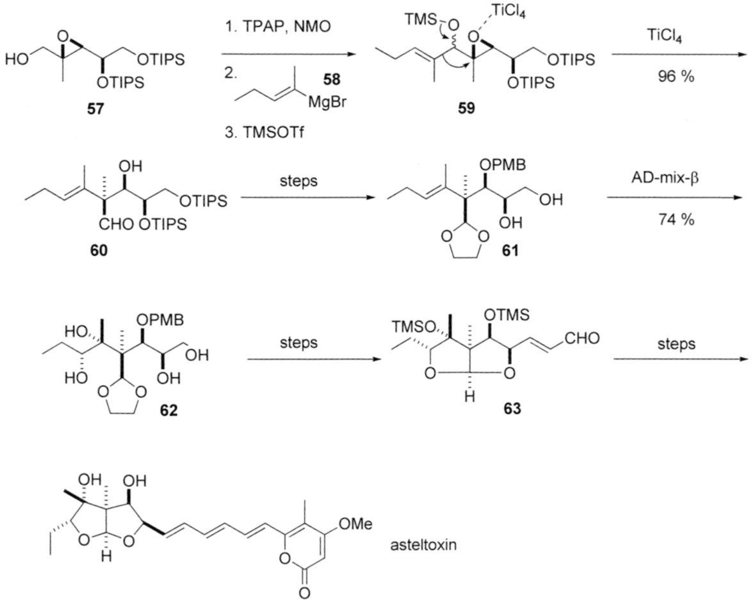

Scheme 1.16 Cha's total synthesis of asteltoxin via stereospecific 1,2-rearrangement.

Suzuki also utilizes his 1,2-rearrangement conditions in the total synthesis of furaquinocin Scheme 1.17) [35]. The starting material was again prepared by Sharpless epoxidation. An alkyne, which is not a good migrating group owing to

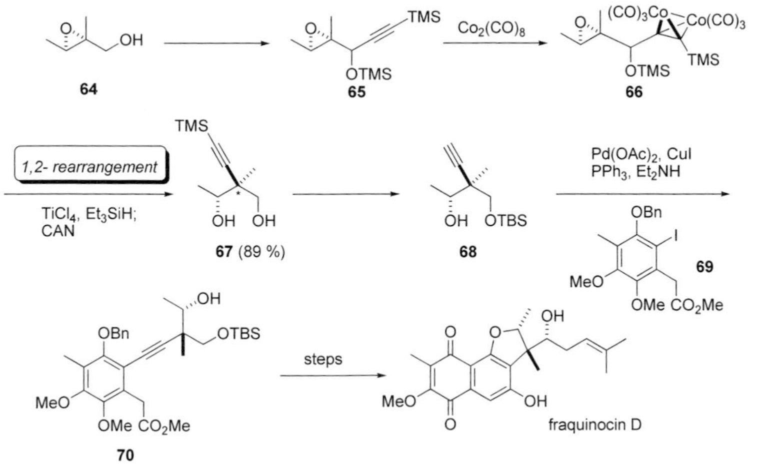

Scheme 1.17 Suzuki's synthesis of furaquinocin D.

its electron-poor nature, was transformed to its Co-complex, and was successfully employed for the stereospecific rearrangement.

These 1,2-rearrangements might be regarded as special cases of intramolecular S$_N$2 displacement at tertiary stereocenters by carbon nucleophiles. For a long time S$_N$2 reactions at tertiary centers have been believed to be quite difficult. Organic chemistry textbooks state that S$_N$2 reactions cannot occur with nucleophiles at tertiary centers owing to steric hindrance at the backside of leaving groups. However, such descriptions are not always true for oxygen nucleophiles. It seems worth mentioning that Mukaiyama (Scheme 1.18) [37] and Shi [38] recently showed that highly stereospecific S$_N$2 replacements are possible when tertiary alcohols are activated by phosphonium salts. In both of these examples, the nucleophiles are so far limited to phenols. Considering that carbon nucleophiles

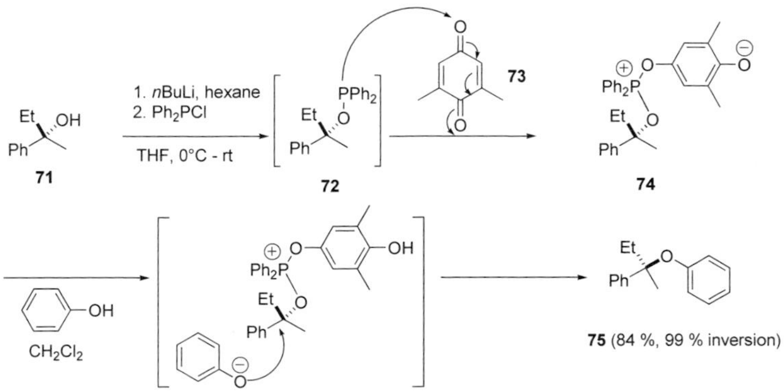

Scheme 1.18 Complete S$_N$2 inversion of a tertiary alcohol by phenol.

has been employed in the modified Mitsunobu reactions of secondary systems [39], hitherto unexplored approaches to quaternary carbons, i.e. the stereospecific inversion of tertiary alcohols by carbon nucleophiles, might be available in the near future.

1.5
Carbometallation Reactions

1.5.1
Addition of a Carbon Nucleophile to a β,β-Disubstituted α,β-Unsaturated Enone

Tetrasubstituted carbon centers, such as tertiary alcohols or amines, can be prepared by the stereoselective nucleophilic addition of carbon nucleophiles to polar C=X bonds. Conjugate addition of a carbon nucleophile to a β,β-disubstituted α,β-unsaturated enone affords a quaternary carbon center.

Mulzer [40] and Ogasawara [41] independently used the conjugate addition of a vinyl cuprate to α,β-unsaturated enones in their synthesis of morphine, where the neighboring chemical environments controlled the stereochemical outcomes.

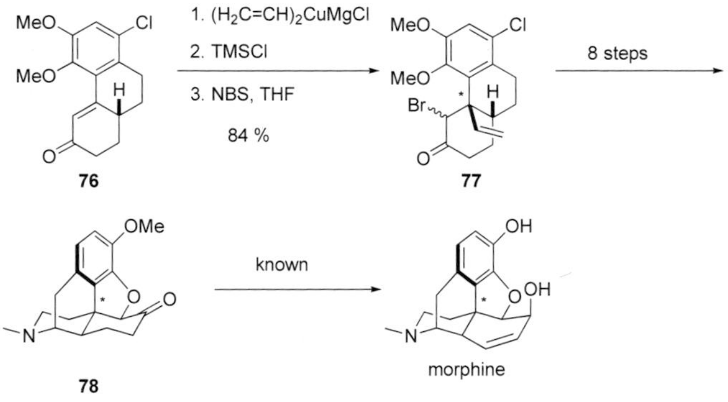

Scheme 1.19 Mulzer's morphine synthesis via conjugate addition of a vinyl cuprate to form a quaternary center.

1.5.2
Asymmetric and Diastereomeric Addition of a Carbon Nucleophile to Unactivated Alkenes Catalyzed by Palladium [42]

The Heck reaction is a palladium-catalyzed coupling of alkenes with organic halides or triflates lacking an sp³-hybridized β-hydrogen. The intramolecular Heck reaction is a powerful tool not only in the assembly of complex natural-product skeletons but also in diastereomeric stereoinductions. When a trisubstituted alkene is employed as the coupling partner, the products include a new quaternary carbon center with excellent diastereomeric control.

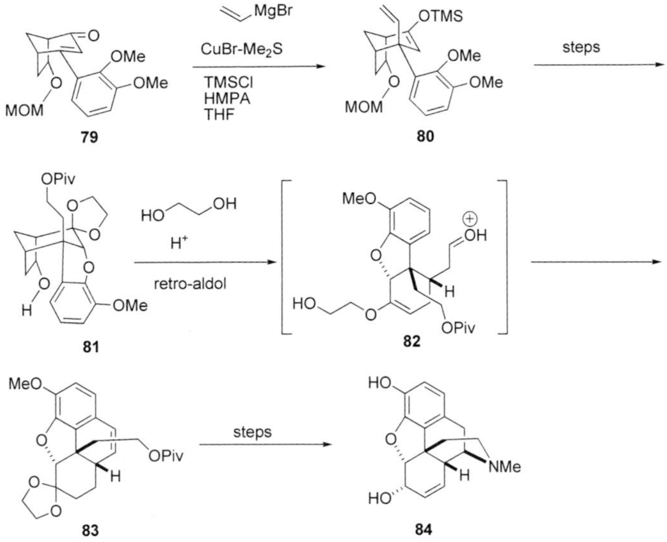

Scheme 1.20 Ogasawara's morphine synthesis.

Configuration information of carbon–heteroatom bonds, which might be constructed by asymmetric catalysts, can be transferred to the new quaternary carbon centers. Scheme 1.21 shows a brief outline of Trost's total synthesis of furaquinocin [36b]. The π-allyl palladium complex intermediate with optically active bisphosphine ligands **87** was generated from an allyl carbonate, and reacted with a bisphenol to gave bisallyl ether **88**. Subsequently, a reductive Heck cyclization of **88** to a furan ring generated a quaternary carbon stereocenter. The authors rationalized the stereochemical outcome by suggesting that steric repulsion between the palladium and the adjacent methyl group in the intermediate **90** played an important role in the diastereoinduction. The enantiomeric purity of the Heck cyclization product **89** was 87% *ee*, which was raised to 99% *ee* by recrystallization.

The enantioselective formation of quaternary carbon centers by a Heck reaction was first reported by Overman in 1989 [43]. This asymmetric Heck (AH) reaction was extensively investigated by Overman and Shibasaki [44, 45]. A number of applications of the AH reaction have been reported. A recent example is the total synthesis of (–)-spirotryprostatin B [46].

Among the three stereocenters in the molecule, quaternary spiro and adjacent centers were stereoselectively constructed via Heck insertion of a conjugated triene. The η³-allylpalladium intermediate was trapped by the nitrogen of a tethered diketopiperazine which proceeds with *anti* stereochemistry (PMP: 1,2,2,6,6-pentamethylpiperidine).

Shibasaki has also reported the synthesis of halenaquinol [47]. Heck cyclization of (*Z*)-trisubstituted alkene **94** in the presence of Pd(OAc)$_2$-(*R*)-BINAP afforded the tetrahydronaphthalene with 99% *ee*. The one-pot Suzuki coupling–asymmetric

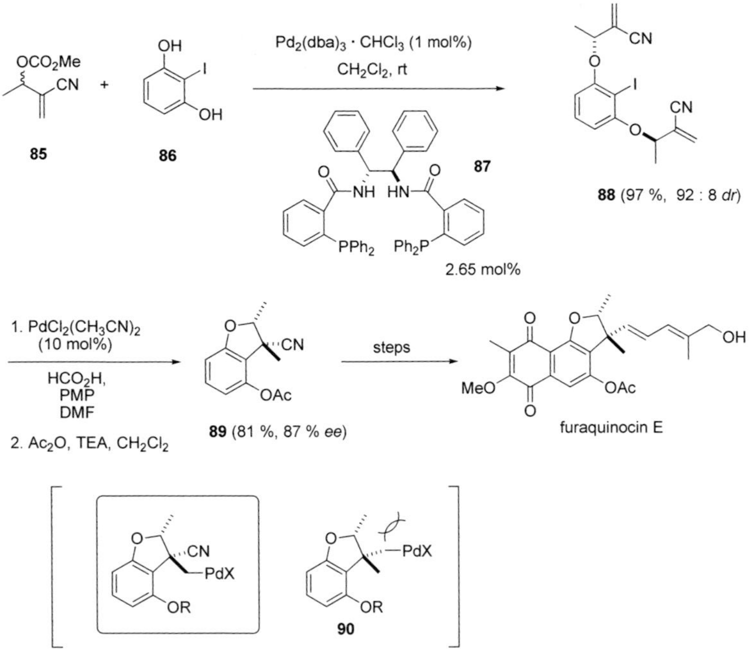

Scheme 1.21 Trost's synthesis of furaquinocin E.

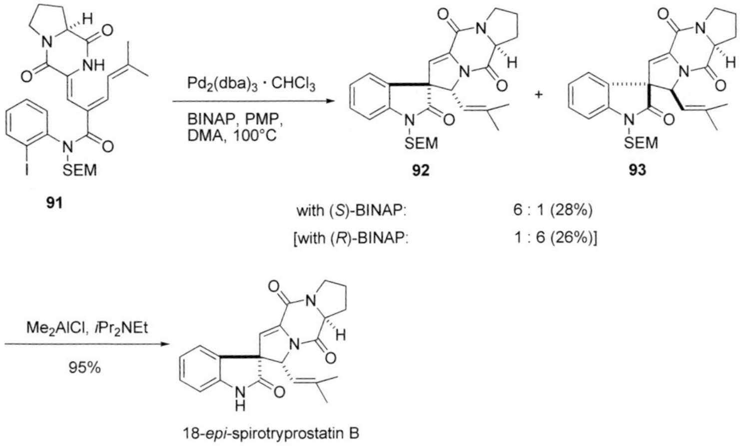

Scheme 1.22 Heck cyclization in the synthesis of
18-*epi*-spirotryprostatin B by chiral palladium catalysts.

Heck cyclization from ditriflate **97** was also achieved in 85% *ee*, albeit the yield was poor (20%).

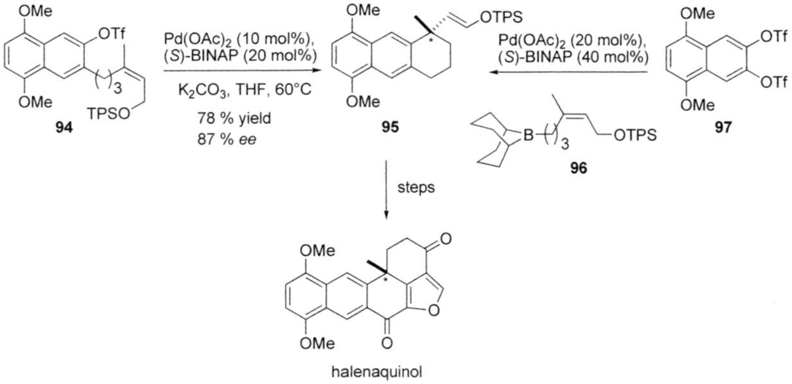

Scheme 1.23 Application of an asymmetric Heck reaction to the total synthesis of halenaquinol.

1.6
C–H Functionalization Reactions

The successful development of practical methods for C–C bond formation via C–H activation would revolutionize natural-product synthesis. However, the use of highly oxidative metal catalysts for oxidative addition to C–H bonds is in essence very difficult. The catalyst needs to be highly reactive and simultaneously selective among the ubiquitous C–H bonds.

On the other hand, C–H insertion reactions mediated by carbenes or by metal–carbenoid complexes are powerful methods to functionalize unactivated C–H bonds [48]. The latter complexes are easily generated from diazo compounds. In general, C–H insertion reactions occur preferentially at tertiary and secondary sites, and often form five-membered rings among other sized rings. An important feature of the reaction is the inherent ability to convert a tertiary stereogenic center into a quaternary center with retention of the absolute configuration.

White utilized a diastereoselective C–H insertion reaction in the total synthesis of (+)-codeine (Scheme 1.24) [49].

The product distribution from decomposition of diazoketone **100** was found to depend markedly on the Rh(II) catalyst employed as well as on small structural variations in the substrate.

Du Bois has reported a beautiful total synthesis of tetrodotoxin, which is the active poison of Japanese *fugu*, with C–H bond functionalization as the key steps [50, 51]. Many functional groups exist densely in the rather small carbon framework. Actually, tetrodotoxin does not contain any quaternary carbon centers but only quaternary-substituted carbons. However, stereoselective construction of the C-6 and C-8a quaternary-substituted carbons was synthetically challenging and Du Bois' elegant synthesis of this difficult target underscores the power of carbene/nitrene insertion reactions.

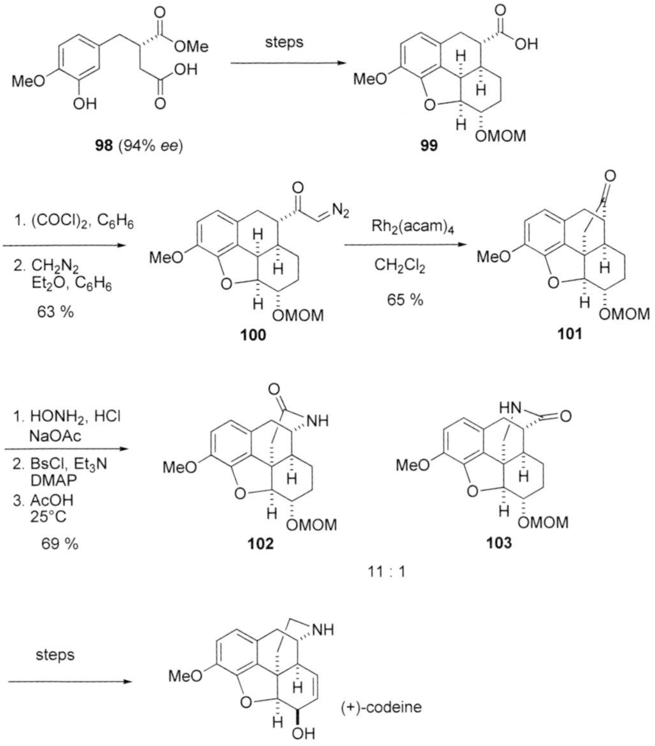

Scheme 1.24 Total synthesis of codeine with C–H insertion as the key step.

Two key C–H functionalizations are incorporated in the scheme. In the first stereospecific Rh-carbene C–H insertion reaction of **106**, the choice of Rh-catalyst was quite important. The first attempts using $Rh_2(OAc)_4$ yielded a complex product mixture, but it was found after screening catalysts that 1.5 mol% $Rh_2(HNCOCPh_3)_4$ gave the cyclic ketone as the sole product. The cyclohexanone was converted to a primary carbamate **109**. Installation of the tetrasubstituted carbinolamine at C-8a was accomplished through stereospecific Rh-catalyzed nitrene insertion using 10 mol% $Rh_2(HNCOCF_3)_4$.

Stereoselective construction of trisubstituted carbons is less difficult, and application of the above C–H functionalization reaction enables them to be converted stereospecifically to tetrasubstituted carbon centers. Steric, electronic, and conformational variations have a great effect on the favored reaction pathway and product distribution. These parameters include catalyst ligands and diazo and substrate substitution patterns. No catalyst has been found to be effective for all substrates. Readers should also be careful to optimize selectivity over C–H insertion, cyclopropanation, Buchner reaction, and ylide generation with the appropriate structural choices.

There are also examples of C–H insertion reactions that do not use diazo-precursors and Rh-catalysts (Scheme 1.26). Taber et al. investigated the use of

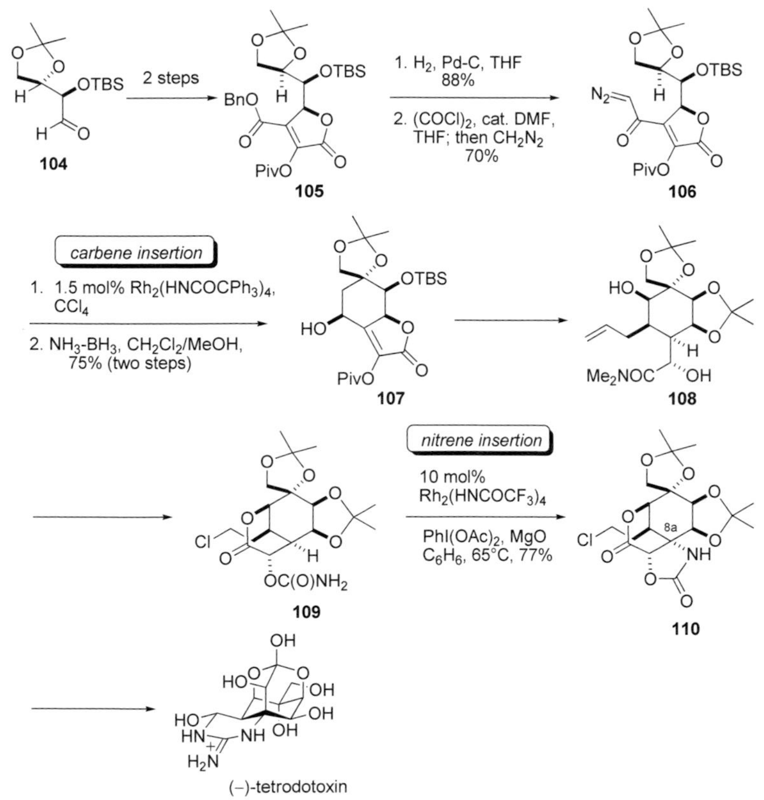

Scheme 1.25 Total synthesis of (–)-tetrodotoxin by Du Bois.

alkylidene carbenes, which can be generated from haloalkenes [52]. Their total synthesis of fumagillin is noteworthy.

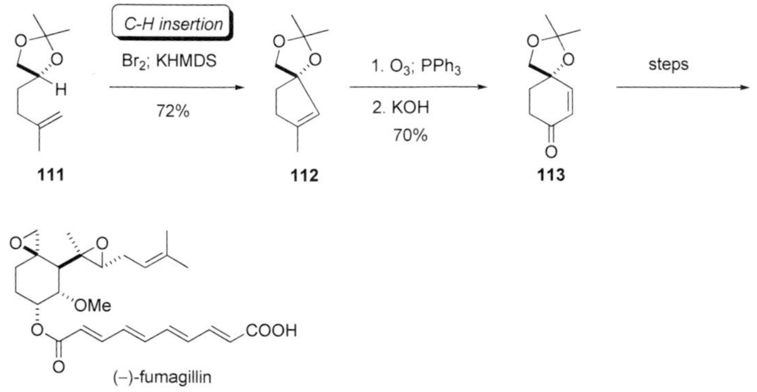

Scheme 1.26 The use of a haloalkene as the source of a carbene intermediate and its application to the total synthesis of fumagillin.

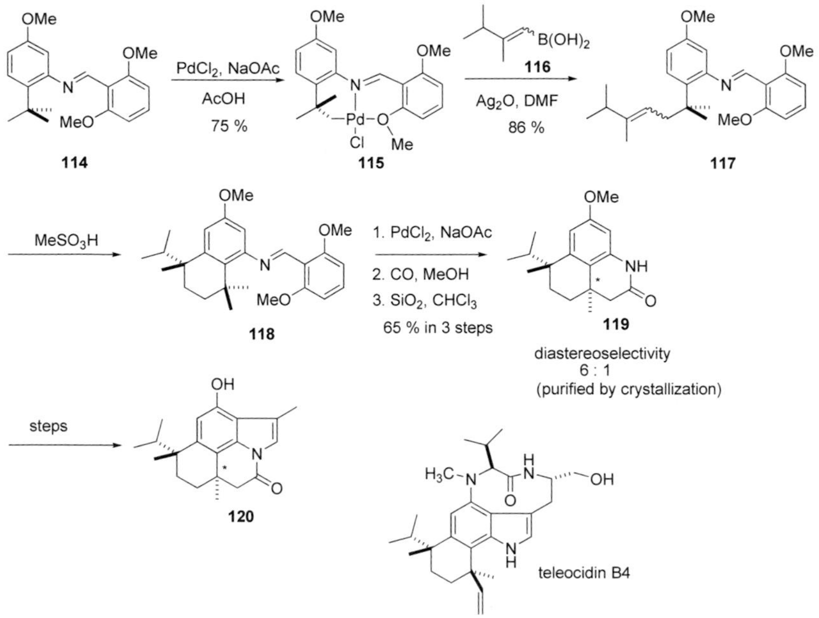

Scheme 1.27 Consecutive functionalizations of three methyls of the *tert*-butyl substituent by diastereoselective C–H activations.

1.7
Asymmetric Modification of Enantiotopic/Diastereotopic Substituents of Quaternary Carbon Centers

Asymmetric quaternary centers can also be generated indirectly via selective modification of diastereotopic or enantiotopic substituents on a symmetric quaternary center. Traditionally, desymmetrizations of enantiotopic substituents have frequently been conducted with enzymes [53]. Sames et al. employed C–H bond activation to desymmetrize *gem*-dimethyl groups in their synthesis of the teloocidin B4 core [54].

This was accomplished via sequential cyclometallation and transmetallation. Stoichiometric PdCl$_2$ in the presence of NaOAc afforded palladacycle **115**. Two methoxy groups on the aryl imine were proved to be important as the directing element. The organopalladium **115** was coupled with vinyl boronic acid to accomplish the first functionalization of a *tert*-butyl group. After the acid-catalyzed cyclization of a cyclohexane ring to **118**, the second C–H activation was conducted. The intermediate palladacycle was treated with CO and methanol to give the methyl ester. Hydrolysis of the Schiff base was accompanied by lactam cyclization. The stereoselectivity in the second C–H bond functionalization was 6:1.

1.8
Summary

This chapter has attempted to present an overview of the stereoselective formation of quaternary carbon centers in reported natural product syntheses. We have tried to choose examples from recently published papers, and did not intend to make this chapter comprehensive. Thus, much outstanding work may have been excluded.

There can be little doubt that the stereoselective construction of quaternary centers is one of the most challenging issues in synthetic natural product chemistry. Catalytic asymmetric alkylations or asymmetric Heck reactions have been employed as powerful tools in the total syntheses, especially for those target compounds with one or a small number of asymmetric centers. For the synthesis of polystereogenic targets, these asymmetric reactions are often used in the very early stage so that substrate-dependent stereoinductions are minimized.

Diastereoselective transformations to quaternary centers will continue to be very important, and these are complementary to those involving asymmetric catalysts. Among them all, the alkylation of chiral enolates is the most popular approach. Intramolecular Diels–Alder reactions will continue to be investigated, in part owing to the rapid progress of biosynthetic studies on Diels–Alderase.

The development of methods applicable for the diastereomeric construction of quaternary carbon centers in the very late stage of total synthesis is still in great demand. Recent attention in the synthetic community to unactivated C–H bond functionalization has also led to the revisiting of transformations such as Rh-carbenoid chemistry.

References

1 Some recent reviews on the synthesis of quaternary centers in natural products: (a) C. J. Douglas, L. E. Overman, *PNAS* **2004**, *101*, 5363; (b) E. A. Peterson, L. E. Overman, *PNAS* **2004**, *101*, 11943.

2 A detailed review on the progress of stereoselective formation of quaternary carbon centers; I. Denissova, L. Barriault, *Tetrahedron* **2003**, *59*, 10105.

3 (a) B. Weber, D. Seebach, *Tetrahedron* **1994**, *50*, 6117; (b) L. Pu, H.-B. Yu, *Chem. Rev.* **2001**, *101*, 757; (c) S. E. Denmark, N. Nakajima, O. J.-C. Nicaise, *J. Am. Chem. Soc.* **1994**, *116*, 8797; (d) S. Itsuno, M. Sasaki, S. Kuroda, K. Ito, *Tetrahedron: Asymmetry* **1995**, *6*, 1507.

4 (a) Sharpless asymmetric epoxidation; R. A. Johnson, K. B. Sharpless, in *Catalytic Asymmetric Synthesis*, ed. I. Ojima, 2nd edn, Wiley-VCH, New York, **2000**, Chapter 6A; (b) Sharpless asymmetric

dihydroxylation; K. B. Sharpless, W. Amberg, Y. L. Bennani, G. A. Crispino, J. Hartung, K.-S. Jeong, H.-L. Kwong, K. Morikawa, , Z.-M. Wang, D. Xu, X.-L. Zhang, *J. Org. Chem.* **1992**, *57*, 2768; (c) Chiral dioxirane epoxidation; M. Frohn, Y. Shi, *Synthesis* **2000**, 1979; (d) Jacobsen and Katsuki asymmetric salen epoxidation; T. Katsuki, in *Catalytic Asymmetric Synthesis*, ed. I. Ojima, 2nd edn, Wiley-VCH, New York, **2000**, Chapter 6B; (e) Jacobsen asymmetric epoxide ring opening reaction; E. N. Jacobsen, *Acc. Chem. Res.* **2000**, *33*, 421.

5 (a) S. Hanessian, *Pure Appl. Chem.* **1993**, *65*, 1189; (b) S. Hanessian, J. Franco, B. Larouche, *Pure Appl. Chem.* **1990**, *62*, 1887; (c) S. Hanessian, J. Franco, G. Gagnon, D. Laramee, B. Larouche, *J. Chem. Inf. Comput. Sci.* **1990**, 30, 413; (d) H.-U. Blaser, *Chem. Rev.* **1992**, *92*, 935.

6   T. Hirose, T. Sunazuka, T. Shirahata,
    D. Yamamoto, Y. Harigaya, I. Kuwajima,
    S. Omura, *Org. Lett.* **2002**, *4*, 501.

7   (a) A. I. Meyers, M. Harre, R. Garland,
    J. Am. Chem. Soc. **1984**, *106*, 1146; (b) a
    recent review by Meyers; M. D. Groaning,
    A. I. Meyers, *Tetrahedron* **2000**, *56*, 9843.

8   A. P. Degnan, A. I. Meyers, *J. Am. Chem.
    Soc.* **1999**, *121*, 2762.

9   (a) Y. Hamashima, D. Hotta, M. Sodeoka,
    J. Am. Chem. Soc. **2002**, *124*, 11240;
    (b) Y. Hamashima, M. Sodeoka, *Chem. Rec.*
    2004, *4*, 231.

10 J. F. Austin, S.-G. Kim, C. J. Sinz, W.-J. Xiao,
    D. W. C. MacMillan, *PNAS* **2004**, *101*, 5482.

11 (a) Some reviews on the progress of asym-
    metric conjugate additions : J. Christoffers,
    A. Baro, *Angew. Chem. Int. Ed.* **2003**, *42*,
    1688; M. P. Sibi, S. Manyem, *Tetrahedron*
    2000, *56*, 8033; (b) Chiral acylating agents by
    Fu are also of interest; however, no applica-
    tion in total synthesis of natural products
    has appeared; I. D. Hills, G. C. Fu, *Angew.
    Chem. Int. Ed.* **2003**, *42*, 3921; A. H.
    Mermerian, G. C. Fu, *J. Am. Chem. Soc.*
    2003, *125*, 4050.

12 M. Bruncko, D. Crich, R. Samy, *J. Org.
    Chem.* **1994**, *59*, 5543.

13 D. L. J. Clive, M. Yu, *Chem. Commun.* **2002**,
    2380.

14 K. C. Nicolaou, G. Vassilikogiannakis,
    W. Mägerlein, R. Kranich, *Angew. Chem. Int.
    Ed.* **2001**, *40*, 2482.

15 (a) H. Oikawa, Y. Suzuki, A. Naya,
    K. Katayama, A. Ichihara, *J. Am. Chem. Soc.*
    1994, *116*, 3605; (b) some reviews on this
    subject; H. Oikawa, *J. Synth. Org. Chem. Jpn.*
    2004, *62*, 778; (c) E. M. Stocking,
    R. M. Williams, *Angew. Chem. Int. Ed.* **2003**,
    42, 3078.

16 K. Auclair, A. Sutherland, J. Kennedy,
    D. J. Witter, J. P. Van den Heever,
    C. R. Hutchinson, J. C. Vederas, *J. Am.
    Chem. Soc.* **2000**, *122*, 11519.

17 (a) H. Oikawa, K. Watanabe, K. Yagi,
    S. Ohashi, T. Mie, A. Ichihara, M. Honma,
    Tetrahedron Lett. **1999**, *40*, 6983; (b)
    K. Watanabe, H. Oikawa, K. Yagi, S. Ohashi,
    T. Mie, A. Ichihara, M. Honma, *J. Biochem.*
    2000, *127*, 467.

18 T. Ose, K. Watanabe, T. Mie, M. Honma,
    H. Watanabe, M. Yao, H. Oikawa, I. Tanaka,
    Nature **2003**, *422*, 185.

19 (a) S. Fukuzawa, Y. Hayashi,
    D. Uemura, A. Nagatsu, K. Yamada,
    Y. Ijyuin, *Heterocyclic Commun.* **1995**, *1*,
    207; (b) M. Kuramoto, K. Hayashi,
    K. Yamaguchi, M. Yada, T. Tsuji,
    D. Uemura, *Bull. Chem. Soc. Jpn.* **1998**,
    71, 771.

20 M. Miyashcta, M. Sasaki, I. Hattori,
    M. Sakai, K. Tanino, *Science* **2004**, *305*,
    495.

21 M. E. Layton, C. A. Morales, M. Shair,
    J. Am. Chem. Soc. **2002**, *124*, 773.

22 J. E. Baldwin, T. D. W. Claridge,
    A. J. Culshaw, F. A. Heupel, V. Lee,
    D. R. Spring, R. C. Whitehead, *Chem.
    Eur. J.* **1999**, *5*, 3154.

23 A comparison of the relative reactivi-
    ties of IMDA and TADA; G. Bérubé,
    P. Deslongchamps, *Tetrahedron Lett.*
    1987, *28*, 5255.

24 K. Araki, K. Saito, H. Arimoto,
    D. Uemura, *Angew. Chem. Int. Ed.*
    2004, *43*, 81; see also the recent total
    synthesis of (+)-maritimol by the
    TADA strategy: A. Toró, P. Nowak,
    P. Deslongchamps, *J. Am. Chem. Soc.*
    2000, *122*, 4526.

25 The 13- and 14-membered trienes have
    been frequently employed as the pre-
    cursors of transannular Diels–Alder
    reactions. The 12-membered trienes
    have rarely been investigated. In
    Roush's synthesis of spinosyn A, stereo-
    selectivity among Diels–Alder products
    was not satisfactory. (a)
    D. J. Mergott, S. A. Frank,
    W. R. Roush, *PNAS* **2004**, *101*, 11955;
    (b) S. A. Frank, W. R. Roush, *J. Org.
    Chem.* **2002**, *67*, 4316; (c) S. A. Frank,
    A. B. Works, W. R. Roush, *Can.
    J. Chem.* **2000**, *78*, 757.

26 unpublished results

27 (a) A recent review on PKR; K. M.
    Brummond, J. L. Kent, *Tetrahedron*,
    2000, *56*, 3263; (b) M. Ishizaki,
    K. Iwahara, Y. Niimi, H. Satoh, O.
    Hoshino, *Tetrahedron* **2001**, *57*, 2729;
    (c) N. E. Shore, M. J. Kundsen, *J. Org.
    Chem.* **1987**, *52*, 569.

28 M. Ishizaki, Y. Niimi, O. Hoshino,
    Tetrahedron Lett. **2003**, *44*, 6029.

29 (a) F. W. Ng, H. Lin, S. J. Danishefsky,
    J. Am. Chem. Soc. **2002**, *124*, 9812; (b) a

review on gelsemine synthetic studies; H. Lin, S. J. Danishefsky, *Angew. Chem. Int. Ed.* **2003**, *42*, 36.

30 S. Yokoshima, H. Tokuyama, T. Fukuyama, *Angew. Chem. Int. Ed.* **2000**, *39*, 4073.

31 C. Clarke, I. Fleming, J. M. D. Fortunak, P. T. Gallather, M. C. Honan, A. Mann, C. O. Nubling, P. R. Raithby, J. J. Wolff, *Tetrahedron* **1988**, *44*, 3931.

32 K. D. Eom, J. V. Raman, H. Kim, J. K. Cha, *J. Am. Chem. Soc.* **2003**, *125*, 5415.

33 (a) M. Shimazaki, H. Hara, K. Suzuki, G.-i. Tsuchihashi, *Tetrahedron Lett.* **1987**, *28*, 5891; (b) other reports on stereospecific 1,2-rearrangements; K. Ooi, K. Maruoka. H. Yamamoto, *Org. Synth.* **1995**, *72*, 95; (c) H. Nemoto, M. Nagamochi, H. Ishibashi, K. Fukumoto, *J. Org. Chem.* **1994**, *59*, 74.

34 (a) G. J. Kruger, P. S. Steyn, R. Vleggaar, C. J. Rabie, *J. Chem. Soc., Chem. Commun.* **1979**, 441; (b) P. S. Steyn, R. Vleggaar, *J. Chem. Soc., Chem. Commun.* **1984**, 977; (c) A. E. de Jesus, P. S. Steyn, R. Vleggaar, *J. Chem. Soc., Chem. Commun.* **1985**, 1633; (d) R. Vleggaar, *Pure Appl. Chem.* **1986**, *58*, 239.

35 T. Saito, T. Suzuki, M. Morimoto, C. Akiyama, T. Ochiai, K. Takeuchi, T. Matsumoto, K. Suzuki, *J. Am. Chem. Soc.* **1998**, *120*, 11633.

36 Total synthesis of furaquinocins: (a) A. B. Smith, III, J. Sestelo, P. G. Dormer, *J. Am. Chem. Soc.* **1995**, *117*, 10755; (b) B. M. Trost, O. R. Thiel, H.-C. Tsui, *J. Am. Chem. Soc.* **2003**, *125*, 13155.

37 T. Shintou, T. Mukaiyama, *J. Am. Chem. Soc.* **2004**, *126*, 7359.

38 Y.-J. Shi, D. L. Hughes, J. M. McNamara, *Tetrahedron Lett.* **2003**, *44*, 3609.

39 (a) T. Tsunoda, Y. Yamamiya, S. Ito, *Tetrahedron Lett.* **1993**, *34*, 3609; (b) J. M. Takacs, Z. R. Xu, X. T. Jiang, A. P. Leonov, G. C. Theriot, *Org. Lett.* **2002**, *4*, 3843; (c) I. Sakamoto, H. Kaku, T. Tsunoda, *Chem. Pharm. Bull.* **2003**, *51*, 474.

40 J. Mulzer, G. Dürner, D. Trauner, *Angew. Chem. Int. Ed. Engl.* **1996**, *35*, 2830.

41 H. Nagata, N. Miyazawa, K. Ogasawara, *Chem. Commun.* **2001**, 1094.

42 L. F. Tietze, H. lla, H. P. Bell, *Chem. Rev.* **2004**, *104*, 3453.

43 (a) N. E. Carpenter, D. J. Kucera, L. E. Overman, *J. Org. Chem.* **1989**, *54*, 5845; see also (b) Y. Sato, M. Sodeoka, M. Shibasaki, *J. Org. Chem.* **1989**, *54*, 4738.

44 The asymmetric intramolecular Heck reaction in natural product total synthesis; A. B. Dounay, L. E. Overman, *Chem. Rev.* **2003**, *103*, 2945.

45 The asymmetric Heck reaction; M. Shibasaki, C. D. J. Boden, A. Kojima, *Tetrahedron* **1997**, *53*, 7371.

46 L. E. Overman, M. D. Rosen, *Angew. Chem. Int. Ed.* **2000**, *39*, 4596.

47 A. Kojima, T. Takemoto, M. Sodeoka, M. Shibasaki, *J. Org. Chem.* **1996**, *61*, 4876.

48 (a) J. C. Gilbert, D. H. Giamalva, M. E. Baze, *J. Org. Chem.* **1985**, *50*, 2557; (b) D. F. Taber, E. M. Petty, K. Raman, *J. Am. Chem. Soc.* **1985**, *107*, 196; reviews on carbene/carbenoids chemistry and their applications of natural product synthesis: (c) D. F. Taber, S.-E. Stiriba, *Chem. Eur. J.* **1998**, *4*, 990; (d) H. M. L. Davies, R. E. J. Beckwith, *Chem. Rev.* **2003**, *103*, 2861; (e) C. A. Merlic, A. L. Zechman, *Synthesis* **2003**, 1137.

49 (a) J. D. White, P. Hrnciar, F. Stappenbeck, *J. Org. Chem.* **1997**, *62*, 5250; (b) J. D. White, P. Hrnciar, F. Stappenbeck, *J. Org. Chem.* **1999**, *64*, 7871.

50 A. Hinman, J. Du Bois, *J. Am. Chem. Soc.* **2003**, *125*, 11510.

51 Tetrodotoxin is a potent voltage-dependent Na channel blocker. Previous total syntheses of tetrodotoxin: (a) Y. Kishi, T. Fukuyama, M. Aratani, F. Nakatsubo, T. Goto, S. Inoue, H. Tanino, S. Sugiura, H. Kakoi, *J. Am. Chem. Soc.* **1972**, *94*, 9219; (b) N. Ohyabu, T. Nishikawa, M. Isobe, *J. Am. Chem. Soc.* **2003**, *125*, 8798; (c) T. Nishikawa, D. Urabe, M. Isobe, *Angew. Chem. Int. Ed.* **2004**, *43*, 4782; (d) T. Nishikawa, M. Asai, N. Ohyabu, N. Yamamoto, M. Isobe, *Angew. Chem. Int. Ed.* **1999**, *38*, 3081.

52 D. F. Taber, T. E. Chritos, A. L. Rheingold, I. A. Guzei, *J. Am. Chem. Soc.* **1999**, *121*, 5589.

53 A lipase-catalyzed enantioselective desymmetrization of prochiral diol was recently employed to construct a spiro-quaternary center in an antitumor antibiotic fredericamycin analogue; A. Akai, T. Tsujino, N. Fukuda, K. Lio, Y. Takeda, K. Kawaguchi, T. Naka, K. Higuchi, Y. Kita, *Org. Lett.* **2001**, *3*, 4015.

54 B. D. Dangel, K. Godula, S. W. Youn, B. Sezen, D. Sames, *J. Am. Chem. Soc.* **2002**, *124*, 11856.

2
Important Pharmaceuticals and Intermediates
Johannes G. de Vries

2.1
The Chirality of Drugs and Agrochemicals

The majority of drugs are chiral. Chiral drugs that have been obtained from natural sources, such as fermentation or agriculture, are obtained in enantiopure form. Until fairly recently, synthesized chiral drugs were racemic. This is rather surprising, as the enzymes and receptors that constitute the targets of these drugs are also chiral and it is thus to be expected that the two enantiomers of a drug will have entirely different interactions with their targets. This message did not register till the disaster with the chiral drug Thalidomide **1a** happened. This drug was prescribed as a treatment for morning sickness for pregnant women. However, it turned out that whereas the (*S*)-enantiomer had indeed this desired property, the other enantiomer was teratogenic and caused deformities in the children born from the women treated with the drug (Fig. 2.1) [1].

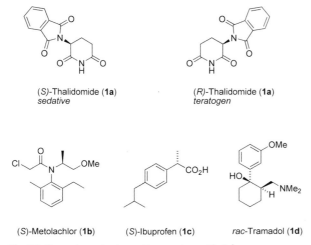

(*S*)-Thalidomide (**1a**)
sedative

(*R*)-Thalidomide (**1a**)
teratogen

(*S*)-Metolachlor (**1b**)   (*S*)-Ibuprofen (**1c**)   *rac*-Tramadol (**1d**)

Fig. 2.1 Some important enantiopure drugs **1b–1d**.
The two enantiomers of Thalidomide **1a** have different
biological effects.

Quaternary Stereocenters: Challenges and Solutions for Organic Synthesis. Edited by Jens Christoffers, Angelika Baro
Copyright © 2005 WILEY-VCH Verlag GmbH & Co. KGaA, Weinheim
ISBN: 3-527-31107-6

This dreadful event has created a greater awareness of the fact that racemates are probably less desirable as drugs than enantiopure forms. For many years this has been a hot topic debated fiercely by many chemists and pharmacologists [2]. In the meantime a number of drugs were "switched" to their enantiopure forms, and in many cases this has been highly successful [3]. This success has rubbed off on agrochemicals, where chiral switches are also actively pursued. A celebrated instance is the herbicide Metolachlor (**1b**), which today is produced by asymmetric hydrogenation [4]. Use of the enantiomerically enriched herbicide not only reduces the cost of production but also decreases the environmental burden to a substantial degree.

On the other hand, a number of studies have shown that the switch of a racemic drug to an enantiopure one is not always successful [5]. Indeed, the (*S*)-enantiomer of the non-steroidal anti-inflammatory drug Ibuprofen **1c** has been struggling for a position on the market [3, 6]. An even more interesting study concerned the painkiller Tramadol **1d**. Here (+)-Tramadol was shown to be more effective than the (–)-enantiomer in clinical trials. However, the (+)-enantiomer was also associated with increased vomiting, thus leading to the conclusion that racemic Tramadol was the best compromise in terms of efficacy and side effects [7]. Nevertheless, in modern drug development, single enantiomers are generally preferred over racemates.

A number of methods exist for the production of enantiopure chemicals (Scheme 2.1) [8]. A common method is the resolution of a racemic mixture, which can be achieved either by a classical resolution via crystallization of diastereomeric salts or by kinetic resolution using enzymes or, occasionally, a chiral catalyst. The problem with these two methods is that the maximum yield is only 50%. Usually it is possible to racemize the unwanted isomer. Better still is to perform the resolution in the presence of a catalyst that causes racemization during the resolution [8f]. This is called a dynamic kinetic resolution in the latter case and an asymmetric transformation in the case of crystallization [9]. Recently, the physical separation of racemates using chromatography on a small scale or a chiral simulated moving bed (SMB) for large-scale separations has become practical [10].

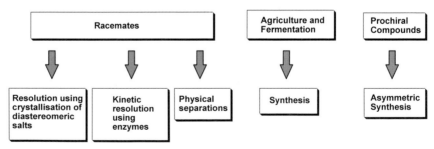

Scheme 2.1 Methods for the production of enantiopure compounds.

A limited number of chemicals is available in enantiopure form from nature. Usually they are obtained via either agriculture or fermentation.

Ideally, a chiral catalyst is used in the synthetic step that introduces chirality into the molecule, to assure formation of only the desired enantiomer. This is called asymmetric catalysis [11]. Enantiopure transition metal catalysts as well as enzymes have been used successfully to achieve this [8a]. Recently, organocatalysis, which uses catalysts based on small chiral organic molecules, has become the subject of intense study [12].

Currently, classical resolution is by far the most important industrial method. This is related to the fact that the final process development of pharmaceuticals is always carried out under extreme time pressure [13]. However, with the introduction of high-throughput experimentation, the use of asymmetric catalysis based on chiral transition metal catalysts is rapidly increasing [14].

Once a chiral molecule has been obtained it can be elaborated into a larger molecule. If new chiral centers are introduced, the existing chirality usually influences the extent of enantiopurity obtained at the new chiral center. Such reactions are called diastereoselective reactions. Chiral auxiliaries are enantiopure fragments that are tethered onto a prochiral molecule only to aid in the introduction of a new chiral center via a diastereoselective reaction and are removed thereafter. This latter method is rarely used for industrial production. A stereoselective synthesis can be either diastereoselective or enantioselective.

Of the existing chiral drugs, a large proportion is still racemic. If there are no safety or efficacy problems there may not be enough incentive to develop an enantiopure form of the drug. Some drugs are obtained enantiopure via fermentation or from extraction of agricultural products, but usually these compounds are further modified to obtain semi-synthetic drugs. Examples are the steroids, the β-lactam antibiotics, the alkaloids, such as the cinchonas, the opiates and the ergot alkaloids, the tetracyclines and several other classes of mycines.

Of the drugs currently on the market roughly 250–300 contain quaternary stereocenters [15]. Most of these are found in semi-synthetic drugs and are either formed during the biosynthesis or during the further elaboration. A much smaller proportion of the drugs containing quaternary stereocenters are entirely synthetic. Many of these are racemic. The small number of fully synthetic enantiopure drugs with quaternary stereocenters is usually obtained by classical resolution. Only in rare cases is chirality induced via asymmetric catalysis followed by a diastereoselective step to form the quaternary stereocenter. Very few examples were found among existing drugs or agrochemicals where asymmetric catalysis was used to form the quaternary stereocenter.

2.2
Steroids

By far the largest class of pharmaceuticals containing a quaternary stereocenter is the steroids. Currently about 170 drugs contain a steroid skeleton and more are still under development [15]. The therapeutic areas are highly diverse and comprise:

- the glucocorticoids [16]: these are C21 steroids related to the pregnane series. They are used to treat allergic reactions, inflammation, arthritis, asthma, gout, eczema and exanthema. In the past they have also been used as immunosuppressants;
- the steroid sex hormones [16]: their most important application is in female contraception;
- calcium-regulating sterols: the best known are Vitamin D and hydroxylated analogues [16].

All steroids contain quaternary stereocenters. The β-methyl-substituted quaternary stereocenters in positions 10 and 13 are a result of the biosynthetic pathway, which invariably proceeds through the cyclization of an isoprene-based triterpene. For the numbering of the steroid carbon atoms see insert in Fig. 2.2. Human steroids are all biosynthetically derived from cholesterol **3**. Several natural raw materials can be used for their industrial production (Fig. 2.2) [17]. In the past, sex hormones have also been extracted from female urine, but this practice is now discontinued.

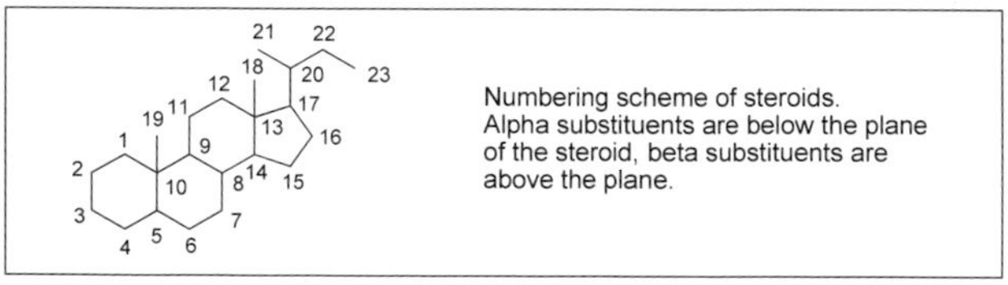

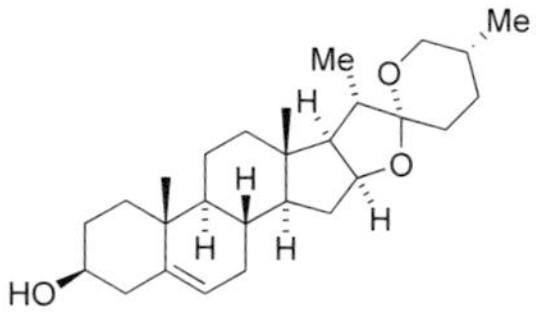

2 Diosgenin
(from Mexican yams)

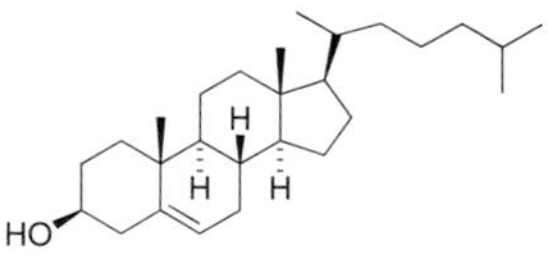

3 Cholesterol
(from wool grease)

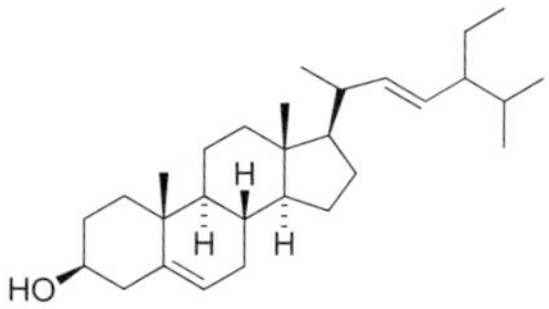

4 Stigmasterol
(fom soybeans)

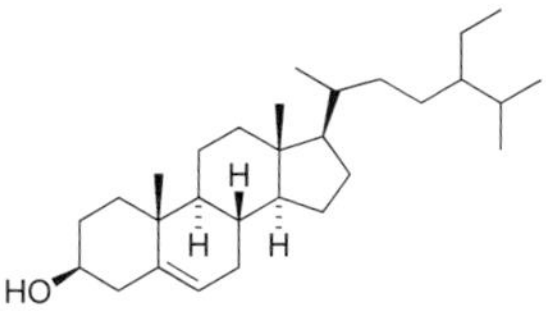

5 β-Sitosterol
(from wheat germ oil,
soy bean oil or
sugar cane wax)

Fig. 2.2 The steroid numbering system and the structures of some steroids.

From these four starting materials the side chain is removed chemically or by microbiological synthesis [18, 19]. Diosgenin and stigmasterol are the preferred precursors (Scheme 2.2) as the removal of their side-chains is easier than from cholesterol or β-sitosterol, which contain saturated side-chains [17].

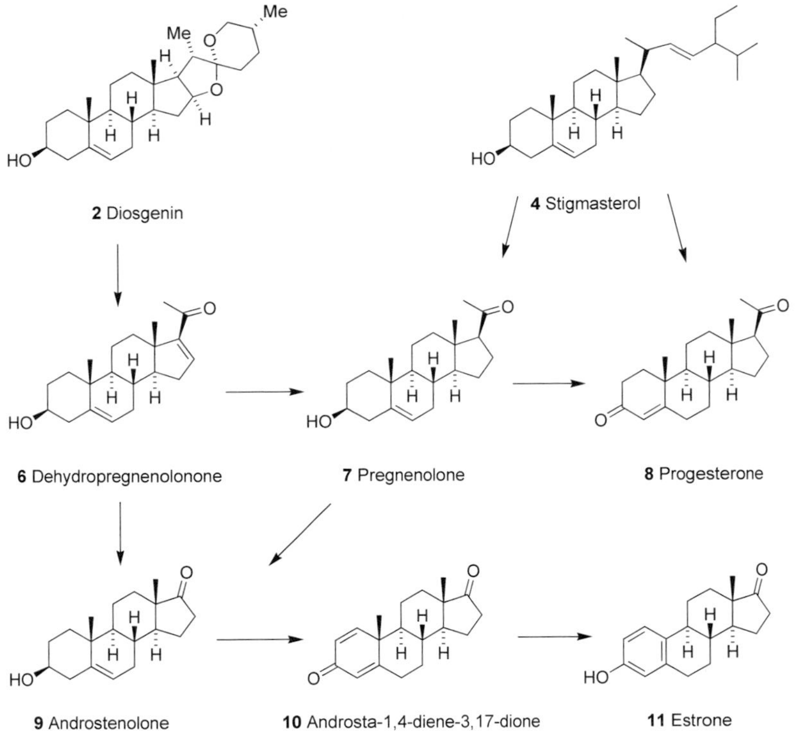

Scheme 2.2 Production of important steroids from readily available raw materials.

In the past, most of the conversions necessary to transform these raw materials into key intermediates for steroid synthesis were carried out using chemical processes. However, both the formation of androstadienedione **10** and its conversion to estrone **11** are now performed using a microbiological synthesis (precursor fermentation) [20].

The glucocorticoids are used extensively to treat skin diseases. In a number of the glucocorticoids new quaternary sterocenters are introduced. Cortisone **17** is prescribed only in rare cases as more potent analogues have been developed. However, it is still important as a raw material for these analogues. The most economic production is from **8**, which is hydroxylated diastereoselectively in the 11-position using precursor fermentation to form **12**. The remaining steps of the synthesis are depicted in Scheme 2.3 [21]. The new stereogenic quaternary center is introduced by diastereoselective osmylation of the trisubstituted double bond in **15** using *N*-methyl-morpholine-*N*-oxide (NMO) as the primary oxidant, followed by oxidation to the acyloin **16**.

The most potent members of the class of glucocorticoids are Dexamethasone **21** and Betamethasone **27**. In Dexamethasone the hydroxyl group in position 17 is introduced by acetoxylation of the enol acetate formed from the C-20 ketone,

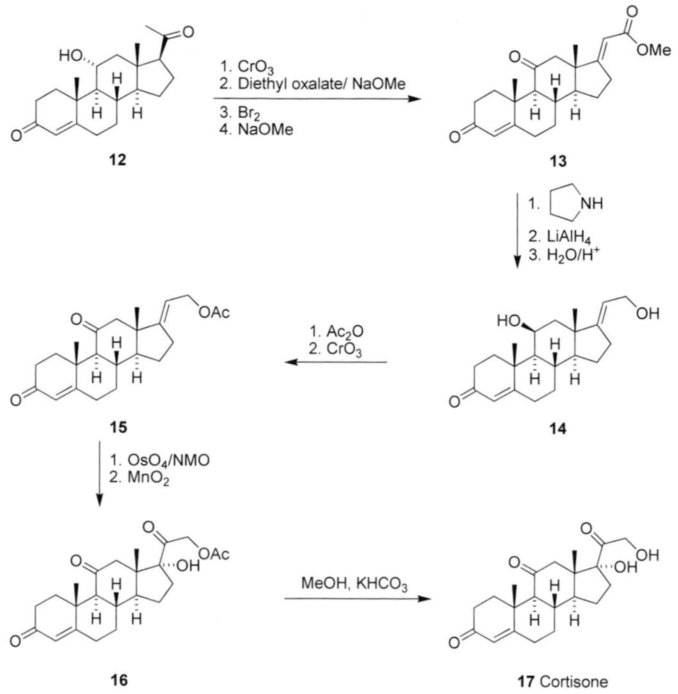

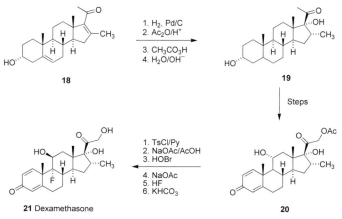

Scheme 2.3 Cortisone from progesterone.

after hydrogenation of the two double bonds (Scheme 2.4). The introduction of the fluorine group is achieved by creating a double bond in the 9,11 position via tosylation of the alcohol and elimination. This bond is then converted into the β-9,11-epoxide, which is further reacted with HF to give the desired fluorohydrin [22].

Scheme 2.4 Introduction of quaternary stereocenters in Dexamethasone.

In Betamethasone **27** the 16-methyl substituent is positioned β. This can be achieved in a number of different ways. One possibility is epoxidation of the 16,17 olefinic bond of **22**, which occurs exclusively in the α-position, followed by the addition of MeMgBr, which occurs from the β-side (Scheme 2.5). Also in this case, the fluorohydrin in position 9,11 is obtained by stereoselective ring opening of the epoxide [23].

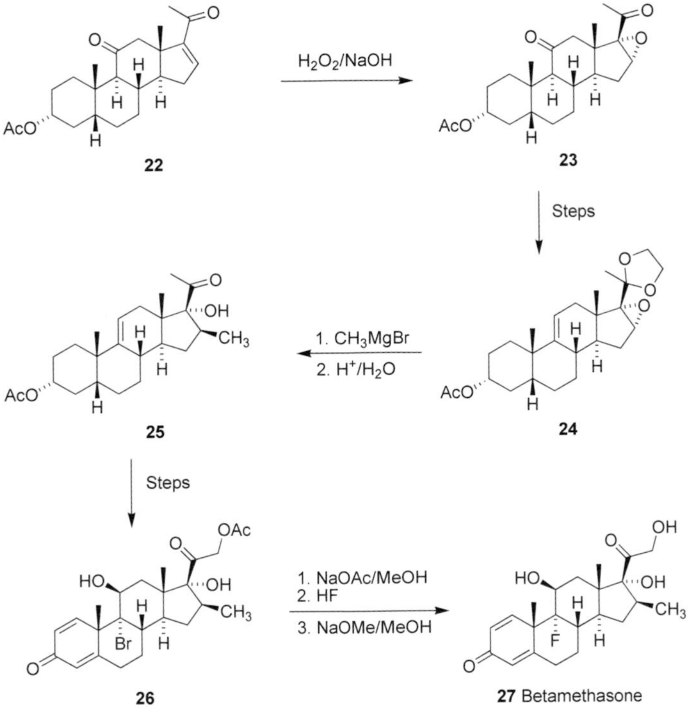

Scheme 2.5 Introduction of quaternary stereocenters in Betamethasone.

The steroid sex hormones are divided into the following subclasses:

- androgens, male sex hormones, such as testosterone
- estrogens, female sex hormones, such as estrone 11 and estradiol
- gestagens, female sex hormones, such as progesterone 8.

For female contraception a gestagen–estrogen combination is used. This mixture leads to inhibition of luteinizing hormone secretion. This hormone in turn is responsible for inducing ovulation.

The introduction of new quaternary stereocenters is important in a number of these compounds, as it not only modulates the biological activity but also increases the bioavailability. In one example, the production of a 17-ethynyl compound is achieved by diastereoselective addition of sodium acetylide to the 17-keto group of

29, which is obtained by Birch reduction of estrone methyl ether **28**. Hydrolysis of the product gives Norethihisterone **31**, a gestagen (Scheme 2.6) [24].

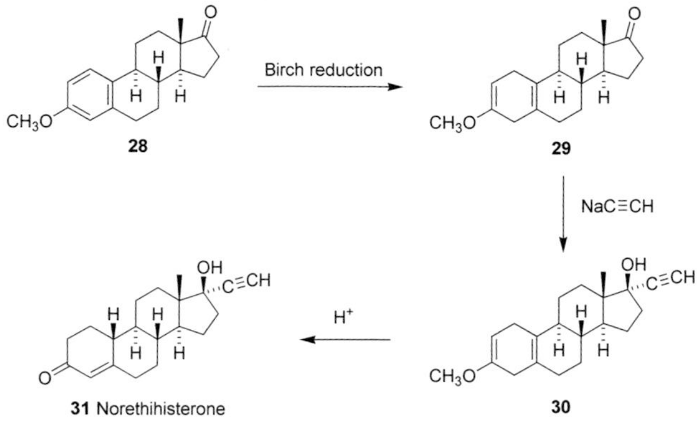

Scheme 2.6 Norethihisterone via ethynylation.

Some modifications of the steroid skeleton cannot be achieved by functionalization of one of the natural products. Thus, occasionally total synthesis becomes necessary as in the case of D-Norgestrel **39**, another gestagen [25]. Here, the skeleton is put together very efficiently by combining 6-methoxy-1-tetralone with vinylmagnesium chloride followed by condensation with 2-ethyl-cyclopenta-1,3-dione. Correct absolute as well as relative stereochemistry is achieved via a simple and elegant yeast reduction of the ketone, which leads to a single diastereomer in 100% *ee* and greater than 75% yield. The rest of the chemistry is straightforward and high yielding. The skeleton is prepared by an acid-catalyzed ring-closure. Hydrogenation of the 14,15 double bond is followed by a Birch reduction of the aromatic ring, which additionally reduces the remaining double bond in position 8,9. Oppenauer oxidation of the alcohol to the ketone followed by addition of lithiumacetylide yields the hydroxyacetylide, and an acid-catalyzed hydrolysis leads to the formation of **39** (Scheme 2.7).

The steroids form probably the most important class of pharmaceuticals containing quaternary stereocenters. The above has only shown a tiny fraction of the total number of different syntheses and production methods, although it has highlighted all the methods used for the introduction of new quaternary stereocenters in steroids.

2.3
Pharmaceuticals and Agrochemicals Based on α-Dialkylated Amino Acids

A number of drugs and agrochemicals contain α-dialkylated amino acid residues. Sometimes they are incorporated into drugs or agrochemicals to increase the biological potency compared to their non-alkylated analogues. Often, the presence of

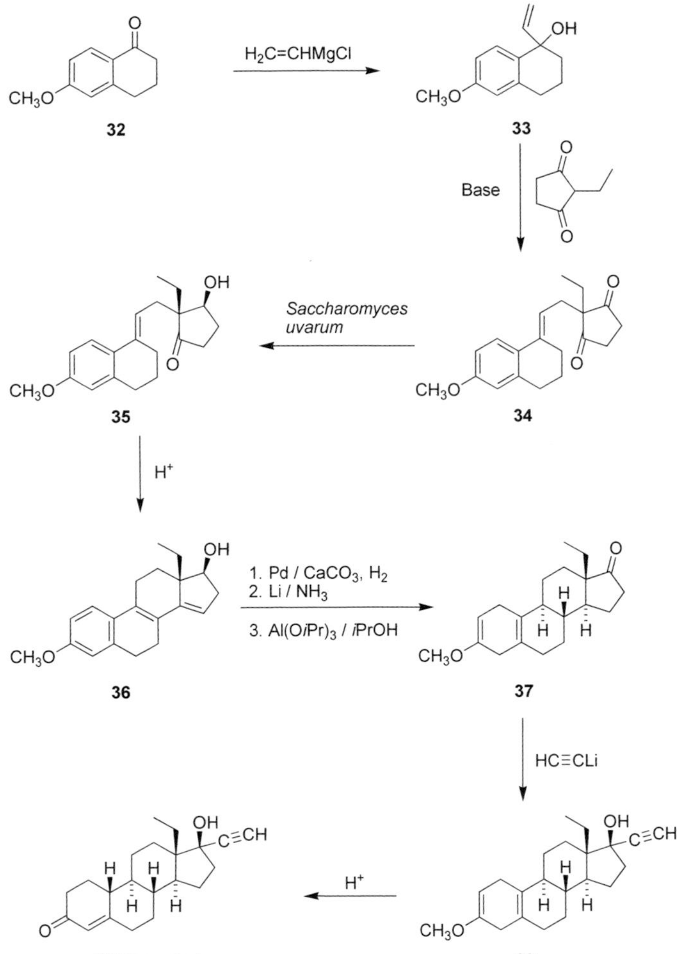

Scheme 2.7 Synthesis of Norgestrel.

the extra alkyl group results in increased resistance towards hydrolysis of the peptide bond, which may be an efficient way to improve the pharmacokinetics of the drug. These are not natural products and thus a number of different synthetic methods have been developed for their production in enantiopure form.

A very general access to this class of compounds was developed by chemists from DSM. They use an amidase enzyme from *Mycobacterium neoaurum* or from an *Ochrobacterium anthropi* strain to hydrolyze racemic α-dialkylated amino acid amides [26]. This is a very efficient resolution, in which only the (S)-enantiomer is hydrolyzed and the (R)-enantiomer remains in the amide form (Scheme 2.8). This reaction is very effective for practically all amino acid amides where R^1 = Me or Et and R^2 is aromatic, aliphatic or allylic.

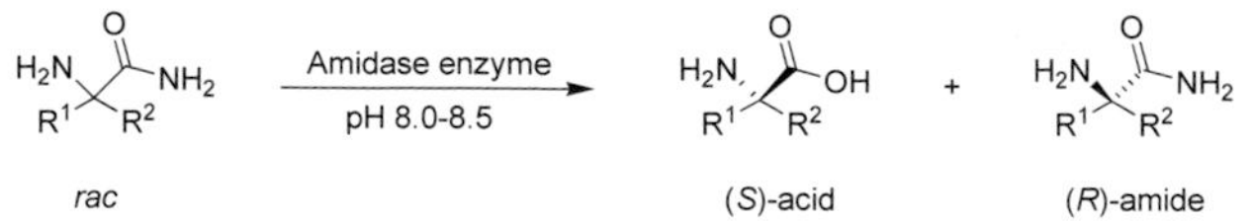

Scheme 2.8 Production of dialkylated amino acids via amidase-catalyzed resolution.

Two general methods exist for the preparation of the racemic amides (Scheme 2.9):

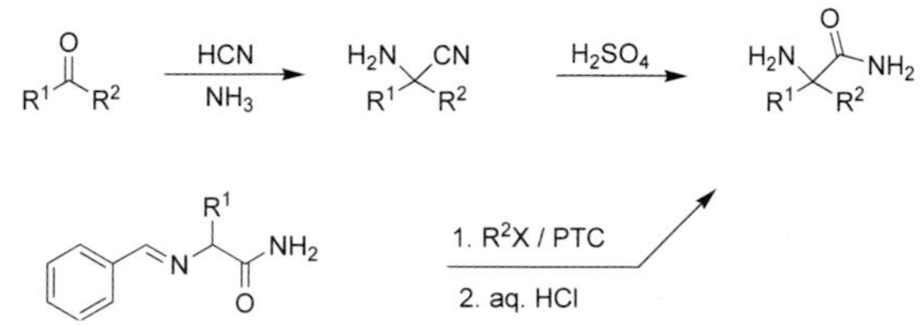

Scheme 2.9 Preparation of racemic dialkylated amino acid amides.

1) a Strecker reaction of a ketone followed by acid-catalyzed selective hydration of the nitrile;
2) phase-transfer-catalyzed alkylation of the benzaldehyde imine of amino acid amides, which works especially well with activated alkylating agents, such as benzyl bromide or allyl bromide.

Other solutions exist for the preparation of enantiopure dialkylated amino acid derivatives and have been used extensively in the laboratory [27]. In particular the methods of Seebach [28] and Schöllkopf [29], based on the diastereoselective alkylation of oxazolinones derived from amino acids and bis-lactim ethers of diketopiperazines, respectively, have proven their utility. More recently, the asymmetric alkylation of imines of amino acids catalyzed by chiral phase-transfer catalysts, such as those based on quaternized cinchona alkaloids has been developed as a useful synthetic method [30]. However, as these methods use 10 mol% of catalyst, they are not yet cost-effective. The quaternary axially chiral bis-1,1′-naphthyl-based quaternary ammonium salts developed by Maruoka are more efficient and can be used at only 1 mol% [31]. However, their synthesis is rather cumbersome, and thus, in this case, the cost of the catalyst forms a barrier to industrial application.

It is possible to use simple non-chiral phase-transfer catalysis (PTC) catalysts to effect the alkylation of an amino acid imine and use a classical resolution for the separation of the racemic α-dialkylated amino acid.

There are several examples of the use of α-alkylated amino acids in drugs [15]. Methyldopa **40**, its ethyl ester Methyldopate **41** and α-methyl-tyrosine (Metirosine, **42**) are antihypertensives (Fig. 2.3). These three compounds are obtained enantiopure by classical resolution [15]. Eflornithine **43**, which is α-difluoromethylated ornithine, was designed as a suicide enzyme inhibitor of ornithine decarboxylase and is used, as the racemate, as an antineoplastic, an antiprotozoal and an antipneumocystis agent [15].

In agrochemicals, large-scale applications of an α-dialkylated amino acid derivative are found in a family of imidazolone antifungal compounds that all contain a pyridine-substituted dihydroimidazolone based on α-methylated-valine [32]. Studies have appeared on the antifungal activity of both enantiomers of **48a** and although the (R)-enantiomer is more effective the difference in activity with the racemate was not sufficiently large to justify a chiral switch [33]. Thus, these compounds are produced and sold as racemates. The synthesis proceeds either via the amino acid amide or via the aminonitrile (Scheme 2.10).

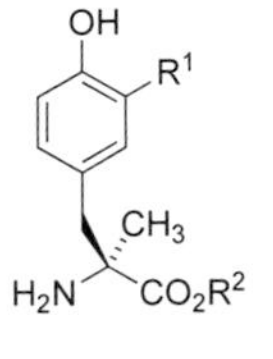

40 R^1 = OH, R^2 = H, Methyldopa
41 R^1 = OH, R^2 = Me, Methyldopate
42 R^1 = H, R^2 = H, Metirosine

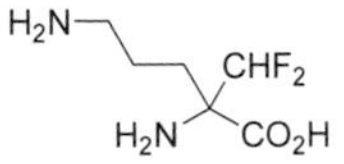

43 Eflornithine

Fig. 2.3 Drugs based on a-dialkylated amino acids.

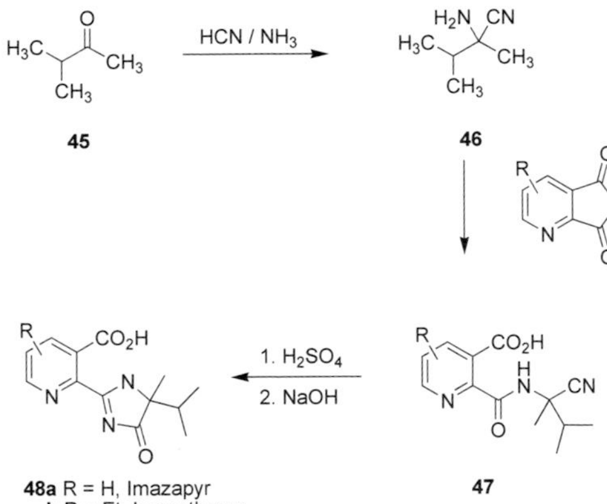

45

46

48a R = H, Imazapyr
  b R = Et, Imazethapyr
  c R = 5-methoxymethyl, Imazamox
  d R = 5,6-benzo, Imazaquin

47

Scheme 2.10 Synthesis of imidazolone antifungals.

Fenamidone **52** is another broad-spectrum antifungal based on the dihydroimidazolone structure. The compound shows excellent activity against fungi of the oocyte class [34]. It finds application among other things in fighting mildew in vineyards. It was developed by chemists from Rhône-Poulenc but is now marketed by Bayer Cropscience. In this case, there was a marked difference in activity between the enantiomers, and so only the (S)-enantiomer is produced. The synthesis (Scheme 2.11) starts with (S)-α-methyl-phenylglycine **49**, which can either be made by the amidase technology described above [35], via resolution of the diastereomeric salt, or via the hydantoine [36].

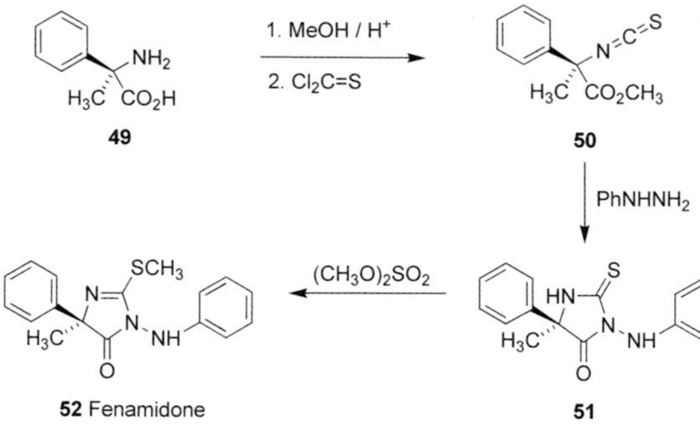

Scheme 2.11 Synthesis of the antifungal Fenamidone.

2.4
Azole Antimycotics

Another very important class of compounds containing a quaternary stereocenter are the azole antimycotics [37, 38]. These compounds all contain a triazole or a histidine group. They are applied both for human use and as antifungals in agriculture. The mechanism of action is via inhibition of ergosterol biosynthesis. More specifically, they inhibit the cytochrome P450-dependent 14-α-demethylation of lanosterol or of 24-methylenedihydrolanosterol. Since ergosterol is an essential constituent of the fungal plasma membrane this inhibition eventually leads to the death of the fungi. In the group of pharmaceutical azole antifungals, three closely related compounds, Ketoconazole **53**, Terconazole **54a** and Itraconazole **54b** all contain a quaternary stereocenter (Fig. 2.4) [15]. All three compounds are marketed as racemate and in the case of Itraconazole even as a mixture of diastereomers. The synthesis of *rac*-Ketoconazole is depicted in Scheme 2.12.

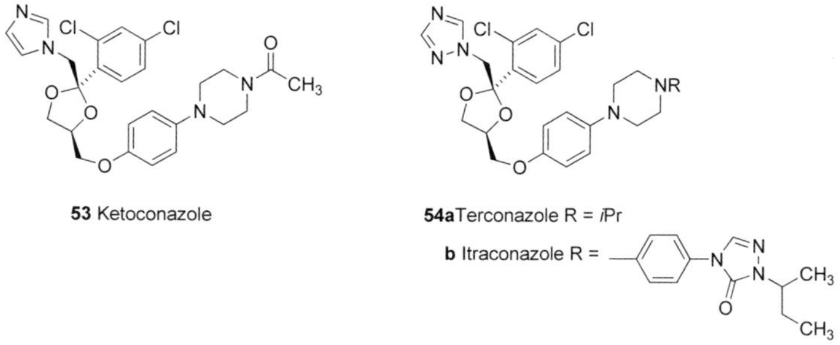

Fig. 2.4 Azole antimycotics containing quaternary stereocenters.

In 2002 a new triazole antifungal named Voriconazole was launched by Pfizer, which seems to be the first enantiopure member of this class of compounds containing a quaternary stereocenter. The compound was synthesized as a racemate and resolved by classical resolution with camphorsulfonic acid in the very last step (Scheme 2.13). Obviously, it is not very efficient to have the resolution at such a

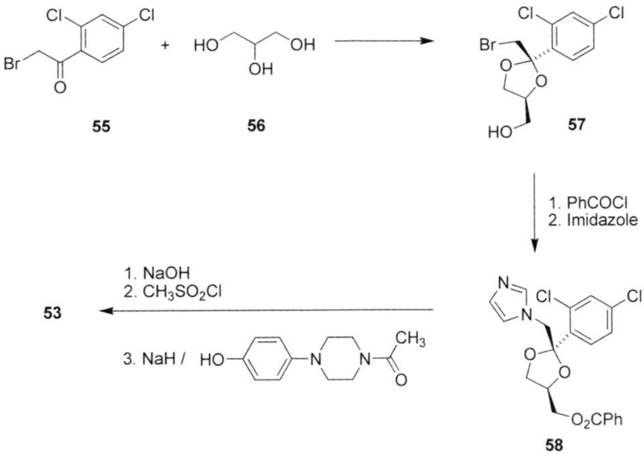

Scheme 2.12 Synthesis of *rac*-Ketoconazole **53**.

late stage of a total synthesis. It underlines the need for a better synthetic arsenal for asymmetric C–C bond formation in general and for the formation of quaternary stereocenters in particular.

A number of azole antifungals containing quaternary stereocenters have been developed for agricultural use [32]. All of these are marketed as racemates and or diastereomers (Fig. 2.5).

2.5
Alkaloids

Most pharmaceutically used alkaloids contain quaternary stereocenters. Important classes are the opioid alkaloids such as morphine, and the ergot alkaloids such as the prolactin inhibitor Bromocryptine (Fig. 2.6). Morphine is isolated from opium, which is harvested from poppies. Bromocryptine is a semi-synthetic drug and, as in most alkaloids, the quaternary stereocenter is introduced during the biosynthesis.

All opioids are centrally acting analgesics [39]. A relatively small number of alkaloids are produced by total synthesis. One of these is Dextromethorphan **74**, an antitussive used in many anticough syrups. This compound is produced by total synthesis and is made enantiopure by classical resolution (Scheme 2.14) [40]. The quaternary stereocenter is formed in the Grewe cyclization of **73**, which is in effect a highly diastereoselective Friedel–Crafts reaction. A slight variation of this route is known, in which **72** is resolved using mandelic acid.

Two approaches have been published in which the amine **72** is obtained in enantiopure form by asymmetric hydrogenation (Scheme 2.15).

The most straightforward method, the selective asymmetric hydrogenation of imine **71** to (*S*)-**72** was developed by Werbitsky from Lonza using iridium in combination with a ferrocene-based bisphosphine ligand developed by Ciba-Geigy

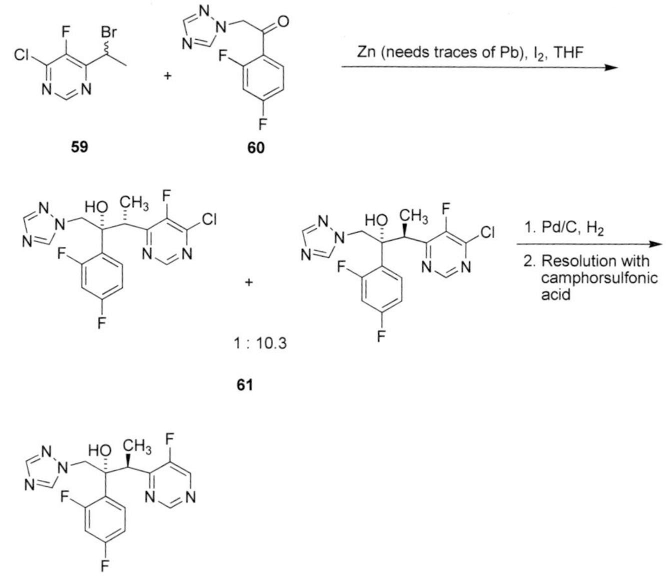

Zn (needs traces of Pb), I$_2$, THF

59   +   **60**

1. Pd/C, H$_2$

2. Resolution with camphorsulfonic acid

1 : 10.3

61

62 Voriconazole

Scheme 2.13 Synthesis and resolution of Voriconazole.

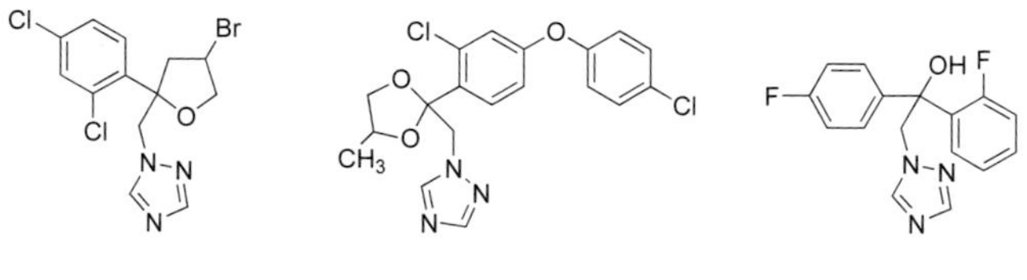

63 Bromuconazole   **64** Difenoconazole   **65** Flutriafol

Fig. 2.5 Fungicides containing quaternary stereocenters.

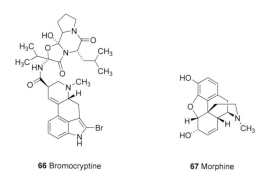

66 Bromocryptine   **67** Morphine

Fig. 2.6 Alkaloid-based drugs containing quaternary stereocenters.

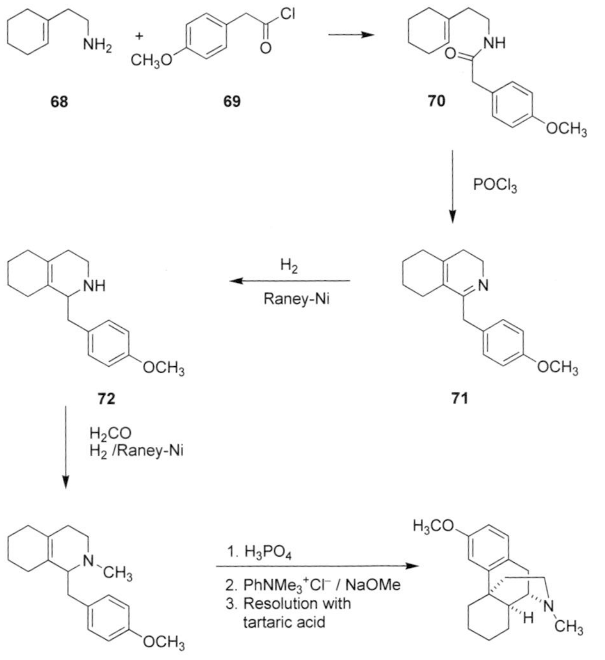

Scheme 2.14 Synthesis and resolution of Dextromethorphan.

(now Solvias) [41]. He was able to obtain (S)-**72** in 89% *ee* using an S/C (Substrate/ Catalyst) ratio of 2000. Less straightforward is the asymmetric hydrogenation of the enamides **75a** [42] and **75b** [43], since two extra synthetic steps are involved. However, this strategy was rewarded with a substantially higher enantioselectivity. It is unclear if any of these processes has ever been introduced into the plant as a replacement for the existing process. In the earlier process the unwanted isomer is racemized [44]. Thus, it remains questionable if the extra costs of registration and product development are offset by the lower cost of the asymmetric hydrogenation process. In addition, there is the additional cost of the rhodium metal and the relatively expensive ligand.

2.6
HIV Inhibitors

The advent of the human immunodeficiency virus (HIV), the cause of acquired immunodeficiency syndrome (AIDS) in humans, has led to a pandemic of tremendous proportions. Untreated, mortality is extremely high and this has spurred massive drug development activity in industrial and academic laboratories worldwide. The starting point of the therapy was the use of known nucleoside analogues such as Zidovudine (azidothymidine, AZT), which act as reverse

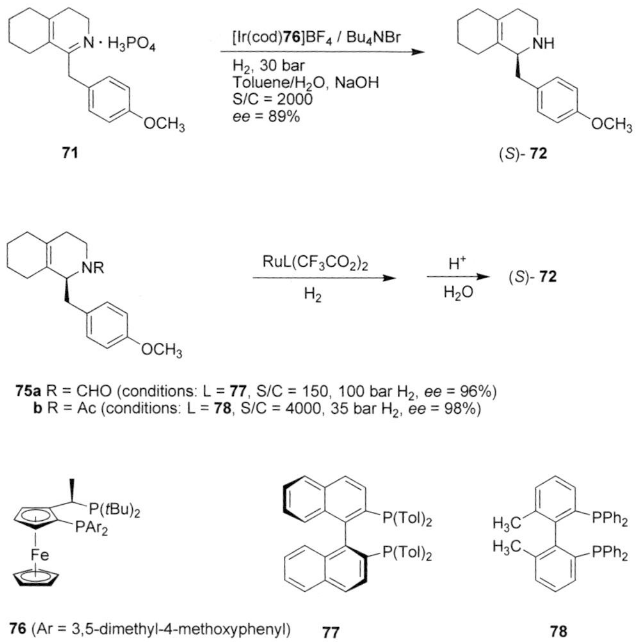

Scheme 2.15 Asymmetric hydrogenation approaches to Dextromethorphan.

transcriptase inhibitors, but this turned out to be insufficient in view of the rapid mutation of the virus. Thus, combination therapy with other drugs became necessary. To this end protease inhibitors and non-nucleoside reverse transcriptase inhibitors have been developed; two of these contain quaternary stereocenters.

The first, Efavirenz **83** is a non-nucleoside reverse transcriptase inhibitor and was developed by the Merck–Dupont combination. The quaternary stereocenter of **83** was established by an asymmetric catalytic C–C bond formation; this is one of the rare examples of the use of such a procedure in production (Scheme 2.16) [45]. Addition of lithium 2-cyclopropylacetylide (2 equivalent) in the presence of the lithium salt of 1-phenyl-2-pyrrolidin-1-yl-propan-1-ol (2 equivalent) to ketone **79** at –55°C gave the tertiary alcohol **80** in 91% crystalline yield and 98% *ee*. The acetylide is delivered to the ketone via a cubic complex **84**, which is preformed by stirring all ingredients except the ketone from –10 to 0°C. The synthesis is completed by oxidative removal of the *p*-methoxybenzyl group using chloranil or DDQ. The oxidation leads to formation of the cyclic aminal, which is hydrolyzed with base, after which the aldehyde side product is reduced with $NaBH_4$ before further purification is possible. The cyclic carbamate **83** is formed by reaction of the amino alcohol **82** with phosgene or alternatively by treatment with methyl chloroformate, followed by LiO*t*Bu.

The main shortcoming of this process is the twofold excess of the chiral ligand and the cyclopropylacetylene that need to be used, presumably because of the presence of the acidic amine proton. A number of improvements to this process

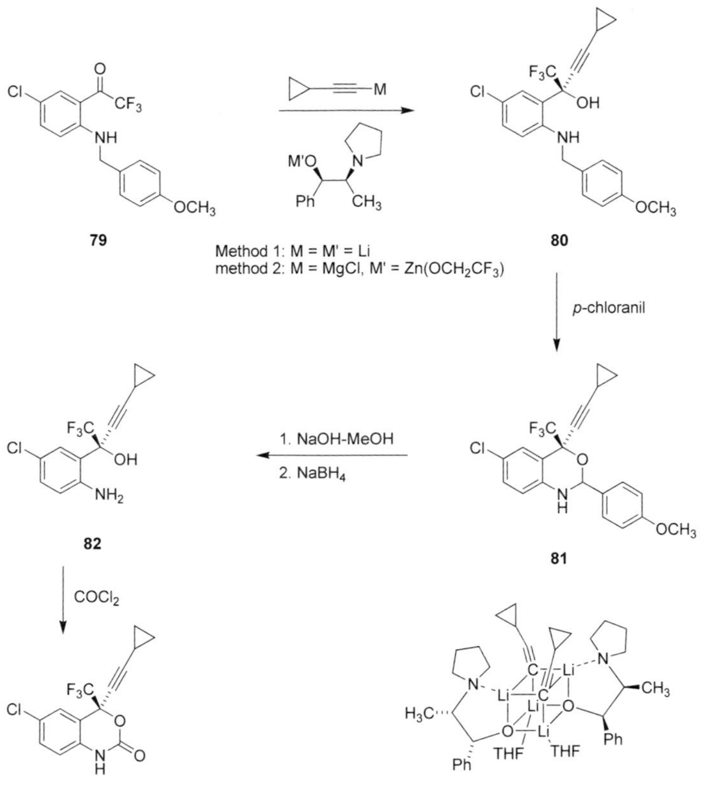

Scheme 2.16 Synthesis of Efavirenz.

have since been reported, such as the *in situ* recycling of the chiral ligand and the surplus nucleophile, which greatly improved the economics of the process [46]. Another interesting improvement has been the use of a zinc instead of a lithium nucleophile. Because of the reduced basicity of the zinc acetylide it is now possible to perform the reaction on the unprotected aniline, which saves a protection and a deprotection step. The enantioselectivity is much influenced by the second achiral auxiliary that is part of the zinc reagent, but also by the metal in the reagent used to deprotonate the acetylene. The best results were obtained by a catalyst that was prepared from Et_2Zn and (1*R*, 2*S*)-*N*-pyrrolidinylnorephedrine in a mixture of trifluoroethanol and THF in combination with the cyclopropylacetylide salt, which was prepared by reacting the acetylene with *n*-BuMgCl. In this fashion **82** was obtained in 95% yield and 98% *ee*. Although this process compares favorably with the older one in many aspects, the use of stoichiometric zinc as well as trifluoroethanol is not desirable from an environmental perspective.

In view of the virus's structural simplicity, antiviral targets for drug development are hard to find. Fortunately, the AIDS virus relies on a protease for maturation to the infectious stage. Initial approaches to HIV protease inhibitors were based on peptidomimetics, which often lead to compounds that may have a high binding affinity but a poor pharmacokinetic profile. Thus, most companies active in this field have screened large libraries of compounds in search of a non-peptidic HIV protease inhibitor. Tipranavir **92** is the offspring of such a screening program and is currently in an advanced stage of development.

The initial synthesis of racemic **92** as published by chemists from Pharmacia & Upjohn is relatively straightforward, in particular the synthesis of dihydropyrone **89**, which is made by condensation of the dianion of methyl acetoacetate with phenylhexanone **85** followed by ester hydrolysis and ring closure (Scheme 2.17) [47]. Resolution at an early stage of the synthesis is extremely favorable for the economics of the process and, indeed, a patent describes the resolution of dihydropyrone **89** [48]. Other methods of obtaining enantiopure **89** have been described. These are based on the formation of **86**, either via addition of acetate anion [49] or via a Reformatsky reaction [50]. After hydrolysis, acid **87** could be resolved with ephedrine. Formation of the acetoacetate using CDI and the magnesium salt of monomethyl malonate followed by ester hydrolysis and ring closure gives **89**. A number of variants have been described for the next steps of the process. A short, clean and high-yielding method is depicted in Scheme 2.17 and is based on the condensation of **89** with *m*-nitroacetophenone to give **90** as a mixture of double bond isomers. Diastereoselective hydrogenation of **90** was accomplished using Rh/duphos at a S/C ratio of 1000 [51]. The synthesis is completed by hydrogenation of the nitro group, followed by sulfonylation with 5-trifluoromethyl-2-chlorosulfonylpyridine.

Several asymmetric approaches to **92** have been published. An approach based on the use of an Evans auxiliary established the stereochemistry at the carbon carrying the ethyl substituent with 100% diastereoselectivity. However, in the reaction creating the other chiral center a 3/2 mixture of diastereomers was obtained [52]. A highly creative route to dihydropyrone **89** employing asymmetric ring-opening ring-closing metathesis was published by Hoveyda and co-workers [53]. However, this route requires more steps to obtain **92**. In addition 5 mol% of an expensive chiral ruthenium catalyst is used. Trost and Anderson have published a synthesis of **92** in which two chiral fragments were developed and coupled at a late stage of the synthesis (Scheme 2.18) [54]. Both asymmetric steps are based on a dynamic kinetic resolution (DKR) that gives a theoretical yield of 100%, which is highly preferable to the classical resolution. The two fragments are coupled and oxidized to ketone **100**. The synthesis is completed by oxidative removal of the 4-methoxybenzyl (PMB) protecting group, base-catalyzed ring closure to the pyrone, hydrogenation of the nitro group to the aniline and coupling with sulfonyl chloride as previously. However, it takes nine synthetic steps to synthesize fragment **96**, four synthetic steps to fragment **99**, and six steps for the coupling of the two fragments and further elaboration. In addition, two relatively expensive catalysts are used at uneconomical S/C ratios. Thus, this route also cannot compete with the ones based on resolution at an early stage of the synthesis.

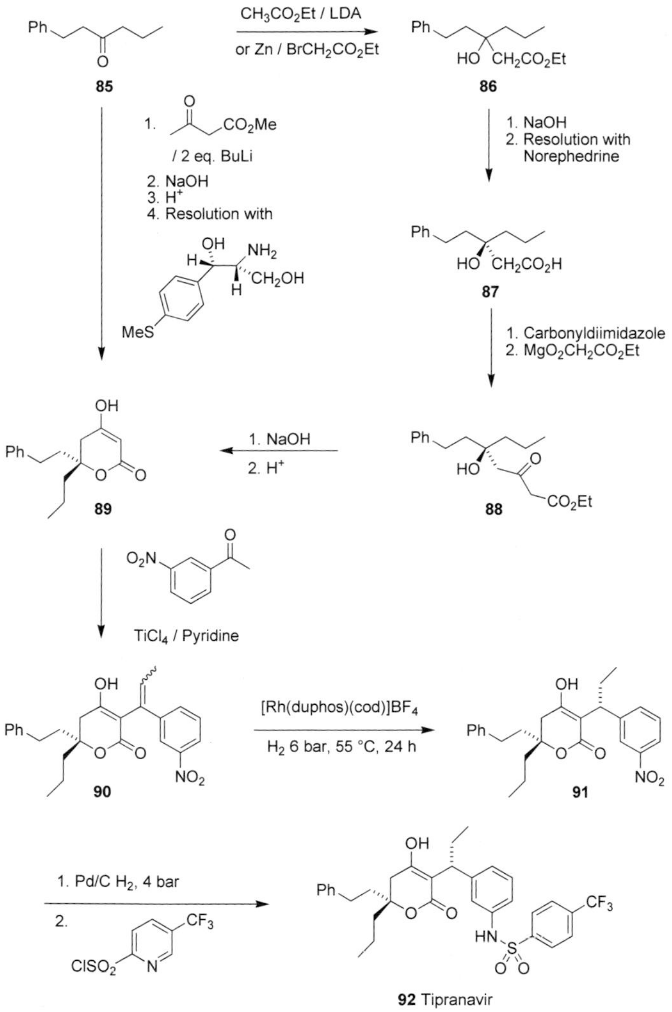

Scheme 2.17 Synthetic routes to Tipranavir.

2.7
β-Lactam Antibiotics

With the exception of the carbapenems and Aztreonam, most classes of β-lactam antibiotics are semi-synthetic, with the nucleus obtained by fermentation. The reason new generations of β-lactam antibiotics are continuously being developed is the emergence of resistance in bacteria. This is largely related to the presence in the bacteria of β-lactamases that are capable of hydrolyzing the β-lactam group, thus defanging the antibiotic. For this reason β-lactamase inhibitors are sometimes co-administered with the antibiotics. Another approach is the addition of

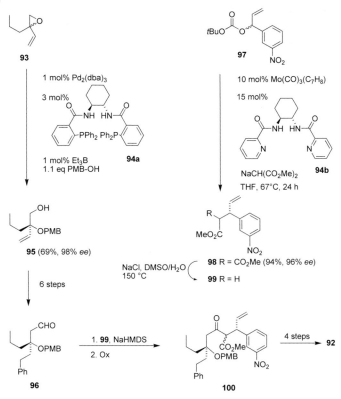

Scheme 2.18 Tipranavir intermediate via asymmetric synthesis.

substituents on the β-lactam ring that reduce the affinity for the β-lactamases, yet leaves the antibiotic potency intact [55].

Temocillin **101** and Latamoxef **102** are two β-lactam antibiotics that carry an additional methoxy substituent adjacent to the amino substituent at the 3-position of the β-lactam ring. Figure 2.7 shows some β-lactam antibiotics that contain quaternary stereocenters.

101 Temocillin   **102** Latamoxef

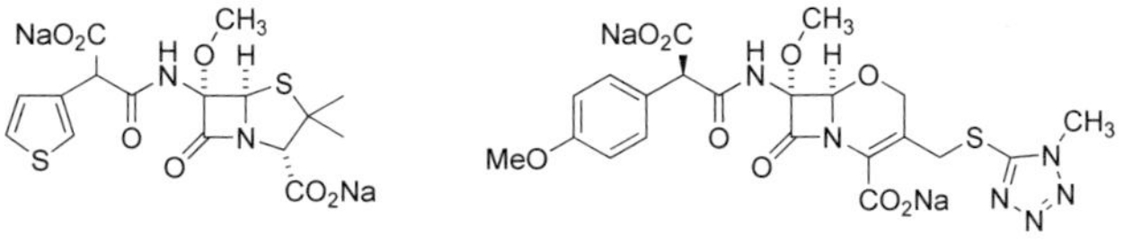

Fig. 2.7 β-Lactam antibiotics with quaternary stereocenters.

A number of different synthetic strategies have been developed. One method that has been used is conversion of the amide functionality in a ketimine by treatment with PCl$_5$/pyridine. Diastereoselective chlorination followed by treatment with LiOCH$_3$ gave the α-methoxylated ketimine, which was hydrolyzed back to the amide by treatment with phosphoric acid. The synthesis was completed by hydrogenolytic removal of the PNB ester group and hydrolysis of the phenyl ester. In this way Temocillin **101** was obtained from the *p*-nitrobenzyl ester of Ticarcillin **103** (Scheme 2.19) [56].

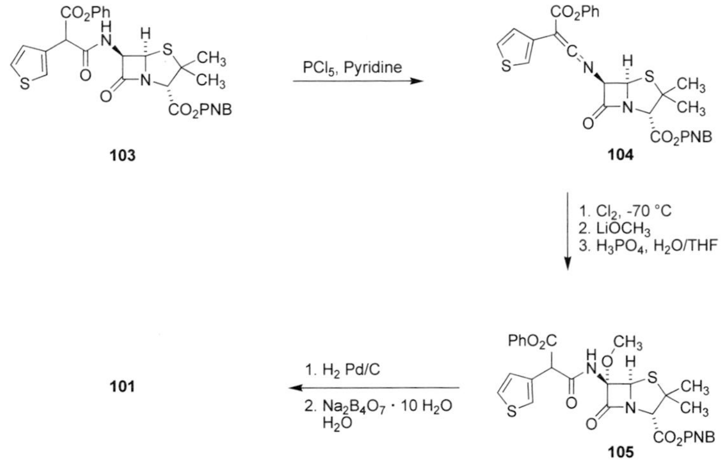

Scheme 2.19 Synthesis of Temocillin from a Ticarcillin derivative.

In the synthesis of Latamoxef the methoxy substituent was introduced in a more direct way by treatment of the amide with *t*BuOCl in the presence of LiOMe (Scheme 2.20) [15]. After introduction of the side chain in the 3-position, the amide substituent was exchanged by hydrolysis of the amide followed by reaction with monobenzhydryl 2-(*p*-methoxyphenyl)malonyl chloride. The two benzhydryl protecting groups were removed by treatment with trifluoroacetic acid in anisole to give **102**.

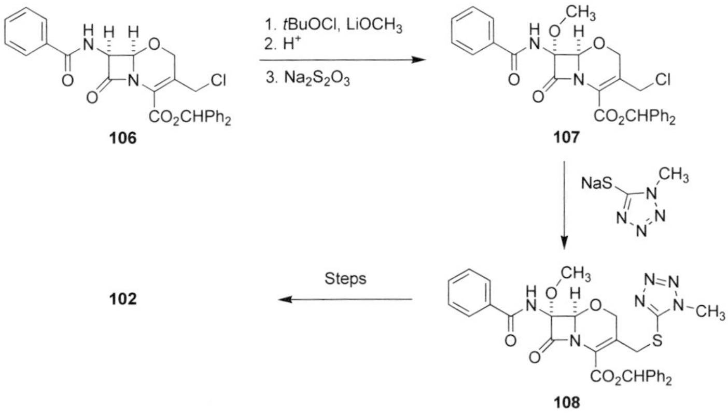

Scheme 2.20 Synthesis of Latamoxef.

2.8
The Tetracyclines

Almost all tetracyclines are semi-synthetic, with the tetracyclic skeleton obtained via fermentation. Recently, Amrubicin hydrochloride, an anti-cancer tetracycline developed by Sumitomo Pharmaceuticals was introduced on the market [57]. This is the first tetracycline to be produced by total synthesis. The compound contains a quaternary stereocenter that was introduced via a Strecker reaction on tetralone **109** (Scheme 2.21). After hydrolysis of the nitrile and esterification, the amino ester was resolved using D-mandelic acid as the resolving agent. A further nine steps took the (*R*)-amino acid ester to Amrubicin hydrochloride **112** [58].

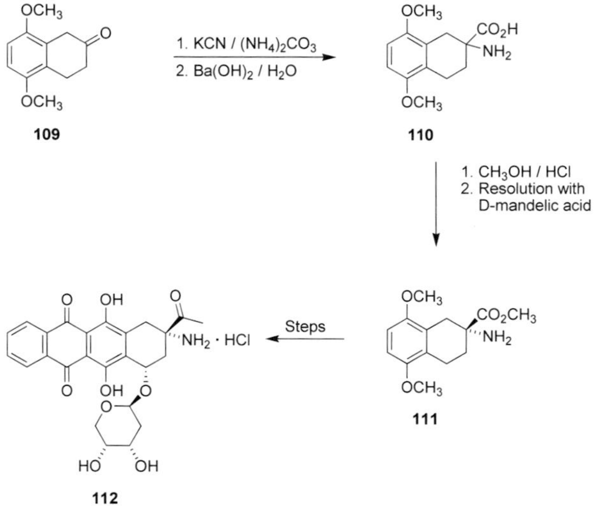

Scheme 2.21 Introduction of a quaternary stereocenter in Amrubicin hydrochloride.

2.9
Summary and Outlook

The above gives an insight into the state of the art of the use of stereoselective synthesis in the industrial production of compounds containing quaternary stereocenters. Disregarding the stereoselectivity aspect for a moment, the synthetic strategies for the establishment of the quaternary centers seem to be fairly limited. We can see the following recurring themes:

1) reactions of reactive organometallic species such as Grignard reagents, alkylzinc reagents or lithium enolates with ketones;
2) reactions of the unhindered "pencil" nucleophiles such as acetylide and cyanide with ketones and imines;
3) nucleophilic or electrophilic substitutions at tetrasubstituted epoxides;

4) electrophilic additions to tri- or tetrasubstituted olefins using reactive reagents such as OsO_4 or peracetic acid, or via Friedel–Crafts addition;
5) radical reactions at CH or NH.

The common denominator of these, appropriate to the hindered nature of the substrates, is reactivity. And indeed, cycloadditions, Michael additions and a host of other reactions are missing for this reason. It is obvious from the above list that further additions would be most welcome.

If we look at the methods used for the introduction of chirality, two methods are used preferentially:

1) resolution of the racemate via crystallization of diastereomeric salts, or occasionally using enzymes;
2) diastereoselective transformations. This is usually a good strategy, as chiral centers stemming from biosynthesis or from well-developed asymmetric synthesis methods can be used.

In only a single case was an asymmetric synthesis used to obtain the compound containing the quaternary stereocenter. At first sight this seems rather disappointing and surprising; however, the scant use of asymmetric catalysis in the production of pharmaceuticals is a well-documented fact and arises from the following reasons [13, 14, 59]:

• Resolution may well be the most economic choice if it is done at an early stage of the total synthesis. In addition, racemisation of the unwanted enantiomer is often possible and increasingly done in situ during the resolution.
• The entire synthesis needs to be economic, not just the step creating the chiral center. If the route using asymmetric catalysis needs more steps it may be less economic for this reason.
• In the development of a process for the production of new pharmaceuticals, time-to-market is of overriding importance in view of the losses a delay in production may cause. Racemate resolution is relatively easy to scale up, whereas catalytic processes may need more time to create a robust process:
• The commercial availability of the catalysts at both gram and kilogram scale at short notice may be a problem.
• The cost of the catalyst may be high.
• The technology may not be sufficiently reliable and robust.
• Licensing may be very costly.

Fortunately, this situation is changing rapidly. The use of high-throughput experimentation in combination with the availability of large libraries of catalysts has provided a solution to the time-to-market problem and, indeed, catalysis has always been one of the possible routes in the process development of the more recent pharmaceuticals.

As for the future outlook, creating economic and robust asymmetric versions of the five reaction types listed above presents an obvious challenge. Both biocatalysis and homogeneous catalysis could play an important role, although the role of biocatalysis would seem to be more limited in view of the non-natural nature of the majority of substrates. In homogeneous catalysis, the development of cost-effective catalysts has only recently taken off. In general, ligands that need to be prepared by multistep syntheses are too expensive, unless the catalyst is active and stable enough to be used at extremely high S/C ratios. We prefer to use ligands that can be prepared in one or two synthetic steps [60]. This is the only way to guarantee that the ligands will be available, both as part of a large library of ligands and on the kilogram scale once production is imminent. Also the use of cheaper metals, such as iron, copper or manganese, is slowly gaining importance. A high reaction rate is the best way to ensure that catalyst costs can be kept at reasonable levels. To improve the rate of a catalytic reaction, mechanistic insight is necessary. Often, the rate is determined by the fact that the catalyst is rapidly deactivated. Other cases are known in which over 99% of the catalyst is in a resting state. These are all interesting challenges for academic research but not many groups have taken up the gauntlet.

This, finally, is the message of this chapter: Much research is needed to bridge the gap between the asymmetric methodologies that have been developed for use in laboratory-scale syntheses, which are well documented in other chapters of this book, and the robust, simple and cost-effective catalysis that is needed to bring asymmetric catalysis into the plant. The economic incentive is obviously there.

References

1 (a) W. G. McBride, *Lancet* **1961**, *2*, 1358; (b) W. Lenz, *Lancet* **1962**, *1*, 45.

2 E. J. Ariëns, *Eur. J. Clin. Pharmacol.* **1984**, *26*, 663.

3 I. Agranat, H. Caner, J. Caldwell, *Nat. Rev. Drug Discov.* **2002**, *1*, 753.

4 H.-U. Blaser, *Adv. Synth. Catal.* **2002**, *344*, 17.

5 P. Mansfield, D. Henry, A. Tonkin, *Clin. Pharmacokinet.* **2004**, *43*, 287.

6 M. J. Cannarsa, *Chimica Oggi* **1999**, *17*, 28.

7 S. Grond, T. Meuser, D. Zech, U. Hennig, K. A. Lehmann, *Pain*, **1995**, *62*, 313.

8 (a) H. U. Blaser, E. Schmidt (eds), *Asymmetric Catalysis on Industrial Scale*, Wiley-VCH, Weinheim, 2004; (b) A. Collins, G. N. Sheldrake, J. Crosby (eds), *Chirality in Industry*, John Wiley & Sons, Chichester, 1992; (c) A. Collins, G. N. Sheldrake, J. Crosby (eds), *Chirality in Industry II*, John Wiley & Sons, Chichester, 1997; (d) D. J. Ager (ed.), *Handbook of Chiral Chemicals*, Marcel Dekker, New York, 1999; (e) R. A. Sheldon, *Chirotechnology*, Marcel Dekker, New York,

1993; (f) R. A Sheldon, *Chem. Ind.* **1990**, 212.

9 J. C. Clark, G. H. Philips, M. R. Steer, *J. Chem. Soc. Perkin Trans. 1* **1986**, 475.

10 (a) G. B. Cox, *Innov. Pharm. Tech.* **2001**, *1*, 131; (b) J. Blehaut, O. Ludemann-Hombourger, S. R. Perrin, *Chimica Oggi* **2001**, *19 (9)*, 24.

11 E.N. Jacobsen, A. Pfaltz, H. Yamamoto (eds), *Comprehensive Asymmetric Catalysis* Vols I–III, Springer, Berlin, 1999.

12 (a) *Adv. Synth. Catal.*, Special Issue on Organic Catalysis, **2004**, *346*, 1007; (b) *Acc. Chem. Res.* Special Issue on Organocatalysis, **2004**, *37*, 487.

13 J. M. Hawkins, T. J. N. Watson, *Angew. Chem. Int. Ed.* **2004**, *43*, 3224.

14 (a) J. G. de Vries, in *Encyclopedia of Catalysis*, ed. I. T. Horvath, John Wiley & Sons, New York, **2003**, Vol. 3, 295; (b) J. G. de Vries and A. H. M. de Vries, *Eur. J. Org. Chem.* **2003**, 799.

15 A. Kleeman, J. Engel, B. Kutscher, D. Reichert (eds), *Pharmaceutical*

Substances. Syntheses, Patents, Applications, 3rd edn, Thieme, Stuttgart, 1999.

16 G. Neef, K. Prezewosky, U. Stache, in *Ullmann's Encyclopedia of Industrial Chemistry*, online version, Wiley-VCH, Weinheim, 2002 (DOI: 10.1002/14356007.a13_089).

17 R. Müller, Steroids, in *Ullmann's Encyclopedia of Industrial Chemistry*, online version, Wiley-VCH, Weinheim, 2002, (DOI: 10.1002/14356007.a25_309).

18 K. Kieslich, *Synthesis* 1969, 120, 147.

19 P. Fernandes, A. Cruz, B. Angelova, H. M. Pinheiro, J. M. S. Cabral, *Enzyme Microb. Technol.* 2003, 32, 688.

20 (a) C. Casas-Campillo, US Patent 3379621, 1968, to Syntex; (b) C. J. Sih, S. S. Lee, Y. Y. Tsong, K. C. Wang, F. N. Chang, *J. Am. Chem. Soc.*, 1965, 87, 2765.

21 N. Appelzweig, *Steroid Drugs*, 1962, 1, 59, 62.

22 (a) G. E. Arth, J. Fried, D. B. R. Johnston, D. R. Hoff, L. H. Sarett, R. H. Silber, H. C. Stoerk, C. A. Winter, *J. Am. Chem. Soc.* 1958, 80, 3161; (b) E. P. Oliveto, R. Rausser, L. Weber, A. L. Nussbaum, W. Gebert, C. T. Coniglio, E. B. Hershberg, S. Tolksdorf, M. Eisler, P. L. Perlman, M. M. Pechet, *J. Am. Chem. Soc.* 1958, 80, 4431.

23 (a) P. L. Julian, W. Cole, E. W. Meyer, B. M. Regan, *J. Am. Chem. Soc.* 1955, 77, 4601; (b) E. P. Oliveto, R. Rausser, H. L. Herzog, E. B. Hershberg, S. Tolksdorf, M. Eisler, P. L. Perlman, M. M. Pechet, *J. Am. Chem. Soc.* 1958, 80, 6687.

24 C. Djerassi, L. Miramontes, G. Rosenkranz, F. Sondheimer, *J. Am. Chem. Soc.* 1954, 76, 4092.

25 C. Rufer, H. Kosmol, E. Schröder, K. Kieslich, H. Gibian, *Liebigs Ann. Chem.* 1967, 702, 141.

26 T. Sonke, B. Kaptein, W. H. J. Boesten, Q. B. Broxterman, J. Kamphuis, F. Formaggio, C. Toniolo, F. P. J. T. Rutjes, in *Stereoselective Biocatalysis*, ed. R. N. Patel, Marcel Dekker, New York, 2000, 23.

27 T. Wirth, *Angew. Chem. Int. Ed. Engl.* 1997, 36, 225.

28 D. Seebach and A. Fadel, *Helv. Chim. Acta* 1985, 68, 1243.

29 U. Schöllkopf, R. Lonsky and P. Lehr, *Liebigs Ann. Chem.* 1985,413.

30 M. J. O'Donnell, *Acc. Chem. Res.* 2004, 37, 506.

31 T. Ooi, M. Takeuchi, M. Kameda, K. Maruoka, *J. Am. Chem. Soc.* 2000, 122, 5228.

32 *The Pesticide Manual*, C.D.S. Tomlin (ed.), British Crop Protection Council, Surrey, 11th edn, 1997.

33 D. H. Johnson, Jr., J. Guerino, Jr., C. Ortlip, Jr., L. S. Quackenbush, US Patent 6060430, 2000, to American Cyanamid Company.

34 P. Genix, J.-L. Guesnet and G. Lacroix, *Pflanzenschutz-Nachrichten Bayer*, 2003, 56, 421.

35 W. H. Kruizinga, J. Bolster, R. M. Kellogg, J. Kamphuis, W. H. J. Boesten, E. M. Meijer, H. E. Schoemaker, *J. Org. Chem.* 1988, 53, 1826.

36 I. Pelta, World Patent WO 00/40545, 2000, to Aventis Cropscience.

37 J. A. Zarn, B. J. Brüschweiler, J. R. Schlatter, *Environmental Health Perspectives* 2003, 111, 255 (doi:10.1289/ehp.5785).

38 A. G. Schmidt, F.-U. Geschke, Antimycotics, in *Ullmann's Encyclopedia of Industrial Chemistry*, On-line version, Wiley-VCH, Weinheim, 2002 (DOI: 10.1002/14356007.a03_077).

39 E. Friederichs, T. Christoph, H. Buschmann, Analgesics and Antipyretics, in *Ullmann's Encyclopedia of Industrial Chemistry*, online version, Wiley-VCH, Weinheim, 2002, (DOI:10.1002/14356007.a02_269).

40 O. Schnider, A. Grüssner, *Helv. Chim. Acta* 1951, 34, 2211.

41 (a) O. Werbitsky, PCT World Patent WO97/03052, 1997, to Lonza; (b) O. Werbitsky, *Chimica Oggi* 1998, 16, (10), 86.

42 M. Kitamura, Y. Hsiao, M. Ohta, M. Tsukamoto, T. Ohta, H. Takayama, R. Noyori, *J. Org. Chem.* 1994, 59, 297.

43 B. Heiser, E. Broger, Y. Crameri, *Tetrahedron: Asymmetry* 1991, 2, 51.

44 O. Schnider, A. Brossi, K. Vogler, *Helv. Chim. Acta* 1954, 37, 710.

45 M. E. Pierce, R. L. Parsons, Jr., L. A. Radesca, Y. S. Lo, S. Silverman, J. R.

Moore, Q. Islam, A. Choudhury, J. M. D. Fortunak, D. Nguyen, C. Luo, S. J. Morgan, W. P. Davis, P. N. Confalone, C. Chen, R. D. Tillyer, L. Frey, L. Tan, F. Xu, D. Zhao, A. S. Thompson, E. G. Corley, E. J. J. Grabowski, R. Reamer, P. J. Reider, *J. Org. Chem.* **1998**, *63*, 8536.

46 A. Choudhury, J. R. Moore, M. E. Pierce, J. M. Fortunak, I. Valvis, P. Confalone, *Org. Process Res. Dev.* **2003**, *7*, 324.

47 S. R. Turner, J. W. Strohbach, R. A. Tommasi, P. A. Aristoff, P. D. Johnson, H. I. Skulnick, L. A. Dolak, E. P. Seest, P. K. Tomich, M. J. Bohanon, M.-M. Horng, J. C. Lynn, K.-T. Chong, R. R. Hinshaw, K. D. Watenpaugh, M. N. Janakiranam, S. Thaisrivongs, *J. Med. Chem.* **1998**, *41*, 3467.

48 B. Jäger, M. Sauter, PCT World Patent WO 02/068404, 2002, to Boehringer Ingelheim Pharma KG.

49 K. S. Fors, J. R. Gage, R. F. Heier, R. C. Kelly, W. R. Perrault, N. Wicnienski, *J. Org. Chem.* **1998**, *63*, 7348.

50 J. Wilken, F. Nerenz, A. Kanschik-Conradsen, US Patent 2004/0110957, **2004**, to Honeywell International, Inc.

51 (a) B. D. Hewitt, M. J. Burk, N. B. Johnson, PCT World Patent WO 00/55150, 2000, to Pharmacia & Upjohn Company; (b) C. J. Cobley, N. B. Johnson, I. C. Lennon, R. McCague, J. A. Ramsden, A. Zanotti-Gerosa, in *Asymmetric Catalysis on Industrial Scale*, ed. H. U. Blaser, E. Schmidt, Wiley-VCH, Weinheim, **2004**, 269.

52 T. M. Judge, G. Phillips, J. K. Morris, K. D. Lovasz, K. R. Romines, G. P. Luke, J. Tulinsky, J. M. Tsitin, R. A. Chrusciel, L. A. Dolak, S. A. Miszak, W. Watt, J.

Morris, S. L. Vander Velde, J. W. Strohbach, R. B. Gammill, *J. Am. Chem. Soc.* **1997**, *119*, 3627.

53 D. R. Cefalo, A. F. Kiely, M. Wuchrer, J. Y. Jamieson, R. R. Schrock, A. H. Hoveyda, *J. Am. Chem. Soc.* **2001**, *123*, 3139.

54 B. M. Trost, N. G. Anderson, *J. Am. Chem. Soc.* **2002**, *124*, 14320.

55 M. Salton, G. D. Shockman (eds), *Beta-Lactam Antibiotics. Mode of action, new developments and future prospects*, Academic Press, New York, **1981**.

56 A. W. Taylor, G. Burton, J. P. Clayton, German Patent DE 2728601, **1978**, to Beecham Group Ltd.

57 J. Li, K. K.-C. Liu, *Mini Rev. Med. Chem.* **2004**, *4*, 207.

58 K. Ishizumi, N. Ohashi, N. Tanno, *J. Org. Chem.* **1987**, *52*, 4477.

59 H.U. Blaser, F. Spindler, M. Studer, *Appl. Catal. A* **2001**, *221*, 119.

60 Some leading references to ligands that can be prepared in 1–2 synthetic steps: (a) C. Claver, E. Fernandez, A. Gillon, K. Heslop, D. J. Hyett, A. Martorell, A. G. Orpen and P. G. Pringle, *Chem. Commun.* **2000**, 961; (b) M. T. Reetz, G. Mehler, *Angew. Chem. Int. Ed.* **2000**, *39*, 3889; (c) M. van den Berg, A. J. Minnaard, E. P. Schudde, J. van Esch, A. H. M. de Vries, J. G. de Vries, B. L. Feringa, *J. Am. Chem. Soc.* **2000**, *122*, 11539; (d) Review: T. Jerphagnon, J.-L. Renaud, C. Bruneau, *Tetrahedron: Asymmetry* **2004**, *15*, 2101; (e) J.G. de Vries, in *Handbook of Chiral Chemicals*, ed. D. J. Ager, S. Laneman, 2nd edn, Marcel Dekker, Inc. New York, **2005**.

3
Aldol Reactions
Bernd Schetter and Rainer Mahrwald

3.1
Introduction

The aldol reaction and the related Claisen reaction are some of the most powerful methods for carbon–carbon bond formation [1]. Particularly since the mid-1980s, intensive research in this field has generated several new routes to stereo- and regiocontrol [2] as well as new catalytic reaction types. Although, overall, many variations of these reactions exist, only a few of them are generally applicable pathways for the creation of quaternary carbon centers.

Formally, there exist two main reaction types to create a quaternary stereocenter via aldol additions (and related reactions). The first pathway is the addition of a carbonyl compound **1** (or a correspondding activated species) with two different α-substituents ($R^2 \neq R^3$) to another carbonyl compound **2**. This method generates a quaternary carbon atom next to the carbonyl group of the aldols **3** and **4** (Scheme 3.1).

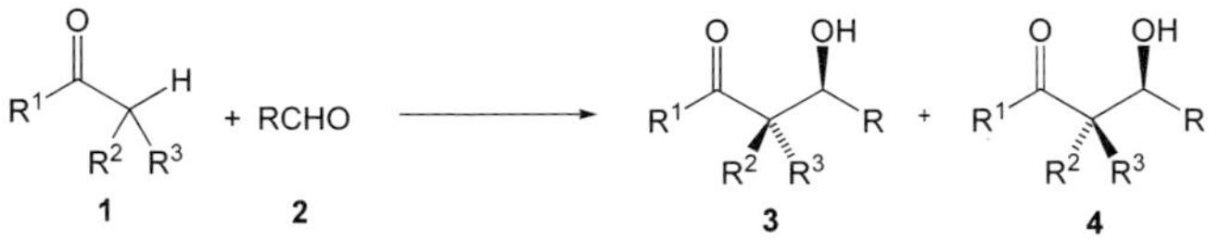

Scheme 3.1

The second pathway (Scheme 3.2) is the aldol addition of an enolizable carbonyl compound **5** to a carbonyl function **6** with two different substituents ($R^2 \neq R^3$), which creates an aldol **7** with an asymmetric tertiary alcohol.

Scheme 3.2

Quaternary Stereocenters: Challenges and Solutions for Organic Synthesis. Edited by Jens Christoffers, Angelika Baro
Copyright © 2005 WILEY-VCH Verlag GmbH & Co. KGaA, Weinheim
ISBN: 3-527-31107-6

The difficulties of these two methods arise from the problems of the formation of defined configured trisubstituted enolates **8** and **9** (Scheme 3.3) and from the lower reactivity of ketones **6** towards nucleophiles.

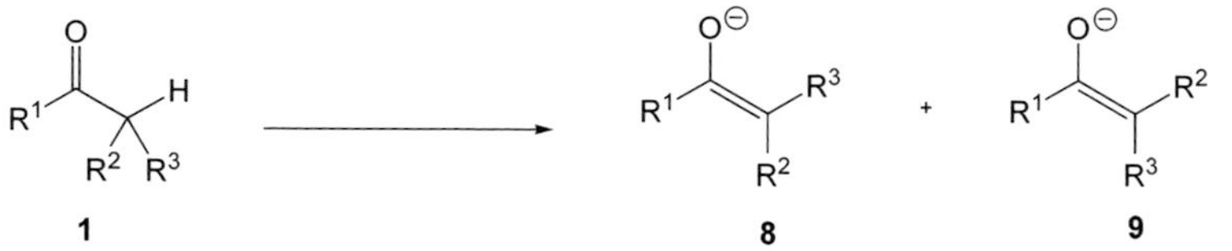

Scheme 3.3

This review will survey equimolar as well as catalytic aldol additions and biochemical methods. Particular attention will be paid to stereoselective variants and the development of stereochemical models to account for enantiomeric enrichment. Applications to natural-product synthesis will be highlighted.

3.2
Metal Enolates

The aldol reaction of di- ($R^1 = H$) or trisubstituted ($R^1 \neq H$) metal enolates **10** to aldehydes **2** creates a quaternary carbon atom in the α-position to the carbonyl group (**13**, **14**; Scheme 3.4) and is expected to be a powerful method for stereo-controlled construction of quaternary carbon centers [3, 4]. The correlation of the enolate geometry and the stereochemistry of the obtained aldols is important for the evaluation of the existing transition-state models, which have been applied so successfully for less-substituted enolates. A chair-like transition state proposed by Zimmerman and Traxler [5] has been well accepted to account for the correlation of the *E/Z* geometry of the enolate to the *syn/anti* stereochemistry in the aldol product. All possible different configurated trisubstituted metal enolates **10** examined by Nakamura et al. (**15–18**; Fig. 3.1) conformed to the Zimmerman–Traxler transition state **11** regardless of the identity of the metal atom [6].

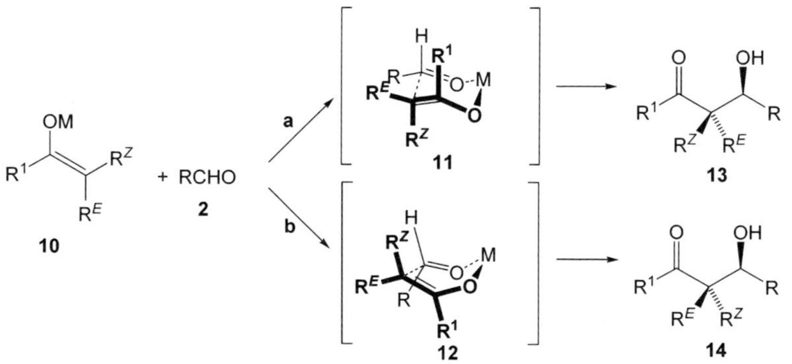

Scheme 3.4

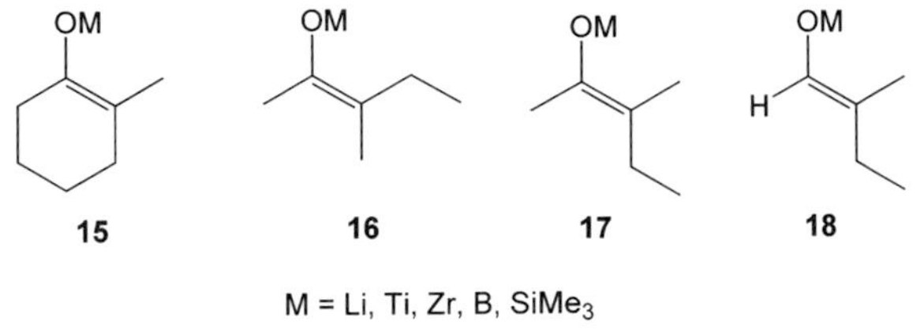

M = Li, Ti, Zr, B, SiMe$_3$

Fig. 3.1 The enolates examined by Nakamura and coworkers [6].

This stands in contrast to the chemistry of less-substituted metal enolates, among which *E* enolates sometimes react via a boat transition state **12** (Scheme 3.4). Destabilization of the latter by the presence of R^Z in the persubstituted enolates is probably the reason behind this experimental observation.

However, the potential of this promising synthesis strategy is often hampered by the selective generation or separation of the di- or trisubstituted enolates. Therefore, some special strategies for the synthesis of enolates with defined *E/Z* configuration were developed, for example the two-electron reduction of thioglycolate lactams **19** (Scheme 3.5) [7].

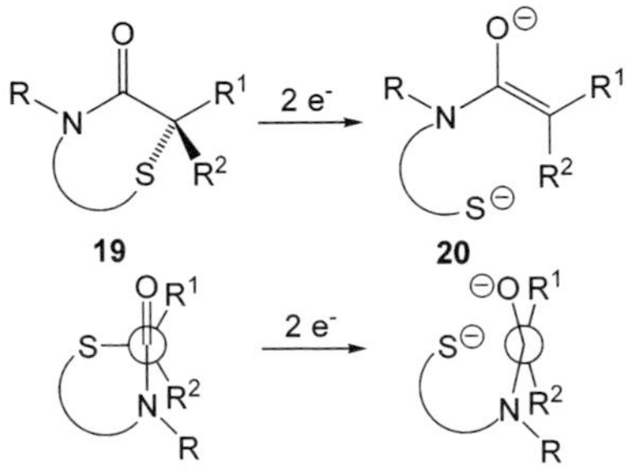

Scheme 3.5

Aldol addition of enolates **21** to unsymmetrically substituted ketones **22** also leads to quaternary carbon centers (**23, 24**; Scheme 3.6). Although aldol additions of aldehydes to enolates are robust methods and the corresponding theory is well developed, aldol additions of ketones to enolates are still largely unexplored, possibly because of the additional complexity of steric differentiation between the two alkyl groups (R^3 and R^4).

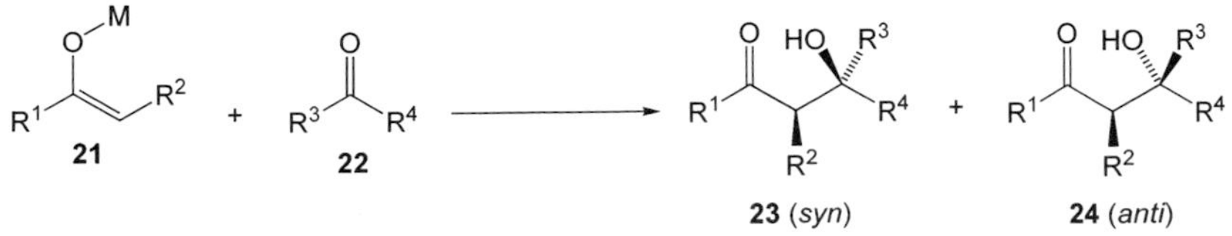

Scheme 3.6

3.2.1
Lithium Enolates

Two systematic studies may serve as examples for the synthetic power of reactions of lithium enolates [8] with aldehydes. The lithium enolates (**15–18**, M = Li; Fig. 3.1) can be generated from the corresponding trimethylsilylenolethers by transmetallation [9]. The results of aldol additions of benzaldehyde with lithium enolates **15–18** are given in Table 3.1 [6]. Not unexpectedly, the trisubstituted enolates were found to react more slowly than their disubstituted counterparts [6].

The use of lithium enolates is limited to those reactions in which a rapid *syn–anti* equilibration of the aldoles via lithium enolates does not take place (entries 1 and 3, Table 3.1), otherwise only poor selectivities are obtained (entries 2 and 4, Table 3.1).

Another interesting method for the formation of lithium enolates is the *in situ* generation and trapping of ketenes **27** from ester enolates **26** (Scheme 3.7) [10].

Reactions investigated by Seebach et al. [10] yielded high diastereoselectivities, but in these cases the authors neither characterized the geometry of the lithium enolate nor explored the configuration of the aldol adduct. The results are summarized in Table 3.2.

The diastereoselectivities are increased by extension of the steric difference between the two alkyl residues of the BHT-ester.

Aldol addition of lithium enolates was successfully employed in the total synthesis of natural products (8-desbromohinckdentine A [11], zaragozic acids [12], the synthesis of natural PI-091, a new platelet inhibitor [13], or in the total synthesis of (–)-verrucarol [14]). Aldol additions of lithium enolates of ketones with butane-2,3-diacetal desymmetrized glycolic acid (5,6-dimethyl-5,6-dimethoxy-1,4-dioxan-2-one) have been described [15]. The obtained aldol adducts can be hydrolyzed to optically pure *anti*-configured 2,3-dihydroxy acids.

3.2.2
Titanium and Zirconium Enolates

Trichlorotitanium enolates can be obtained by transmetallation of the corresponding trimethylsilylenolethers with titanium tetrachloride [16, 17] and triisopropoxytitanium enolates by conversion of lithium enolates with $ClTi(OiPr)_3$.

Table 3.1 Results of aldol additions of benzaldehyde with lithium enolates.

Entry	Enolate	Conditions solvent, *T* (°C), time (s)	*syn:anti*	Yield (%)
1	**15**	THF/hexane, –72, 5	9:91	86
2	**16** (97 % *Z*)	THF/Et$_2$O, –72, 10	45:55	76
3	**17** (86 % *E*)	THF/Et$_2$O, –72, 5	17:83	71
4	**18** (94 % *E*)	THF/Et$_2$O, –72, 5	40:60	30

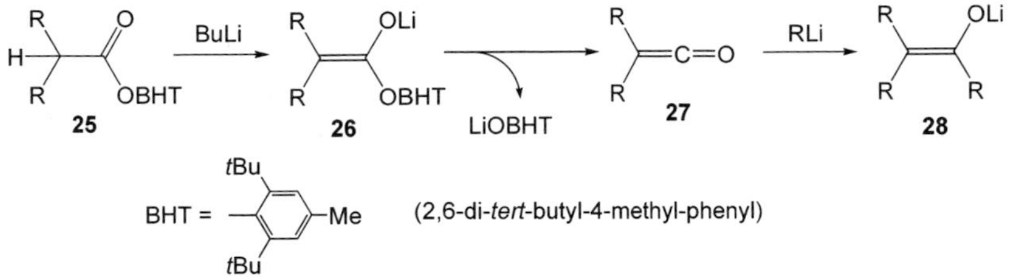

Scheme 3.7

There are only a few examples of the creation of quaternary carbon centers via titanium or zirconium enolates. Nakamura et al. [6] investigated aldol additions of aldehydes to titanium and zirconium enolates (**15–18**; Fig. 3.1). Good to moderate stereoselectivities were obtained, depending on the stereochemistry of the enolates (Table 3.3).

Table 3.2 Aldol adducts of BHT-esters and aldehydes using a lithium nucleophile.

Entry	BHT-ester 25, lithium nucleophile, aldehyde	Aldol product	Yield (%)	Major diastereomer (%)
1	2-methylbutanoate, BnLi, PhCHO	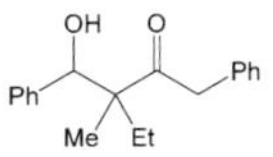 **29**	77	60
2	2,3-dimethylbutanoate, BuLi, PhCHO	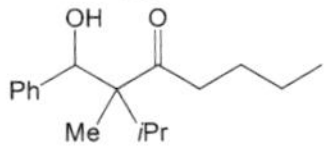 **30**	78	84
3	2,3-dimethylbutanoate, BnLi, C_3H_7CHO	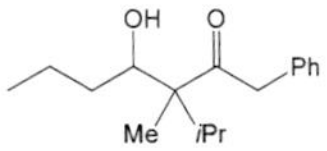 **31**	54	83
4	2,3-dimethylbutanoate, BnLi, PhCHO	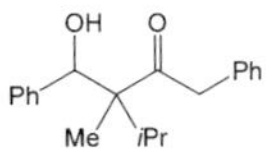 **32**	65	97
5	2,3,3-trimethylbutanoate, BuLi, PhCHO	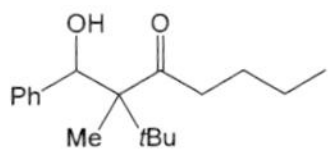 **33**	60	99
6	2,3,3-trimethylbutanoate, BnLi, PhCHO	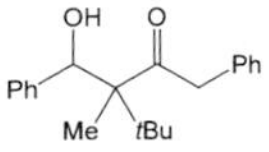 **34**	52	99
7	2,3,3-trimethylbutanoate, BnLi, C_3H_7CHO	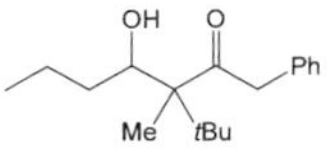 **35**	27	84

Table 3.3 Results of aldol additions to Ti/Zr enolates.

Entry	Enolate	M	R	Conditions Solvent, T (°C), t (h)[a]	syn:anti	Yield (%)
1	15	Ti(OiPr)$_3$	Ph	hexane, –72 , 1	5:95	94
2	15	TiCl$_3$	Ph	CH$_2$Cl$_2$, –72, 5 s	12:88	29
3	15	TiCl$_3$	nPr	CH$_2$Cl$_2$, –72, 5 min	9:91	56
4	15	Cp$_2$ZrCl	Ph	THF/hexane, –72, 1	17:83	56
5	16 (97% Z)	Cp$_2$ZrCl	Ph	THF/Et$_2$O, –72, 3	86:14	68
6	16 (97% Z)	Cp$_2$ZrCl	nPr	THF/Et$_2$O, –72, 1	86:14	34
7	17 (86% E)	Cp$_2$ZrCl	Ph	THF/Et$_2$O, –72, 3	9:91	56
8	17 (86% E)	Cp$_2$ZrCl	nPr	THF/Et$_2$O, –72, 1	15:85	19
9	18 (94% Z)	Cp$_2$ZrCl	Ph	THF/Et$_2$O, –72, 3	31:69	35

a Unless stated otherwise.

Kobayashi et al. [18, 19] investigated the stereochemistry of the aldol addition of lactate derivatives and found a remarkable improvement in the *anti*-selectivity when titanium enolates were employed (Scheme 3.8, Table 3.4).

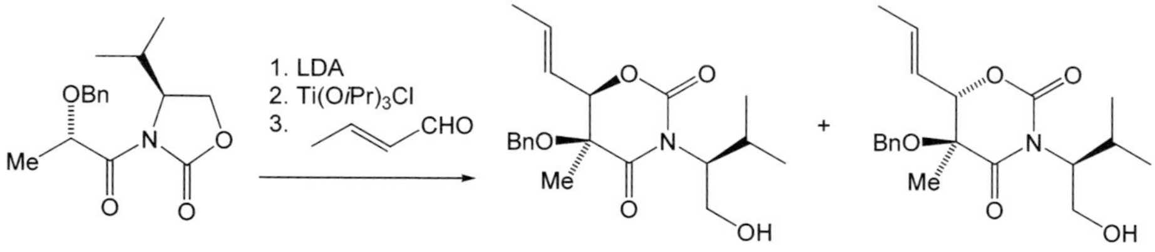

Scheme 3.8

Evans et al. [20] used the aldol addition of titanium enolates in their pathway to zaragozic acids. From the chiral building block tartaric acid, first the trimethylsilylenolether **39** was generated (Scheme 3.9), which was then transformed into the corresponding titanium enolate. The latter could be reacted with several carbonyl compounds (aldehydes **44**, **46**, **48**, **50**, **54** and keto esters **52**). The systematic results reported in Table 3.5 may be rationalized by invoking the proposed transition-state model (**41–43**; Scheme 3.9). The chair-like transition

Table 3.4 Changes in *anti*-selectivity of aldol addition of lactate derivatives using Ti enolate.

Entry	Equiv. Ti(OiPr)$_3$Cl	Temperature (°C)	Yield (%)	anti:syn
1	0	–78	69	1:1
2	1.1	–78 to –40	52	8:1
3	3	–78 to –40	63	20:1

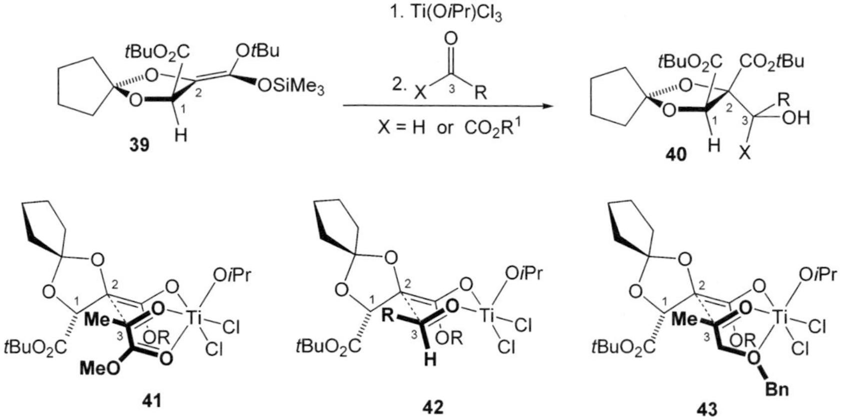

Scheme 3.9

Table 3.5 Stereochemical inductions involving Ti enolates.

Entry	Carbonyl compound	Aldol adduct	C-3 selectivity	Yield (%)
1	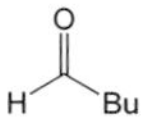 **44**	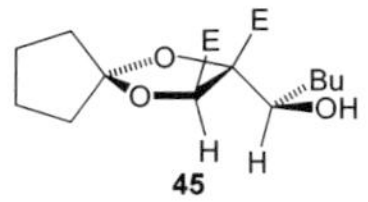 **45**	91:9	48
2	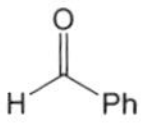 **46**	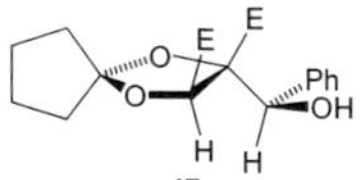 **47**	88:12	67
3	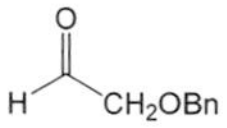 **48**	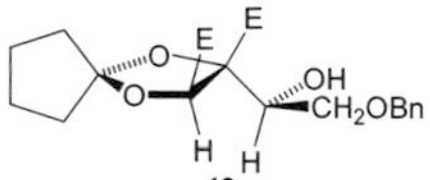 **49**	91:9	64
4	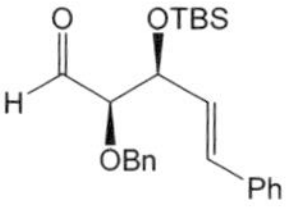 **50**	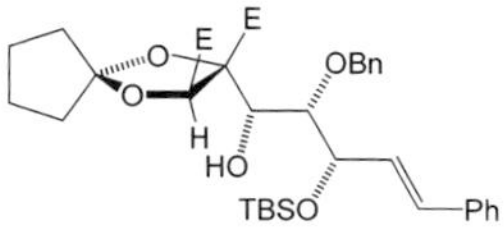 **51**	95:5	80
5	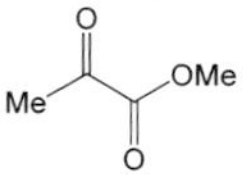 **52**	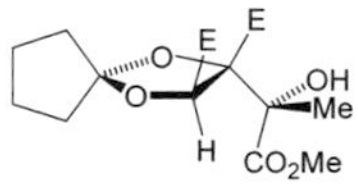 **53**	95:5	66
6	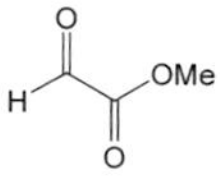 **54**	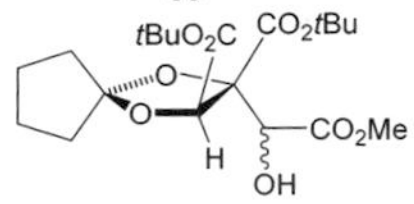 **55**	50:50	44

E = CO$_2$*t*Bu

state **41** orients the α-keto ester on the face of the tartrate enolate opposite the *tert*-butyl ester. Thus, the methyl ester of the electrophile occupies a pseudo-axial orientation, allowing chelation to titanium. The stereoinduction observed for unfunctionalized aldehydes (entries 1 and 2) may be rationalized when these substrates are in an orientation such that the alkyl or aryl group occupies a pseudo-equatorial position in the closed transition state **42**.

On the other hand, benzyloxyacetaldehyde realizes its chelate potential via pseudo-axial orientation of the benzyl ether **43**. An identical arrangement leads to the selectivity observed with a more highly functionalized aldehyde (entry 4). Perhaps the highly activated nature of the aldehyde in entry 6 may lead to a less-rigid transition state and thus the loss of stereochemical induction.

3.2.3
Boron Enolates

Systematic investigations of the reaction behavior of borinates (M = BR$_2$; Fig. 3.1) and borates (M = B(OR)$_2$; Fig. 3.1) carried out by Nakamura et al. are summarized in Table 3.6 [6].

In reactions of the BBu$_2$-enolate of 2-methylcyclohexanone **15** excellent *anti*-selectivity is observed (entries 1 and 2), whereas the corresponding cyclohexanone BBu$_2$-enolate is only 67% *anti*-selective [21]. The reactivity of borate enolates was found to be dependent on the nature of the ligand. Thus, dimethylborate enolate is almost inert to benzaldehyde at room temperature (entry 4), while a cyclic borate (entry 3) reacts even at –72°C.

The lithium enolate obtained by the method of Gleason et al. [7] (Schemes 3.5 and 3.10) can be easily transmetallated with various boron species. The effect on the *syn/anti* selectivity and the diastereomeric excess (*de*) (two diastereomeric *syn*-**57** and two diastereomeric *anti*-**58** products are possible owing to the already existing stereocenter in the enolate) of the aldol reactions is summarized in Table 3.7 [22].

If one considers the envelope conformation of the pyrrolidine ring (**59**, **60**; Scheme 3.10), two possible transition-state models can be discussed. In both

Table 3.6 The reaction behaviors of borinates and borates.

Entry	Enolate	M	R	Solvent, *Temp.* (°C), time (h)	*syn:anti*	Yield (%)
1	**15**	BBu$_2$	Ph	Et$_2$O, –72, 2	0.5:99.5	60
2	**15**	BBu$_2$	*n*Pr	Et$_2$O, –72, 2	3:97	31
3	**15**	B(OCH$_2$CH$_2$O)	Ph	THF/hexane/CH$_2$Cl$_2$, –72, 2	12:88	57
4	**15**	B(OMe)$_2$	Ph	THF/C$_6$H$_{12}$, 4 kbar, 3	3:97	50
5	**16** (97% *Z*)	BBu$_2$	Ph	Et$_2$O, –72, 1.5; 0, 0.5	94:6	30
6	**17** (86% *E*)	BBu$_2$	Ph	Et$_2$O, –72, 1.5; 0, 0.5	3:97	41
7	**18** (94% *E*)	BBu$_2$	Ph	Et$_2$O, –72, 1.5; 0, 0.5	33:67	34

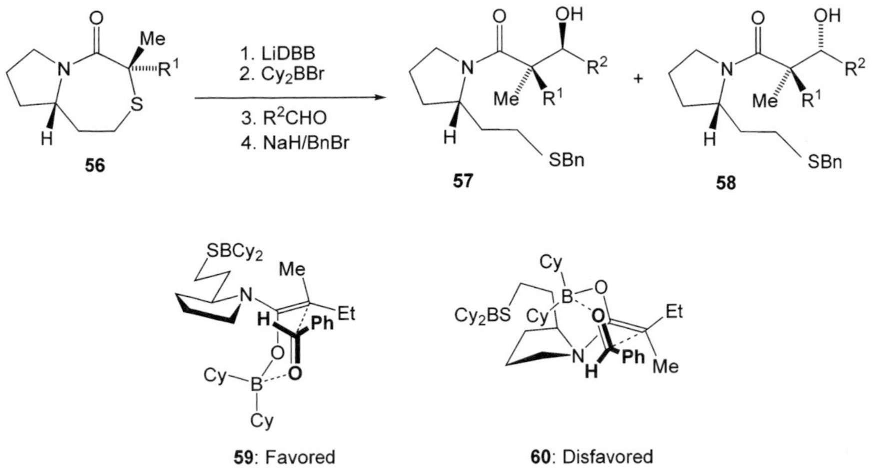

Scheme 3.10

these structures the enolate is held in a pseudo-equatorial position to minimize steric interactions of the enolate with the pyrrolidine ring. Transition state **59** is favored over **60**, as the latter would have significant *syn*-pentane interactions between the enolate oxygen and the pseudo-axial thioethylene chain.

Recently, Eilbracht and Keränen reported a method for the preparation of five-, six- and seven-membered rings by a regioselective hydroformylation of species

Table 3.7 Results of aldol reactions with boron enolates.

Entry	R^1	R^2	Additive (equiv.)	syn:anti	de (%)	Yield (%)
1	Et	Ph	none	52:48	23	71
2	Et	Ph	Bu$_2$BOTf (2.2)	79:21	95	48
3	Et	Ph	Bu$_2$BOTf (3)	83:17	92	39
4	Et	Ph	Cy$_2$BOTf (2.2)	69:31	30	46
5	Et	Ph	Cy$_2$BOTf (3)	91:9	93	53
6	Et	Ph	Bu$_2$BCl (2.1)	75:25	81	63
7	Et	Ph	Cy$_2$BCl (1.1)	73:27	50	n.d.
8	Et	Ph	Cy$_2$BCl (2.1)	92:8	96	31
9	Et	Ph	Cy$_2$BBr (1.1)	55:45	93	n.d.
10	Et	Ph	Cy$_2$BBr (2.1)	91:9	94	80
11	Et	4-(MeO)C$_6$H$_4$-	Cy$_2$BBr (2.1)	91:9	98	83
12	Et	4-Br C$_6$H$_4$-	Cy$_2$BBr (2.1)	93:7	99	81
13	Et	(E)-PhCH=CH-	Cy$_2$BBr (2.1)	91:9	99	63
14	Et	CH$_2$=C(Me)-	Cy$_2$BBr (2.1)	98:2	95	44
15	Et	(E)-MeCH=C(Me)-	Cy$_2$BBr (2.1)	91:9	91	95
16	allyl	Ph	Cy$_2$BBr (2.1)	92:8	91	71
17	Bn	Ph	Cy$_2$BBr (2.1)	91:9	99	91

n.d., not determined

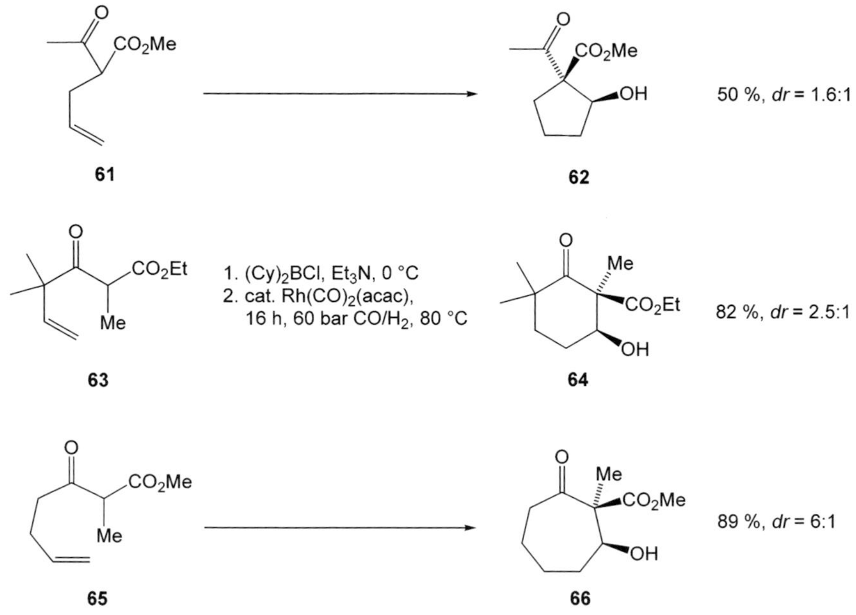

Scheme 3.11

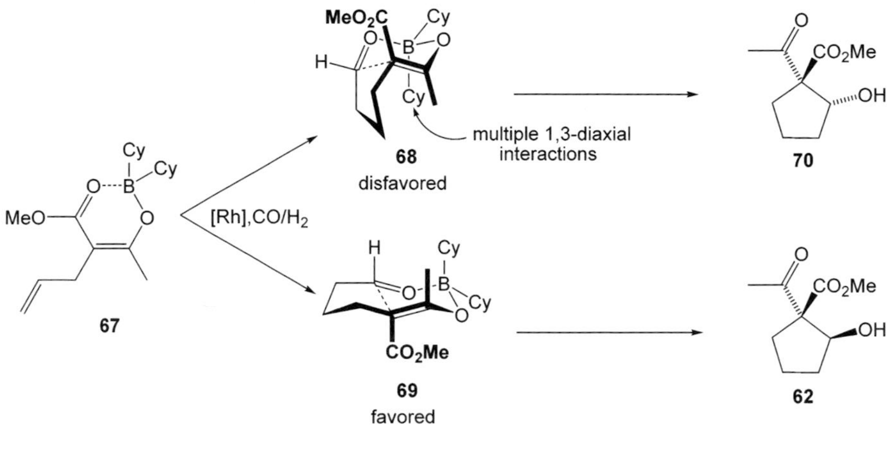

Scheme 3.12

61, **63** and **65** followed by an intramolecular aldol addition (Scheme 3.11) [23]. This reaction is carried out as a one-pot reaction. For the formation of the five-membered ring adduct **62**, one can assume a ring-flipped transition state **69** that minimizes the 1,3-diaxial interactions of the cyclohexyl residues, as depicted in Scheme 3.12. For six- and seven-membered ring formation, the chelate transition state **72** should be considered (Scheme 3.13). After the hydroformylation,

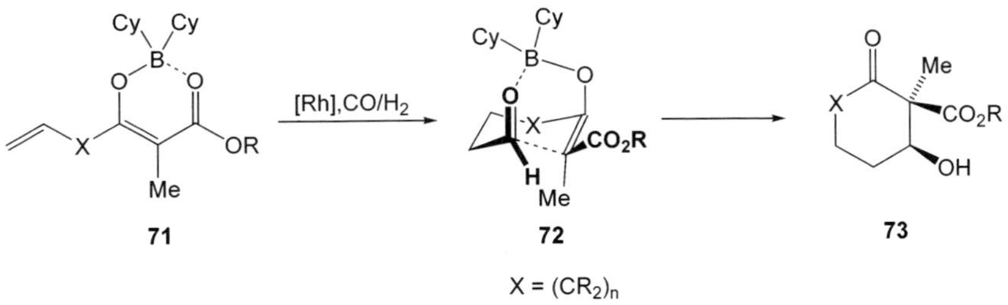

Scheme 3.13

chelation switches from the ester group to the aldehyde, resulting in a rigid bicyclic transition state for the aldol addition.

Further examples of aldol reactions via boron enolates are the addition of aldehydes to trisubstituted 2-alkoxycarbonyl allylboronates to yield α-methylene -γ-lactones [24] and the preparation of chromane carboxylate [25].

3.3
Catalytic Aldol Additions

Catalytic aldol reactions [4] can be divided into two main types: metal catalysis and organocatalysis. Metal catalysis can be divided into two main parts. One is the conversion of trimethylsilylenolether and related activated species with carbonyl compounds in the presence of catalytic amounts of a Lewis acid (Mukaiyama reaction) or fluoride ions (Scheme 3.14). The other type is the direct aldol addition of nonactivated carbonyl compounds by a catalytic species. The latter is of special interest owing to the benefit of omitting the preparation of enolethers. This avoids waste and makes this pathway particularly interesting for "green chemistry".

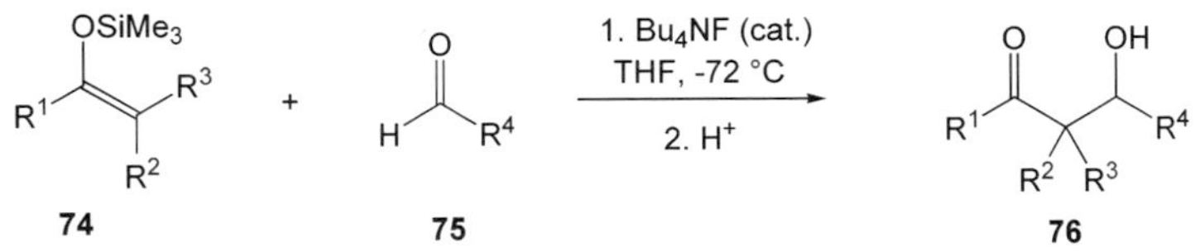

Scheme 3.14

3.3.1
Fluoride-ion-mediated Aldol Addition

A variety of silyl enol ethers **74** and carbonyl compounds **75** undergo a fluoride-catalyzed reaction to give aldols **76** (Scheme 3.14, Table 3.8) [26].

The reaction with tetrabutylammonium fluoride (TBAF) proceeds smoothly even at –78°C. There was found to be an optimum range in the amount of the catalyst. A stoichiometric amount of TBAF caused a considerable decrease in the

Table 3.8 Results of fluoride-catalyzed reactions of silyl enolethers with carbonyl compounds.

Entry	Silyl enol ether (geometrical purity)	Aldehyde/ ketone	Aldol	Temp. (°C)	Time (h)	Yield (%)	*syn:anti*
1	**77**	PhCOCOPh	**78**	−25	20	31	n.d.
2	**79**	PhCHO	**80**	−78	2.5	69	3:7 or 7:3 (not assigned)
3	**81**	R^4CHO	**82**	−30 −72	2 2	R^4 = Ph 62 R^4 = nPr 49	7:3 86:14
4	**83**	PhCHO	**84**	−35	1.5	68	4:6
5	**85** (83 %)	R^4CHO	**86**	−72	0.5 2	R^4 = Ph 68 R^4 = nPr 65	29:71 22:78
6	**87** (97 %)	R^4CHO	**88**	−72	0.5 2	R^4 = Ph 97 R^4 = nPr 60	64:36 68:32
7	**89** (93 %)	PhCHO	**90**	−72	2	68	35:65

n.d., not determined.

product yield. Too small amounts caused the formation of unwelcome by-products. Thus, good results were obtained by using 5–20 mol% of TBAF. It is notable that, in contrast to lithium enolates, the reaction tolerates a wide range of reaction conditions. For example, the reaction was carried out with equal success from −30 to −70°C and over times ranging from 2 to 4 h. The fluoride ion is only moderately basic [27, 28] and did not act as a base under these reaction conditions.

The fluoride-catalyzed aldol reaction of silyl enol ethers is considered not to proceed via chelate (cyclic) transition states because of the presumed intermediacy of metal-free enolates [29]. The existence of free enolate anions under these reaction conditions has been demonstrated [30, 31] and an extended-transition-state model has been proposed to account for the observed diastereoselectivity (Fig. 3.2) [32–34].

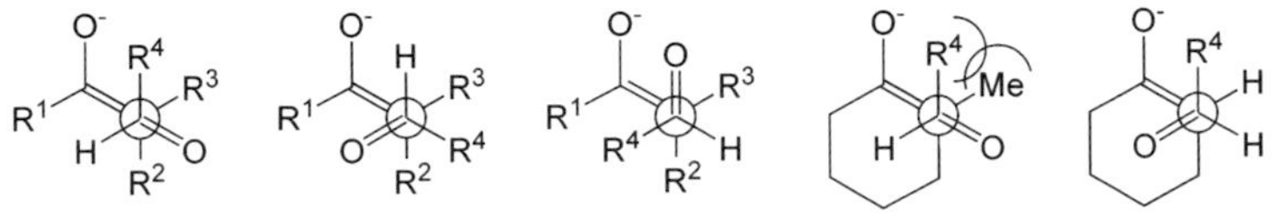

91   **92**   **93**   **94**: Disfavored *syn*   **95**: Disfavored *anti*

Fig. 3.2 Open transition states in fluoride-mediated aldol addition.

Transition state model **91** is only applicable for enolates with sterically bulky substituents, whereas model **92** should be considered for less-hindered enolates. The *syn*-producing behavior of the cyclic enolate (entry 3, Table 3.8) was found to be consistent with transition state model **91**. The lower degree of selectivity compared with the simple cyclohexanone enolate **95** is probably due to the additional gauche interaction caused by the 2-methyl group shown in **94** [29].

On the contrary, the *E*-acyclic enolates (entries 5 and 7, Table 3.8) were found to be *anti*-selective, while the *Z*-enolate (entry 6) is still *syn*-selective. This *anti*-selectivity sharply contradicts with the transition state model **91** and therefore transition state **92** had to be proposed. *Ab initio* calculations resulted in equal energies for transition states **91** and **92**, whereas the third possible transition state **93** is disfavored by about 3.3 kcal mol^{-1} [35].

3.3.2
Lewis-acid-mediated Mukaiyama-type Aldol Reactions

The original Mukaiyama reaction [36] is the titanium(IV) chloride-mediated reaction between silylenolethers **97** and carbonyl compounds **96** (Scheme 3.15). Despite its remarkable power as a method for carbon–carbon bond formation, the level and sense of its stereoselectivity often vary. The *syn/anti* ratio of the aldol products **98** is affected by the nature of the aldehyde and silylenolate, and the character of the Lewis acid catalyst. One of the few examples of this original reaction leading to a quaternary stereocenter is given in Scheme 3.16. The obtained aldol derivative **101** can easily be transformed into substituted furanes **102** [37].

Several other Lewis acids may also be employed instead of titanium(IV) chloride, among them chiral ones that achieve remarkable enantioselectivities in aldol additions.

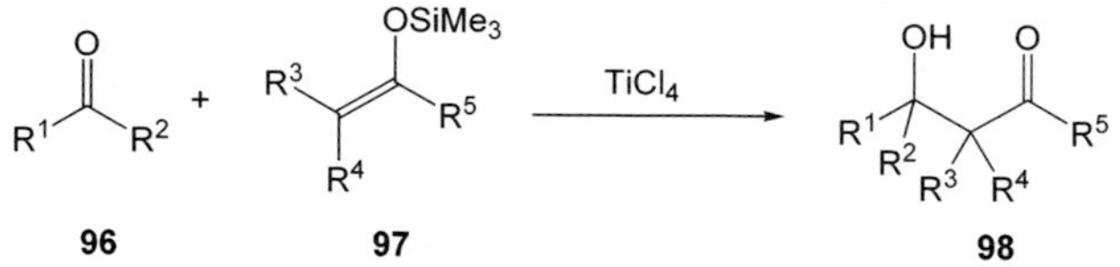

Scheme 3.15

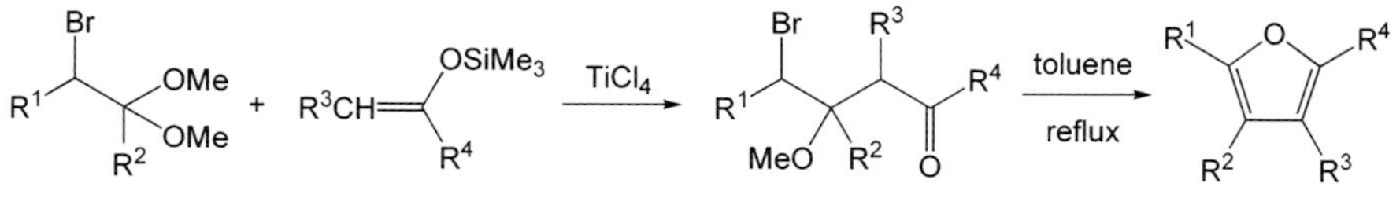

Scheme 3.16

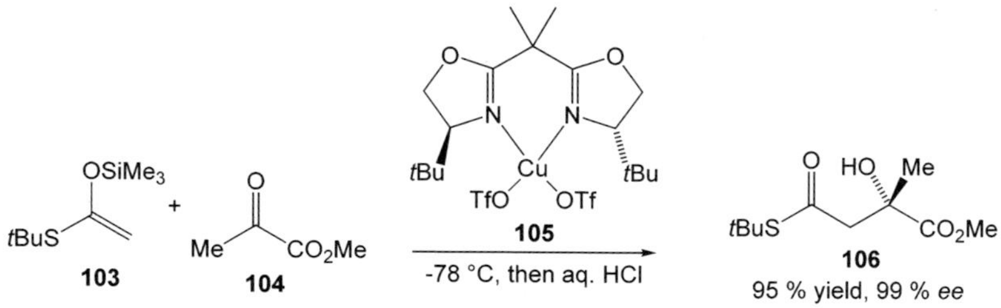

Scheme 3.17

Evans et al. describe the first catalytic, enantioselective addition of the silylenolether **103** to pyruvate esters **104** (Scheme 3.17) mediated by copper Lewis acid [38].

The most effective catalyst for this reaction with regard to yield and control of the enantioselectivity is the (*t*Bu-box)Cu(OTf)$_2$ complex **105**. Some more product examples for this reaction are given in Fig. 3.3.

The enantioselectivities of the pyruvate ester additions are all consistent with the stereochemical model shown in Fig. 3.4. The bulky *tert*-butyl group effectively shields the *re* face of the keto-functionality in the copper complex **115**, directing nucleophile addition to the *si* face.

The diastereoselectivity can be rationalized by an open, antiperiplanar transition state structure **118** (Scheme 3.18) that minimizes steric interactions between the enolether substituent and the pendent ligand substituent. Therefore, the disposition of the –OSiMe$_3$ and the –SR group is less important.

An example of tin-promoted aldol reactions, which lead to stereotriades **122** containing one quaternary stereocenter, is the reaction of ketene silyl acetals **121** with benzyloxypropionaldehyde **120** (Scheme 3.19).

Chiral diamine–Sn(II) complexes (for example **125**) are prepared *in situ* by coordination of chiral pyrrolidine derivatives to Sn(OTf)$_2$. They make it possible to achieve high stereo- and enantioselectivities in the aldol reaction of ketene silyl

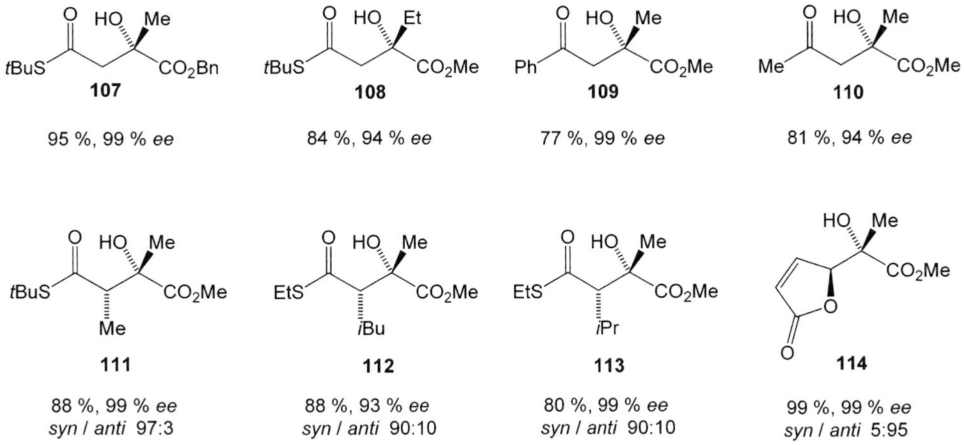

Fig. 3.3 Examples of products obtained by the (*t*Bu-box)Cu(OTf)₂ complex **105**-mediated aldol reaction.

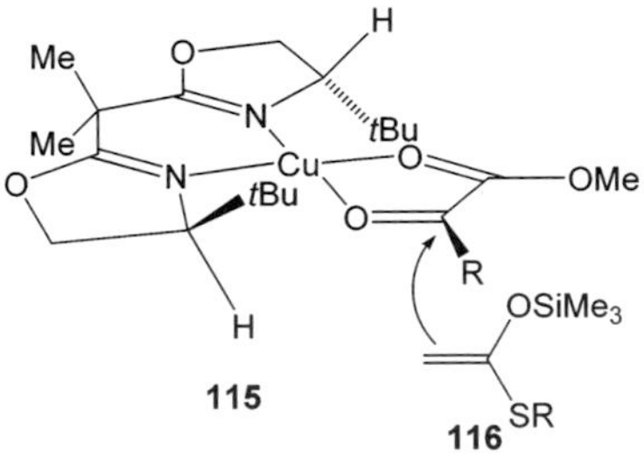

Fig. 3.4 Explanation of the enantioselectivity of the (*t*Bu-box)Cu(OTf)₂ complex **105**-mediated aldol reaction.

acetals **124** [39, 40]. The synthetic power of this reaction is applied in the total synthesis of buergerinine F **129** (Scheme 3.20) [41].

3.3.3
Direct Aldol Additions

Since the mid-1990s, various methods for direct aldol addition have been developed, to avoid the need for preconversion of the carbonyl compounds into more reactive species. Early examples of the creation of quaternary carbon centers were given by Miyoshi et al. [42] and Mahrwald and Gündogan [43]. Miyoshi et al. described the BiCl₃-Zn mediated addition of aldehydes to α-diketones **130** to afford α,β-dihydroxyketones **131** and **132** (Scheme 3.21, Table 3.9).

Mahrwald et al. described the regioselective TiCl₄-mediated aldol addition [43, 44] at the more encumbered side of unsymmetrical ketones **133** (Scheme 3.22). The asterisk marks the more encumbered side.

The results leading to quaternary carbon centers are listed in Table 3.10.

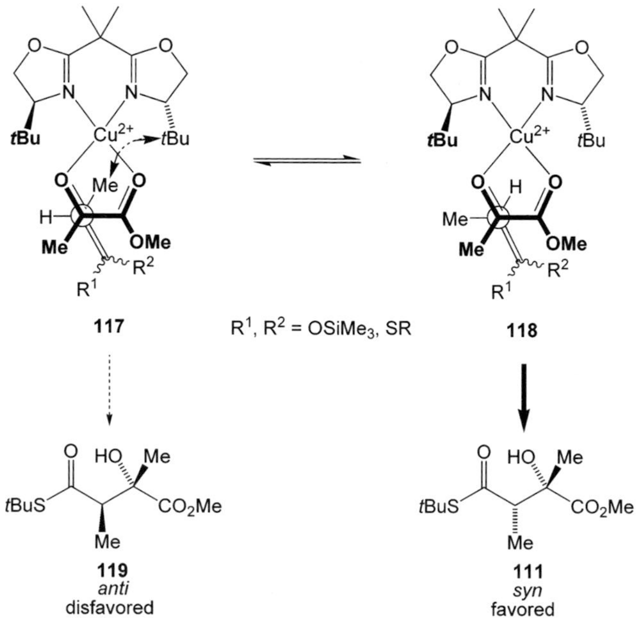

117 ⇌ **118**

R^1, R^2 = OSiMe₃, SR

119
anti
disfavored

111
syn
favored

Scheme 3.18

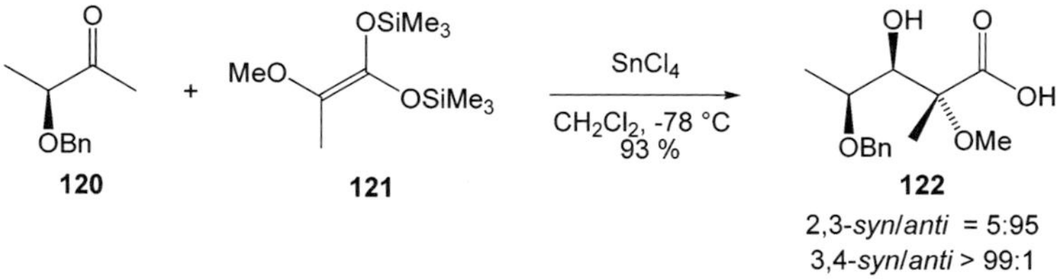

120 + **121** $\xrightarrow[\substack{CH_2Cl_2, -78\ ^\circ C \\ 93\ \%}]{SnCl_4}$ **122**

2,3-*syn*/*anti* = 5:95
3,4-*syn*/*anti* > 99:1

Scheme 3.19

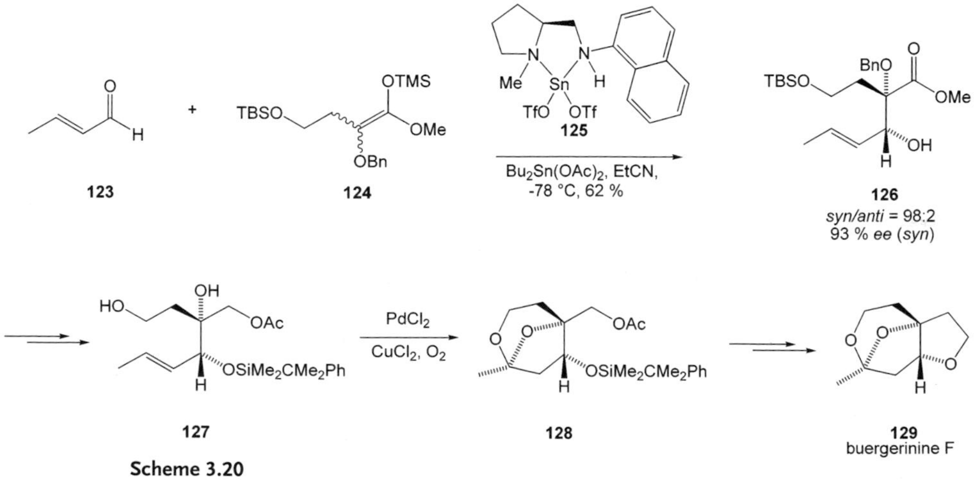

123 + **124** $\xrightarrow[\substack{Bu_2Sn(OAc)_2,\ EtCN, \\ -78\ ^\circ C,\ 62\ \%}]{125}$ **126**

syn/*anti* = 98:2
93 % *ee* (*syn*)

127 $\xrightarrow[CuCl_2,\ O_2]{PdCl_2}$ **128** → **129**
buergerinine F

Scheme 3.20

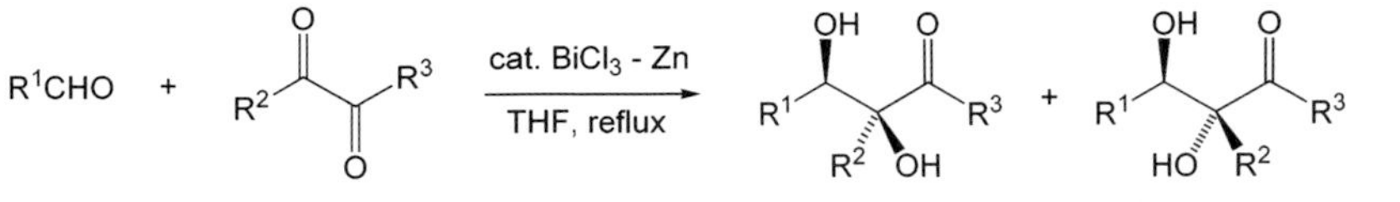

Scheme 3.21

Table 3.9 Results of direct aldol additions of aldehydes to α-diketones.

Entry	R^1	R^2	R^3	Yield (%)	syn:anti
1	PhCH$_2$CH$_2$-	Me	Me	85	26:74
2	Ph	Me	Me	63	32:68
3	H$_3$C(CH$_2$)$_7$-	Me	Me	61	32:68
4	PhCH$_2$CH$_2$-	Et	Et	63	15:85
5	PhCH$_2$CH$_2$-	Me	Et	93[a]	26:74[b]
6	PhCH$_2$CH$_2$-	Me	MeO	54	33:67

a Two regioisomeres 71:29; b *syn:anti* ratio for the major regioisomer.

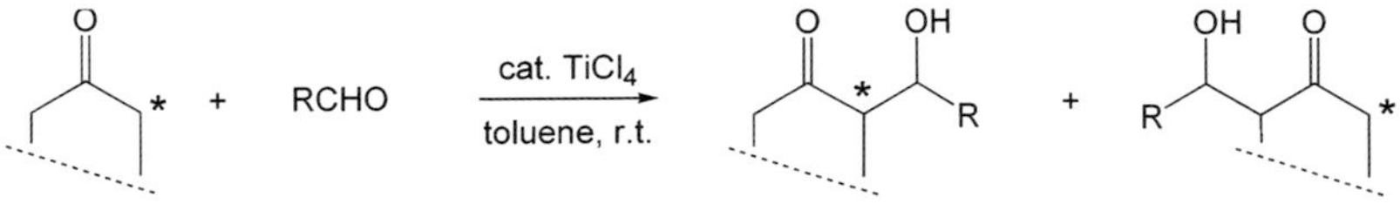

Scheme 3.22

Related to the Mukaiyama reaction is the direct Me$_3$SiCl-mediated aldol addition with TiCl$_4$-R$_3$N-reagent or ZrCl$_4$, and related reactions (Scheme 3.23) [45–48].

In this case, oxygenated ketones react with ketones, ketones or α-chloroketones with aldehydes and α-oxygenated ketones with ketones or aldehydes. Exemplary reactions leading to quaternary carbon centers (Scheme 3.24) are given in Table 3.11. In contrast to the original Mukaiyama reaction, even sterically overcrowded aldehydes and ketones give acceptable yields (entries 9–11).

The conversion of cyclohexanone **143** and β-keto ester **142** may serve as an example of this aldol-type reaction, as illustrated in Scheme 3.25.

An asymmetric, aldol-type reaction leading to β-hydroxy-α-alkylamino acids was presented by Ito et al. [49], who found that gold(I) complexes with chiral ferrocenylphosphanyl ligands **145** (Fig. 3.5) are effective catalysts for an asymmetric aldol reaction of isocyanocarboxylates **146** with aldehydes forming optically active 2-oxazoline-4-carboxylates **147** and **148** (Scheme 3.26). These can easily be transformed to the corresponding α-amino acids.

This reaction is mostly *anti*-selective, but the obtained diastereoselectivity decreases with the steric demand of R^1 and is dependent on the nature of the ligand (NR$_2$). Formation of the *anti*-configured reaction products **147** is favored by high enantiomeric excess (Table 3.12). The reaction time increases with the steric demand of R^1.

Table 3.10 Regioselective aldol additions to unsymmetrical ketones.

Entry	Ketone	Aldehyde	Yield	Ratio 134:135	*syn:anti* ratio of the main product
1	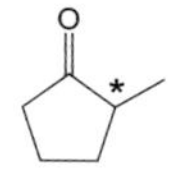 **136**	PhCHO	88	99:1	98:2
2		iPrCHO	81	95:5	84:16
3	**137**	PhCHO	67	89:11	75:25
4	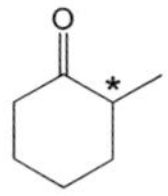	iPrCHO	71	88:12	67:33
5		nPrCHO	62	86:14	73:27

Scheme 3.23

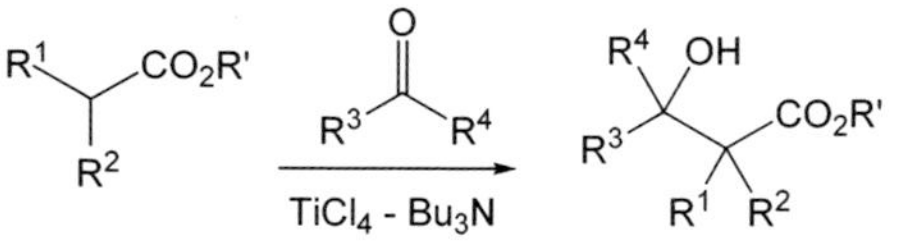

Scheme 3.24

Table 3.11 Results of Me$_3$SiCl-mediated aldol additions.

Entry	R^1	R^2	R^3	R^4	R^5	Yield (%)	*syn:anti*
1	Ph	H	H	Ph	Et	86	–
2	Ph	Me	H	Ph	Me	95	99:1
3	Ph	Me	H	Et	Et	86	n.d.
4	Ph	Me	H	nPr	nPr	81	n.d.
5	Ph	Me	H	Ph	CH$_2$Cl	91	99:1
6	Et	Me	H	Ph	Me	84	84:16
7	Et	Me	H	Ph	Et	92	77:28
8	nPr	Et	H	Me	nHex	60	60:40
9	iPr	Me	Me	iPr	H	87	n.d.
10	iPr	Me	Me	tBu	H	51	n.d.
11	iPr	Me	Me	Me	nHex	45	n.d.
12	Ph	Cl	Me	iPr	H	63	92:8
13	Me	TBSO	H	Ph	CH$_2$Cl	78	–

n.d., not determined.

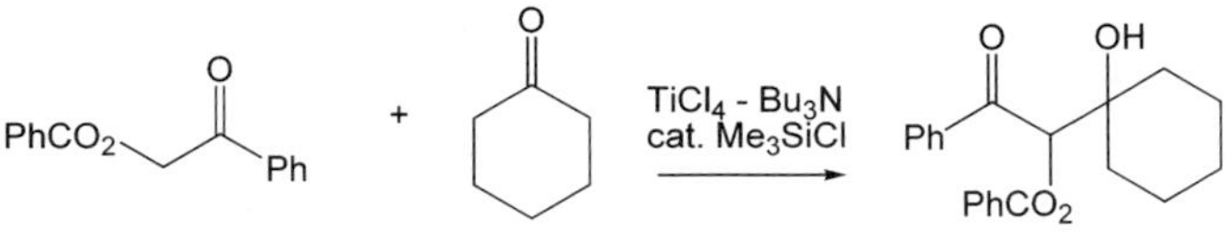

142 **143** **144**

Scheme 3.25

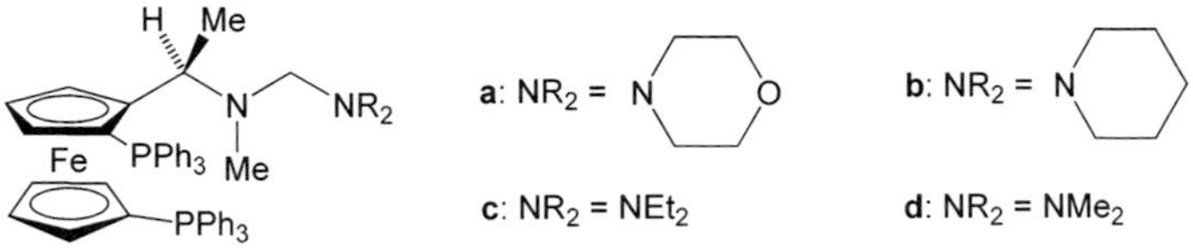

a: NR₂ = morpholine
b: NR₂ = piperidine
c: NR₂ = NEt₂
d: NR₂ = NMe₂

145

Fig. 3.5 Chiral ferrocenylphosphanyl ligand for enantioselective gold(I)-mediated aldol additions.

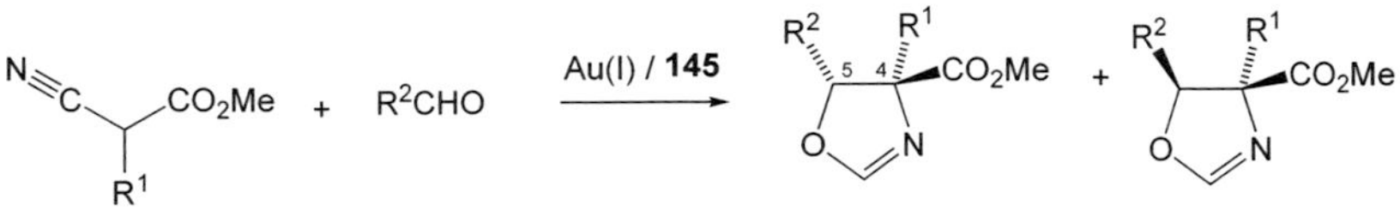

146 **147** **148**

Scheme 3.26

The formation of the *anti*-oxazolines **147** can be explained by the intermediate **149** and the *syn*-species by the intermediate **150** (Fig. 3.6). In both **149** and **150** attack takes place preferentially on the *si* face of the donor center of the enolate.

Table 3.12 The effect of the ligand on the diastereoselectivity of asymmetric aldol-type reactions.

Entry	R¹	R²	Ligand (NR2)	Reaction time (h)	Yield (%)	Ratio anti:syn	ee anti	ee syn
1	Me	Ph	a	67	97	93:7	94	53
2	Me	Ph	b	43	92	88:12	90	5
3	Me	Ph	c	96	90	77:23	82	26
4	Me	Ph	d	65	95	82:18	92	44
5	iPr	Ph	a	330	86	62:38	88	17
6	iPr	Ph	b	280	86	54:46	92	28
7	iPr	Ph	c	100	87	52:48	85	42
8	iPr	Ph	d	200	95	50:50	88	48
9	Me	Me	a	41	86	56:44	86	54
10	Me	Me	b	65	94	44:56	44	6
11	Me	Me	d	94	100	38:62	46	49
12	Et	Me	a	62	92	54:46	87	66
13	iPr	Me	a	260	100	24:76	26	51
14	iPr	Me	b	290	100	22:78	35	23

Nucleophilic attack on the *si* face of aldehydes, which is expected for small substituents R^1, leads to *anti*-oxazolines **147**. Bulky substituents R^1 such as isopropyl are likely to make **149** less favorable, owing to steric interactions between R^1 and R^2. Thus, the nucleophilic attack takes place on the *re* face, **150** (Fig. 3.6), to produce *syn*-oxazolines **148**.

Related to the reaction above is the enantioselective aldol addition of aldehydes and 2-cyanopropionates **151**, catalyzed by a chiral diphosphanyl-rhodium(I) complex to give quaternary chiral carbon centers at the α-position of nitriles (Scheme 3.27) [50]. Complete conversion was obtained with 1 mol % of a rhodium catalyst containing the chiral (S,S)-(R,R)-TRAP-ligand **154** (Fig. 3.7).

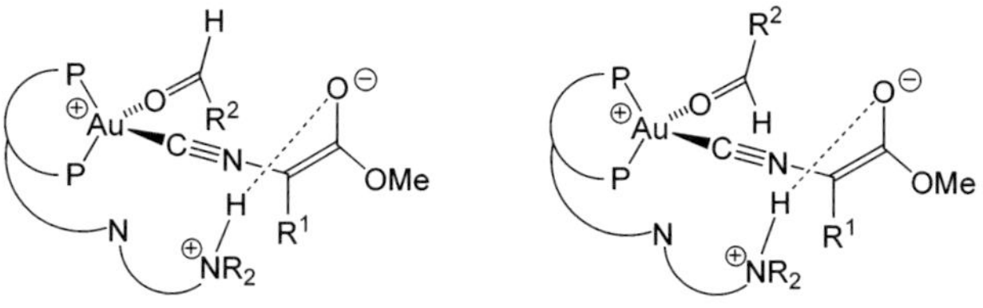

149   **150**

Fig. 3.6 Transition states proposed to explain the obtained enantioselectivities in the gold(I)-mediated aldol addition.

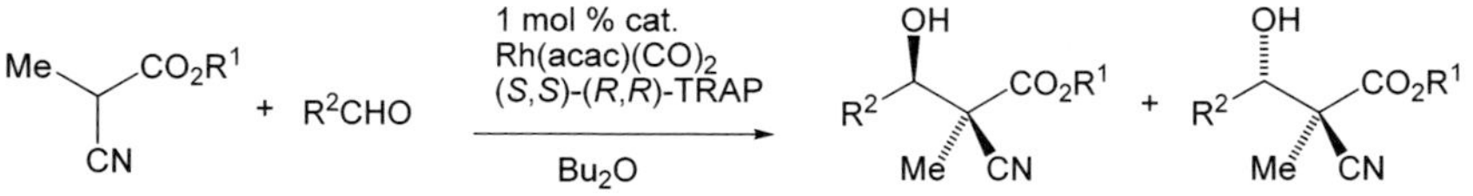

151   **152** (*syn*)   **153** (*anti*)

Scheme 3.27

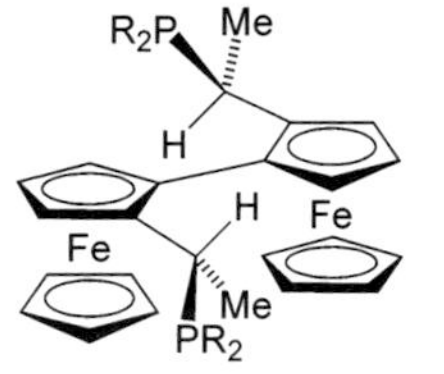

(S,S)-(R,R)-TRAP
((R,R)-2,2''-bis[(S)-1-(dialkylphos-
phanyl)ethyl]-1,1''-biferrocene)

154

Fig. 3.7 Chiral ferrocenylphosphine ligand for enantioselective rhodium(I)-mediated aldol additions.

The observed configuration at the α-position of the aldol product suggests that (S,S)-(R,R)-TRAP **154** on the rhodium complex differentiates between the steric bulkiness of the α-methyl and the ester group of **151**. One of the phenyl substituents blocks the approach of an aldehyde to the *si* face of the enolate coordinated to the rhodium atom. The preferential formation of *anti*-configured aldols

Table 3.13 Results of enantioselective aldol additions of aldehydes and 2-cyanopropionates.

Entry	R^1	R^2	Temperature (°C)	Time (h)	Yield (%)	anti:syn	ee (%) ee anti	ee syn
1	Et	H	0	3	84	–		60
2	Me	H	−30	100	67	–		35
3	Et	H	−30	42	85	–		74
4	iPr	H	−30	90	86	–		78
5	nBu	H	−30	70	80	–		82
6	CHiPr$_2$	H	−10	24	82	–		91
7	CHnBu$_2$	H	−10	24	86	–		93
8	CHPh$_2$	H	−10	24	96	–		87
9	Et	Me	–	24	63	45:55	31	23
10	iPr	Me	–	24	61	47:53	55	50
11	CHiPr$_2$	Me	–	24	67	81:19	86 (2S,3S)	33
12	CHiPr$_2$	Me	–	24	80	77:23	78 (2S,3S)	28
13	CHiPr$_2$	Me	–	72	80	84:16	78 (2S,3S)	43
14	CHiPr$_2$	Et	–	48	76	75:25	57 (2S,3S)	10
15	CHiPr$_2$	EtO$_2$C	–	88	88	68:32	91 (2S,3R)	63

in entries 11–15 (Table 3.13) may suggest that this reaction proceeds through the antiperiplanar transition state **155** (Scheme 3.28). Compared with transition state **157** giving *syn*-aldol, **155** avoids the steric repulsion between the aldehyde substituent R and the bulky CHiPr$_2$-ester. The synclinal transition state **156** giving an *anti*-aldol may be less favorable than **158**, owing to the steric interaction between R and one of the phenyl groups of the (*S,S*)-(*R,R*)-TRAP. The low diastereoselectivity of the reactions of 2-cyanopropionates with sterically less-demanding ester groups (entries 9 and 10, Table 3.13) may be due to the lower steric repulsion between R and the ester group.

An example of a dia- and enantioselective intramolecular aldol addition, catalyzed by [Rh(cod)(OMe)]$_2$ and (*S*)-BINAP was published by Krische et al. [51]. This method was employed in the synthesis of *cis*-fused C-D rings of alkaloids.

An interesting attempt to create a mimic of a class II zinc-containing aldolase is the Et$_2$Zn/(*S,S*)-linked-BINOL complex [52]. The linked BINOL **159**, **160** (Fig. 3.8) was reacted with diethyl zinc (Et$_2$Zn) to give a catalytically active complex **161** (Fig. 3.9). This complex was able to promote the aldol reaction shown in Scheme 3.29. The results of this reaction are summarized in Table 3.14.

With (*S,S*)-oxygen linked-BINOL **159** the reaction turned out to be *syn*-selective.

It was found that the heteroatom in the linker part of linked BINOL affected the diastereoselectivity. With (*S,S*)-sulfur-linked BINOL **160** the reaction proceeded *anti*-selectively (entries 7–9). The major *anti*-aldols were obtained in high *ee*, whereas the *ee* of the minor *syn*-isomers was rather low.

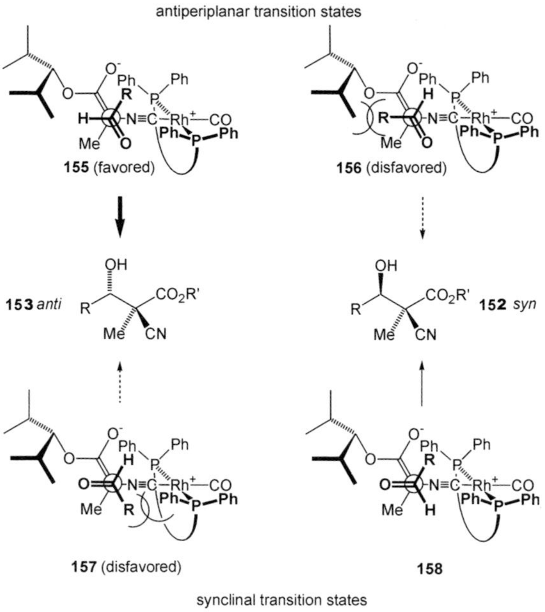

antiperiplanar transition states

155 (favored)   156 (disfavored)

153 *anti* R OH CO₂R'
         Me CN

OH CO₂R' 152 *syn*
R Me CN

157 (disfavored)   158

synclinal transition states

Scheme 3.28

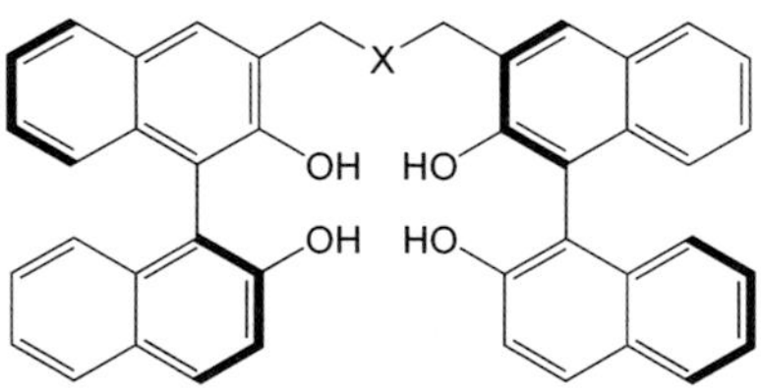

X = O: (*S,S*)-linked BINOL **159**
X = S: (*S,S*)-sulfur-linked BINOL **160**

Fig. 3.8 Chiral BINOL ligands for the formation of chiral zinc complexes for direct, catalytic aldol reactions.

3.3.4
Organocatalysis

Natural aldolases use combinations of acids and bases in their active sites to accomplish direct asymmetric aldol reactions of unmodified carbonyl compounds. To make enolization possible under essentially neutral, aqueous conditions, these enzymes decrease the pKa value of the carbonyl donor by converting it into a cationic species **166**, typically an iminium ion. A relative weak Brønsted-base cocatalyst then generates the nucleophilic species, an enamine enolate **167** for example, via deprotonation (Scheme 3.30).

Although the reaction imitates that of the enzymes, aldolase-like direct catalytic aldol addition remained an elusive challenge for a long time.

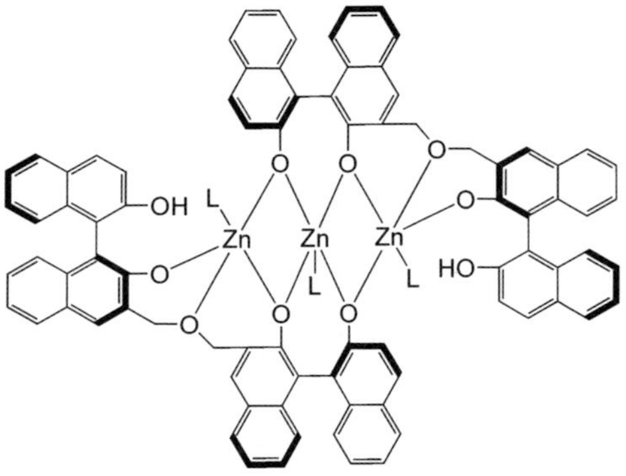

161

Fig. 3.9 Structure of the chiral zinc complex proved by X-ray crystal structure analysis.

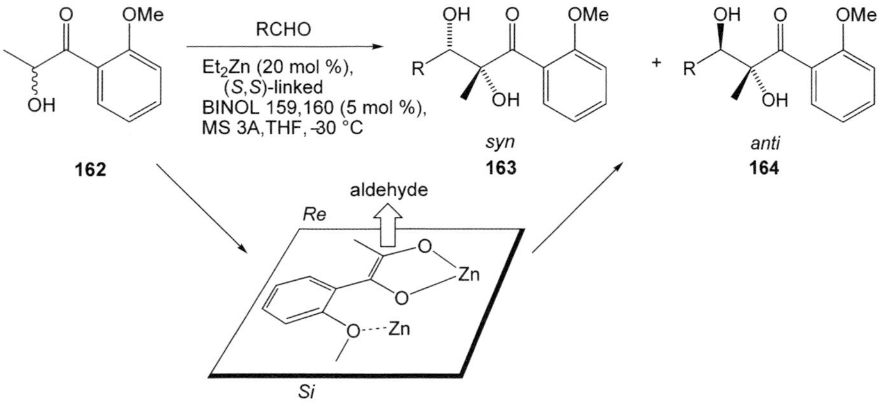

Scheme 3.29

Table 3.14 Reactions of the catalytically active $Et_2Zn/(S,S)$-linked-BINOL complex.

Entry	R	X	Time (h)	Yield (%)	syn:anti	ee syn (%)	ee anti (%)
1	$Ph(CH_2)_2$	O	16	97	62:38	87	96
2	$Ph(CH_2)_3$	O	19	72	64:36	78	90
3	Et	O	12	88	71:29	68	86
4	2-methylpropyl	O	10	80	68:32	72	87
5	$BnOCH_2$	O	18	80	65:35	85	92
6	cC_6H_{11}	O	24	0	–	–	–
7	$Ph(CH_2)_2$	S	45	62	35:65	60	92
8	$Ph(CH_2)_3$	S	24	63	41:59	45	86
9	Et	S	24	56	41:59	48	87

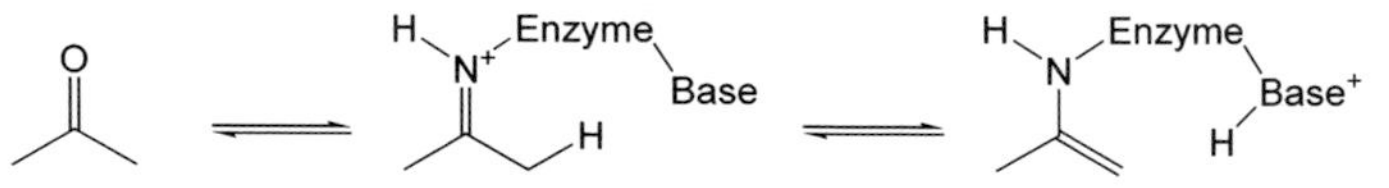

165 166 167

Scheme 3.30

In the early 1970s, Eder, Sauer and Wiechert at Schering AG in Germany observed that chiral amino acids were able to promote an intramolecular aldol condensation with high enantiomeric excess (Scheme 3.31) [53]. This reaction is of special interest for the stereoselective synthesis of steroids. It was soon found that proline is particularly effective as a catalyst for this reaction.

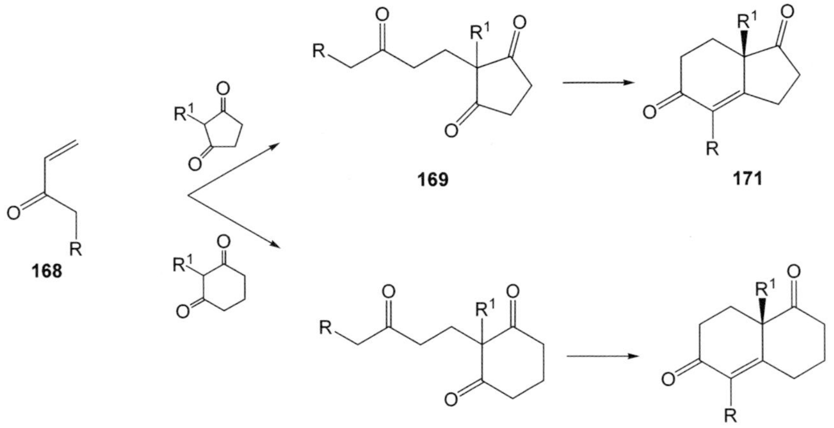

168 169 170 171 172

Scheme 3.31

Only three years later, Hajos and Parrish made the corresponding proline-catalyzed aldol **174** addition products accessible (Scheme 3.32) [54], and proposed a reaction mechanism (**176**, Fig. 3.10). Two hydrogen bonds provide a 6,6,7-membered ring conformation and the rigidity necessary to achieve the stereoselectivity observed in these reaction. The vicinal hydrogens indicated in the pyrrolidine ring would have to be on the same side of the molecule to allow the two hydrogen bonds to form. The bulky proline molecule would have to be opposite to the β-angular methyl group.

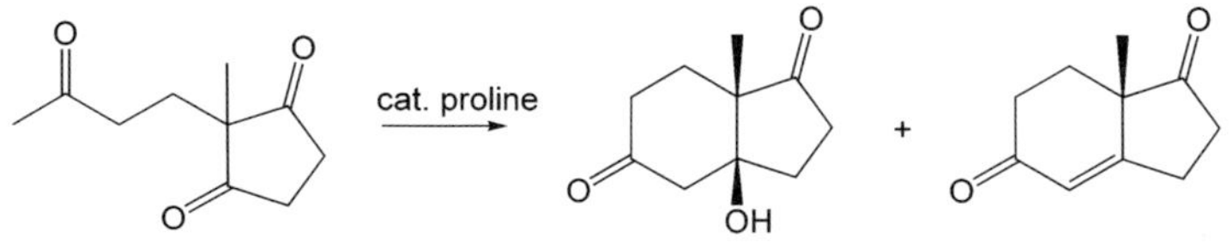

173 174 175

Scheme 3.32

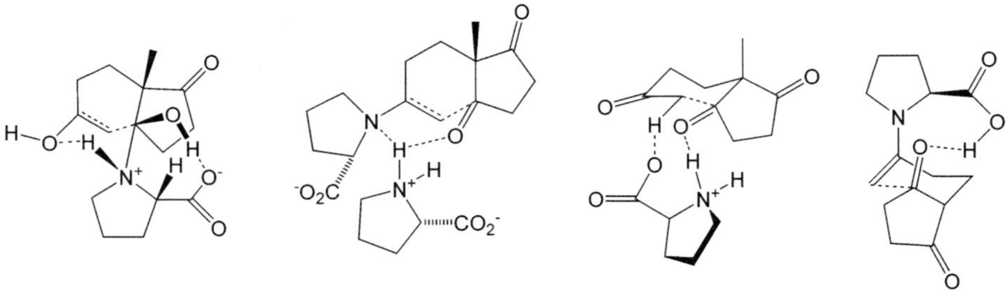

176 Hajos Model | **177** Agami Model | **178** Swaminathan Model | **179** Houk Model

Fig. 3.10 Proposed transition states for the proline-mediated enantioselective catalytic aldol addition.

The C–C bond formation would have to occur from the side opposite to the methyl group to give a *cis*-fused product. Some other mechanisms have been proposed by Agami [55], Swaminathan [56] and Bahmanyar and Houk [57]. The Houk model **179** has obtained some experimental support [58]. Applications of this reaction are steps in the total synthesis of the protosterol 3β,20-dihydroxyprotoster-24-ene [59], the total synthesis of D-Norgestrel [60] and the synthesis of (4a*S*)-(+) or (4a*R*)-(–)-dimethyl-4,4a,7,8-tetrahydronaphthalene-2,5(3*H*,6*H*)-dione [61].

A more common application of organocatalysis with pyrrolidine as catalyst has been reported, making some intermolecular aldol adducts **185** accessible (Scheme 3.33) [62]. This reaction is carried out in the presence of acetic acid (Scheme 3.34), and DMSO as solvent gave the highest yields. Some examples are listed in Table 3.15. An enantioselective version of this reaction using chiral amines has been described [63].

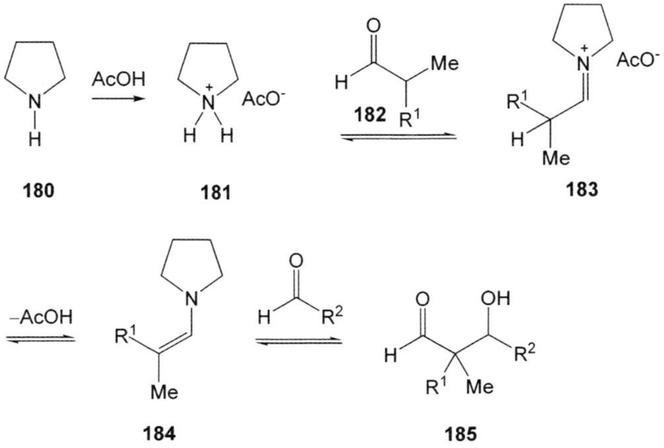

Scheme 3.33

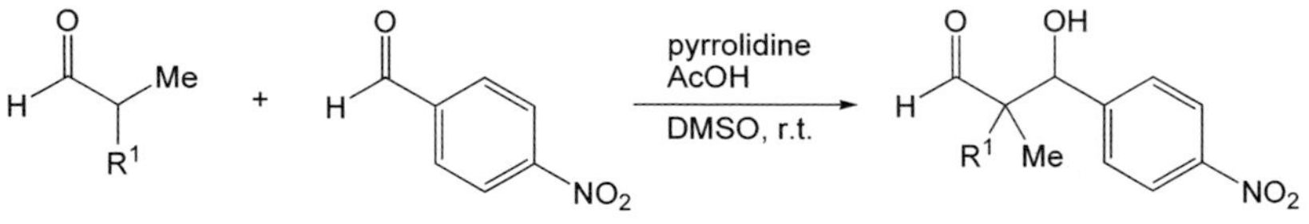

Scheme 3.34

Table 3.15 Results of organocatalysis using pyrrolidine as catalyst in acetic acid.

Entry	R^1	Pyrrolidine (equiv.)	AcOH (equiv.)	Time (h)	Yield (%)
1	Me	0.3	none	48	74
2	Me	0.05	0.25	2	96
3	Et	0.3	1.5	3	89
4	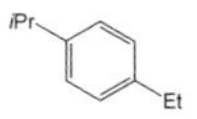	0.3	1.5	3	87

A further two examples of amine catalysis are given in Scheme 3.35 [64].

3.3.5
Enzyme and Antibody Catalysis

As a supplement to classic chemical methods, enzymes are finding increasing acceptance as chiral catalysts for the *in vitro* synthesis of asymmetric species [65]. Enzymes have been optimized by evolution for high selectivity and catalytic efficiency. Several dozen aldolases have been identified in nature, and many of them are commercially available. Class I aldolases bind their substrates covalently via imine–enamine formation (Scheme 3.30) with an active-site lysine residue to initiate bond cleavage or formation. Class II aldolases utilize zinc ions as a Lewis-acid cofactor, which facilitates deprotonation by bidentate coordination of the donor to give the enediolate nucleophile **197** (Scheme 3.36).

The synthesis of Zanamivir **202**, an inhibitor of viral sialidases that is marketed against influenza, may serve as an example of the synthetic application on the multitonne scale of enzyme-catalyzed aldol reactions (Scheme 3.37) [66].

Natural aldolases have substrate specificities that are predetermined by nature. Antibody catalysts designed by synthetic chemists can have specificities different from those of natural enzymes. An aldolase antibody 38C2-catalyzed resolution of tertiary aldols leads to enantiomerically enriched aldols [67]. The results of enantiomeric resolutions are summarized in Table 3.16.

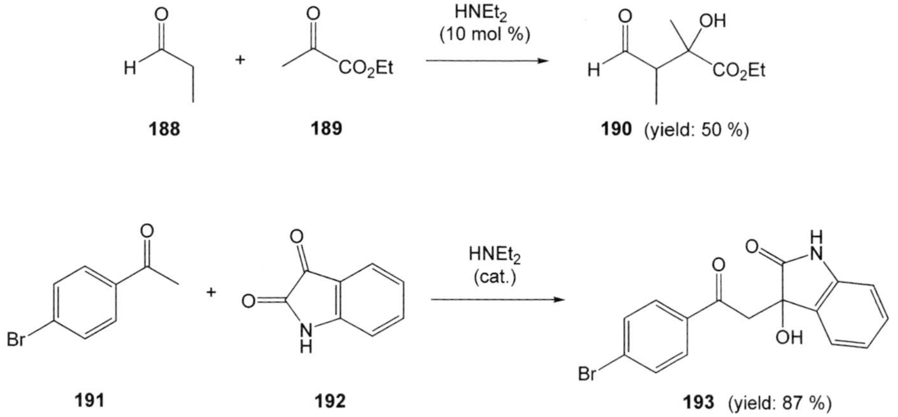

Scheme 3.35

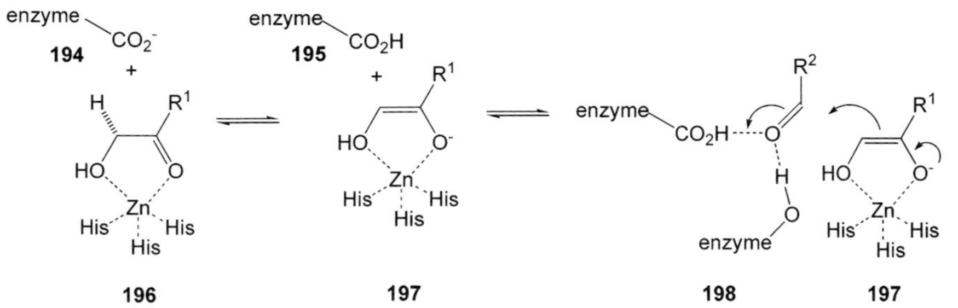

Scheme 3.36

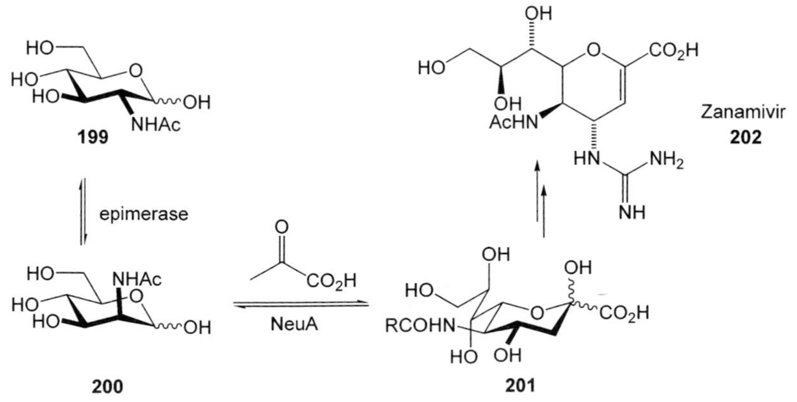

Scheme 3.37

Table 3.16 Results of enantiomeric resolutions using an aldolase antibody.

Entry	Aldol (product)	Conversion (%)	*ee* (%)
1	**203**	50	>99
2	**204**	52	80
3	**205**	50	>99
4	**206**	50	94
5	**207**	50	96
6	**208**	50	95
7	**209**	40	75
8	**210**	50	95

3.4
Conclusions

The chemistry discussed here, a "progress report" of developments, provides additional testimony to the already well-recognized value that the aldol addition brings to the art of organic synthesis.

Recent accomplishments based on stereoselective methods in aldol additions indicate the potential utility of this reaction. So long as nature continues to supply a wealth of aldol-containing molecules that possess interest to all kind of life-sciences, so long will chemists continue to refine existing methods and explore novel transformations for realizing all the demands of such a structure.

Although much has been achieved, there is still a need of control of stereoselectivity. This applies mainly to the development of methods that allow the synthesis of trisubstituted enolates with a defined configuration. On the other hand, the development and results of direct aldol additions and enzymatic methods indicate that these methods are also powerful tools in stereoselective aldol additions.

3.5
Note Added in Proof

After submission of the original manuscript, relevant papers were published or came to the attention of the authors.

These deal with the concept of catalytic enantioselective aldol additions with chiral Lewis bases: Scott E. Denmark, Y. Fan, *J. Am. Chem. Soc.* **2002**, *124*, 4233; Scott E. Denmark, S. Fujimori **2004**, in *Modern Aldol Reactions*, ed. R. Mahrwald, Wiley-VCH, Weinheim, pp. 229–326.

References

1 R. Mahrwald (ed.) *Modern Aldol Reactions*, Wiley-VCH, Weinheim, 2004.
2 C. Palomo, M. Oiarbide, J. M. García, *Chem. Soc. Rev.* 2004, *33*, 65–75.
3 S. F. Martin, *Tetrahedron* 1980, *36*, 419–460.
4 For a general review of the catalytic creation of quaternary carbon centers see: (a) E. J. Corey, A. Guzman-Perez, *Angew. Chem.* 1998, *110*, 402–415; (b) J. Christoffers, A. Mann, *Angew. Chem.* 2001, *113*, 4725–4732.
5 H. E. Zimmerman, M. D. Traxler, *J. Am Chem. Soc.* 1957, *79*, 1920–1923.
6 S. Yamago, D. Machii, E. Nakamura, *J. Org. Chem.* 1991, *56*, 2098–2106.

7 J. M. Manthorpe, J. L. Gleason, *J. Am. Chem. Soc.* 2001, *123*, 2091–2092.
8 D. Seebach, *Angew. Chem.* 1988, *100*, 1685–1715.
9 G. Stork, P. F. Hudrlik, *J. Am. Chem. Soc.* 1968, *90*, 4464–4465.
10 R. Häner, T. Laube, D. Seebach, *J. Am. Chem. Soc.* 1985, *107*, 5396–5403.
11 Y. Liu, W. W. McWorther, Jr., *J. Am. Chem. Soc.* 2003, *125*, 4240–4252.
12 P. Fraisse, I. Hanna, J.-Y. Lallemand, T. Prangé, L. Ricard, *Tetrahedron* 1999, *55*, 11819–11832.
13 R. Shiraki, A. Sumino, K. Tadano, S. Ogawa, *J. Org. Chem.* 1996, *61*, 2845–2852.

14 J. Ishihara, R. Nonaka, Y. Terasawa, R. Shiraki, K. Yabu, H. Kataoka, Y. Ochiai, K. Tadano, *J. Org. Chem.* **1998**, *63*, 2679–2688.

15 D. J. Dixon, A. Guarna, S. V. Ley, A. Polara, F. Rodríguez, *Synthesis* **2002**, 1973–1978.

16 E. Nakamura, J. Shimada, Y. Horiguchi, I. Kuwajima, *Tetrahedron Lett.* **1983**, *24*, 3341–3342.

17 P. J. Murphy, G. Procter, A. T. Russel, *Tetrahedron Lett.* **1987**, *28*, 2037–2040.

18 T. Kamino, Y. Murata, N. Kwai, S. Hosokawa, S. Kobayashi, *Tetrahedron Lett.* **2001**, *42*, 5249–5252.

19 Y. Murata, T. Kamino, S. Hosokawa, S. Kobayashi, *Tetrahedron Lett.* **2002**, *43*, 8121–8123.

20 D. A. Evans, B. W. Trotter, J. C. Barrow, *Tetrahedron* **1997**, *53*, 8779–8794.

21 D. A. Evans, J. V. Nelson, E. Vogel, T. R. Taber, *J. Am. Chem. Soc.* **1981**, *103*, 3099–3111.

22 E. D. Burke, J. L. Gleason, *Org. Lett.* **2004**, *6*, 405–407.

23 M. D. Keränen, P. Eilbracht, *Org. Biomol. Chem.* **2004**, *2*, 1688–1690.

24 J. W. J. Kennedy, D. G. Hall, *J. Am. Chem. Soc.* **2002**, *124*, 898–899.

25 F. Lang, D. Zewge, Z. J. Song, M. Biba, P. Dormer, D. Tschaen, R. P. Volante, P. J. Reider, *Tetrahedron Lett.* **2003**, *44*, 5285–5288.

26 E. Nakamura, M. Shimizu, I. Kuwajima, J. Sakata, K. Yokoyama, R. Noyori, *J. Org. Chem.* **1983**, *48*, 932–945.

27 A. J. Parker, *Adv. Org. Chem.* **1965**, *5*, 1–39.

28 E. Buncel, B. Menon, *J. Am. Chem. Soc.* **1977**, *99*, 4457–4461.

29 I. Kuwajima, E. Nakamura, *Acc. Chem. Res.* **1985**, *18*, 181–187.

30 R. Noyori, I. Nishida, J. Sakata, M. Nishizawa, *J. Am. Chem. Soc.* **1980**, *102*, 1223–1225.

31 R. Noyori, I. Nishida, J. Sakata, *J. Am. Chem. Soc.* **1983**, *105*, 1598–1608.

32 R. Noyori, I. Nishida, J. Sakata, *J. Am. Chem. Soc.* **1981**, *103*, 2106–2108.

33 E. Nakamura, S. Yamago, D. Machii, I. Kuwajima, *Tetrahedron Lett.* **1988**, *29*, 2207–2210.

34 R. Mahrwald, *Chem. Rev.* **1999**, *99*, 1095–1120.

35 Y. Li, M. N. Paddon-Row, K. N. Houk, *J. Org. Chem.* **1990**, *55*, 481–493.

36 (a) T. Mukaiyama, K. Narasaka, K. Banno, *Chem. Lett.* **1973**, 1011–1014; (b) T. Mukaiyama, K. Banno, K. Narasaka, *J. Am. Chem. Soc.* **1974**, *96*, 7503–7509.

37 T. Mukaiyama, M. Hayashi, *Chem. Lett.* **1974**, 15–16.

38 (a) D. A. Evans, M. C. Kozlovski, C. S. Burgey, D. W. C. MacMillan, *J. Am. Chem. Soc.* **1997**, *119*, 7893–7894; (b) D. A. Evans, C. S. Burgey, M. C. Kozlovski, S. W. Tregay, *J. Am. Chem. Soc.* **1999**, *121*, 686–699.

39 S. Kobayashi, M. Horibe, Y. Saito, *Tetrahedron* **1994**, *50*, 9629–9642.

40 S. Kobayashi, Y. Fujishita, T. Mukaiyama, *Chem. Lett.* **1989**, 2069–2072.

41 I. Shiina, J. Kawakita, R. Ibuka, Abstracts of papers, 83rd National Meeting of the Chemical Society of Japan, Tokyo, **2003**, Vol. 2, 2C401.

42 N. Miyoshi, T. Fukuma, M. Wada, *Chem. Lett.* **1995**, 999–1000.

43 R. Mahrwald, B. Gündogan, *J. Am. Chem. Soc.* **1998**, *120*, 413–414.

44 R. Mahrwald, B. Ziemer, S. Troyanov, *Tetrahedron Lett.* **2001**, *42*, 6843–6845.

45 Y. Yoshida, N. Matsumoto, R. Hamasaki, Y. Tanabe, *Tetrahedron Lett.* **1999**, *40*, 4227–4230.

46 Y. Tanabe, N. Matsumoto, T. Higashi, T. Misaki, T. Itoh, M. Yamamoto, K. Mitarai, Y. Nishii, *Tetrahedron* **2002**, *58*, 8269–8280.

47 Y. Yoshida, R. Hayashi, H. Sumihara, Y. Tanabe, *Tetrahedron Lett.* **1997**, *38*, 8727–8730.

48 Y. Tanabe, N. Matsumoto, S. Funakoshi, N. Manta, *Synlett* **2001**, 1959–1961.

49 Y. Ito, M. Sawamura, E. Shirawaka, K. Hayashizaki, T. Hayashi, *Tetrahedron* **1988**, *44*, 5253–5262.

50 R. Kuwano, H. Miyazaki, Y. Ito, *J. Organomet. Chem.* **2000**, *603*, 18–29.

51 (a) B. M. Bocknack, L.-C. Wang, M. J. Krische, *Proc. Acad. Nat. Sci.* **2004**, *101*, 4421–5424; (b) R. R. Huddleston, M. J. Krische, *Org. Lett.* **2003**, *5*, 1143–1146.

52 N. Kumagai, S. Matsunaga, T. Kinoshita, S. Harada, S. Okada, S. Sakamoto, K. Yamaguchi, M. Shibasaki, *J. Am. Chem. Soc.* **2003**, *125*, 2169–2178.

53 U. Eder, G. Sauer, R. Wiechert, *Angew. Chem.* **1971**, *83*, 492–493.

54 Z. G. Hajos, D. R. Parrish, *J. Org. Chem.* **1974**, *39*, 1615–1621.

55 C. Agami, *Bull. Soc. Chim. Fr.* **1988**, *3*, 499–507.

56 D. Rajagopal, M. S. Moni, S. Subramanian, S. Swaminathan, *Tetrahedron: Asymmetry* **1999**, *10*, 1631–1634.

57 S. Bahmanyar, K. N. Houk, *J. Am. Chem. Soc.* **2001**, *123*, 12911–12912.

58 L. Hoang, S. Bahmanyar, K. N. Houk, B. List, *J. Am. Chem. Soc.* **2003**, *125*, 16–17.

59 E. J. Corey, S. C. Virgil, *J. Am. Chem. Soc.* **1990**, *112*, 6429–6431.

60 G. Sauer, U. Eder, G. Haffer, G. Neef, R. Wiechert, *Angew. Chem.* **1975**, *87*, 413–414.

61 H. Hagiwara, H. Uda, *J. Org. Chem.* **1988**, *53*, 2308–2311.

62 N. Mase, F. Tanaka, C. F. Barbas III, *Org. Lett.* **2003**, *5*, 4369–4372.

63 N. Mase, F. Tanaka, C. F. Barbas III, *Angew. Chem. Int. Ed.* **2004**, *43*, 2420–2423.

64 (a) H. G. Lingwall, J. S. MacLennan, *J. Am. Chem. Soc.* **1932**, *54*, 4739–4744; (b) J. Schreiber, C. G. Wermuth, *Bull. Soc. Chim. Fr.* **1965**, *8*, 2242–2249.

65 (a) K. Drauz, H. Waldmann, *Enzyme Catalysis in Organic Synthesis*, 2nd edn, Wiley-VCH, Weinheim **2002**; (b) W. D. Fessner, in *Modern Aldol Reactions*, Vol. 1, ed. R. Mahrwald, Wiley-VCH Verlag, Weinheim **2004**, pp. 201–272.

66 T. Sugai, K. Kuboki, S. Hiramatsu, H. Okazaki, H. Ohta, *Bull. Chem. Soc. Jpn.* **1995**, *68*, 3581–3589.

67 B. List, D. Shabat, G. Zhong, J. Turner, A. Li, T. Bui, J. Anderson, R. A. Lerner, C. F. Barbas III, *J. Am. Chem. Soc.* **1999**, *121*, 7283–7291.

4

Michael Reactions and Conjugate Additions

Angelika Baro and Jens Christoffers

4.1
Introduction

The conjugate 1,4-addition of carbon nucleophiles **1** to acceptor activated olefins **2** is one of the most important C–C bond forming reactions in organic synthesis. This is mainly due to the variety of acceptors and donors (for example, organometallic reagents, **A**, enolates, **B**, or enamines, **C**) applicable in this reaction (Scheme 4.1)[1].

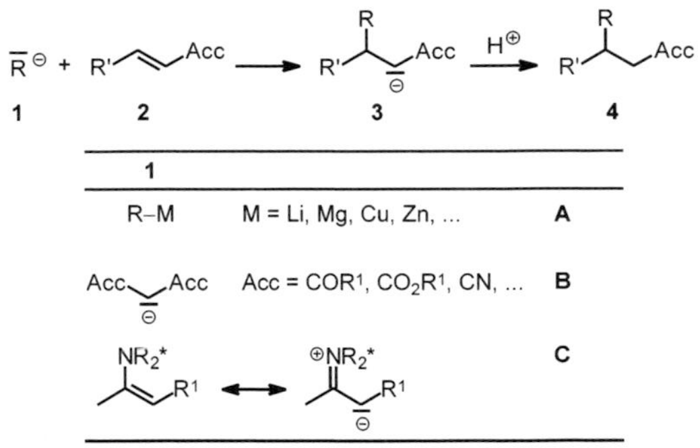

Scheme 4.1 Conjugate addition of a carbon nucleophile to an activated olefin acceptor.

Carbanionic organolithium, Grignard, copper, zinc, or other reagents are often utilized carbon nucleophiles (Scheme 4.1, **A**). However, competing 1,2-addition to the acceptor is observed in the case of hard nucleophiles, especially when the acceptor is a carbonyl group. According to the hard and soft acid and base (HSAB) principle [2], organocuprates [3] are the classic reagents of choice for 1,4-reactivity although conjugate addition is also achieved with "hard" lithium or Grignard reagents in certain cases. Whereas numerous publications deal with the formation of tertiary stereocenters by asymmetric cuprate additions, this method is rarely reported for the generation of quaternary carbon atoms. In particular, the copper-catalyzed addition of organozinc reagents has evolved into a fruitful field of research since 2000 [4].

Quaternary Stereocenters: Challenges and Solutions for Organic Synthesis. Edited by Jens Christoffers, Angelika Baro
Copyright © 2005 WILEY-VCH Verlag GmbH & Co. KGaA, Weinheim
ISBN: 3-527-31107-6

In contrast to organometallic compounds as nucleophiles, the application of derivatives with active methylene moieties (so-called Michael donors, **B**) as typical soft nucleophiles requires only catalytic amounts of a Brönstedt base, since the next equivalent of the donor is deprotonated by the enolate **3** formed as intermediate immediately after conjugate addition of the donor **1** to olefin **2**. This reaction was first observed and reported by Komnenos and Claisen [5], and was later named after Arthur Michael in order to honor his early systematic investigations [6]. Asymmetric Michael reactions have been developed by several groups and are well suited to the highly selective construction of quaternary stereocenters [7]. From a historical point of view, chiral Brönstedt-base catalysis has to be mentioned first. This topic has become a promising area of increasing interest and nowadays is known as "organocatalysis" [8]. Thus, the first part of this chapter is a survey of chiral bases including organocatalysts. The second section will focus on the use of chiral metal complexes as catalysts for asymmetric Michael reactions, and, finally, the third part deals with the utilization of chiral auxiliaries. The application of auxiliaries is attractive, since these methods are often reliable, with high yields and selectivities, and (depending, of course, on the type of auxiliary) can even be scaled up at low cost as in most cases the auxiliary can be recovered. In particular, enamines derived from optically active primary or secondary amines (Scheme 4.1, **C**) are the dominating class of soft nucleophiles for conjugate additions to activated olefins as acceptors.

4.2
Chiral Brönstedt Bases

4.2.1
Cinchona Alkaloids

The pioneer of alkaloid-catalyzed asymmetric Michael reactions is Hans Wynberg, who published the conversion of ten Michael donors **5** with methyl vinyl ketone **6a** in the presence of quinine **7a** [9]. The enantiomeric excess (*ee*) could be determined in only one case: product **8a** was obtained in 87% yield and 68% *ee* in toluene as the solvent (Scheme 4.2). Detailed information on stoichiometry and other reaction parameters was not given in this communication. The Wynberg group extended their studies to other Michael donors and other cinchona alkaloid catalysts and derivatives thereof, and in 1979, the first full paper was published [10]. Interestingly, among all alkaloids investigated, only *N*-methylquininium hydroxide **7b** gave significant conversion rates. This finding was attributed to the phase-transfer catalytic activity of the quaternary ammonium salt, as similarly observed with benzyl trimethylammonium hydroxide. The enantioselectivities, however, were rather low, for example 25% *ee* for the cyclohexanone derivative **8b**, while the initial results for indanone derivative **8a** have now been slightly improved to 76% *ee*.

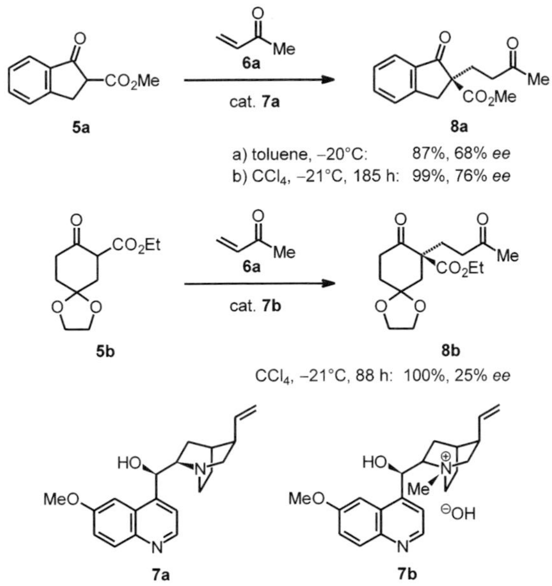

Scheme 4.2 Pioneering work by Wynberg and coworkers.

Further investigations with *N*-alkyl quininium, ephedrinium, and amphetaminium salts did not raise optical yields above 36% *ee* [11]. Cinchona alkaloids have been rediscovered under the heading of organocatalysis. These examples will therefore be discussed below [12, 13].

4.2.2
Polymer-bound Alkaloids

A heterogeneous system is distinguished by simple separation and recovery of the catalyst, while homogeneous catalysts are commonly well-defined unique low-molecular-mass species that are easy to characterize spectroscopically, which is, of course, advantageous for optimization procedures. To take advantage of both concepts, the immobilization of low-molecular-mass catalysts by binding them to a solid polymeric support has gained significantly in importance [14]. Again, Wynberg was the first who used polymer-bound cinchona alkaloid catalysts in Michael reactions: quinine and its derivatives were immobilized by ether and ester linkages of the secondary alcohol or a quinoline alkoxy group to styrene–divinylbenzene copolymers. In the model reaction of carboxylate **5a** to give indanone derivative **8a** enantiomeric excesses with these catalysts did not exceed 53% [15]. Kobayashi and coworkers used the vinyl moiety of quinine for copolymerization with acrylonitrile; however, with maximum 57% *ee* for **8a** they could not significantly improve Wynberg's results [16]. Oda et al. prepared acrylonitrile copolymers by using radical addition of thiols to the vinyl group of quinine to introduce a spacer group of variable length between the catalytically active moiety and the polymer matrix. This strategy was rewarded with 65% *ee* for product **8a** [17].

Relying on the rate-accelerating effect of quaternary ammonium salts, which has already been observed by Wynberg [10], Hodge and coworkers converted several cinchona and ephedra alkaloids with chloromethylated cross-linked poly-styrene–divinylbenzene (PS-DVB) resins to obtain *N*-benzyl-modified catalysts. Unfortunately, the optimal selectivity achieved was 27% *ee* for model compound **8a** [18]. Finally, the group led by d'Angelo was successful using the polymer-supported catalysts shown in Scheme 4.3. First, primary alcohols were prepared by anti-Markownikov hydroxylation of the vinyl moiety in quinines and quinidines. The new catalyst was then synthesized by linking ω-hydroxyalkanoic acids by ether formation to a Merrifield resin, followed by esterification of the carboxyl group with the alkaloid primary alcohol function. The only substrate investigated in this study was indanone carboxylate **5a**, which furnished product **8a** with 85% yield and 87% *ee* [19].

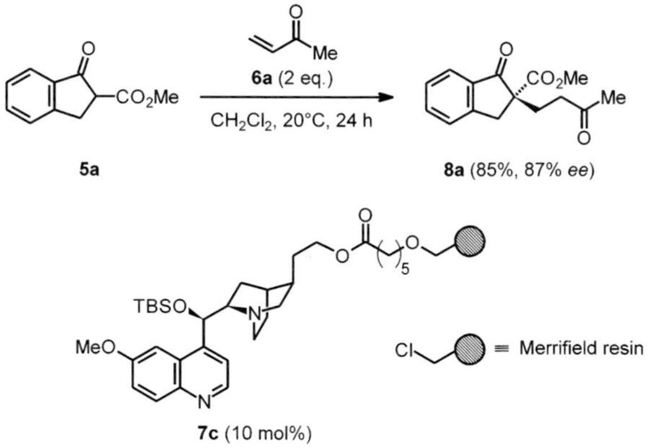

Scheme 4.3 Asymmetric Michael reaction according to d'Angelo and coworkers [19].

4.2.3
Organocatalysis

The development of the Hajos–Parrish–Eder–Sauer–Wiechert reaction in the early 1970s marks the beginning not only of organocatalysis but of modern asymmetric catalysis in general [20]. Achiral Michael adduct **8c** undergoes L-proline-catalyzed Robinson annulation giving the Wieland–Miescher ketone **9**, which is an important optically active intermediate for industrial steroid synthesis (Scheme 4.4) [21]. Actually, the conjugate addition step yields an achiral product, **8c**. The quaternary stereocenter is formed in an asymmetric aldol reaction of an *in situ* formed enamine. The catalysis of enantioselective Michael reactions by proline and derivatives thereof has been elaborated for several substrates since then [22]. However, with the exception of diketone **9** and its homologue with two six-membered rings, *quaternary* stereocenters cannot be obtained with high selectivity [23].

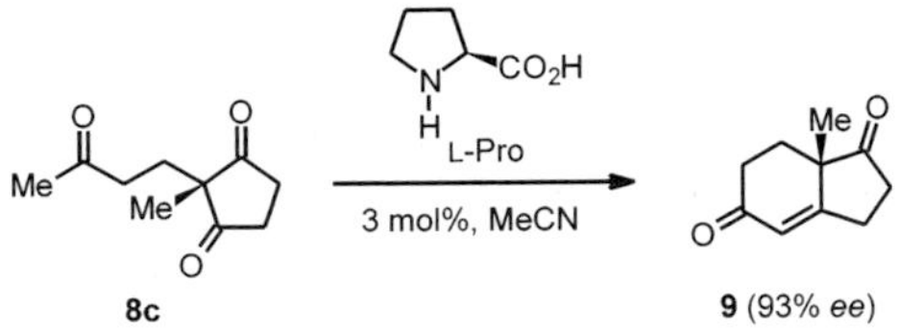

Scheme 4.4 The Hajos–Parrish–Eder–Sauer–Wiechert reaction.

Important contributions to organocatalytic Michael reactions have been made by Jørgensen and coworkers [24]. With respect to quaternary stereocenters the enantioselective conjugate addition of β-diketones **5c–e** to alkynones **10** is to be noted (Scheme 4.5) [12]. This process is catalyzed by bis(dihydroquinine)phthalazine [(DHQ)$_2$PHAL] **11**, initially introduced for asymmetric Sharpless dihydroxylation [25]. Addition products **8d–8i** are primarily formed as mixtures of *E*- and *Z*-isomers in respect of the C–C double bond in the side chain. With catalytic amounts of PBu$_3$, however, these *E/Z*-mixtures can be equilibrated in a two-step one-pot procedure to optically active *E*-isomers without affecting the configuration of the newly formed quaternary center. The yields and selectivities summarized in Scheme 4.5 refer to the isolated *E*-isomers of **8d–8i**.

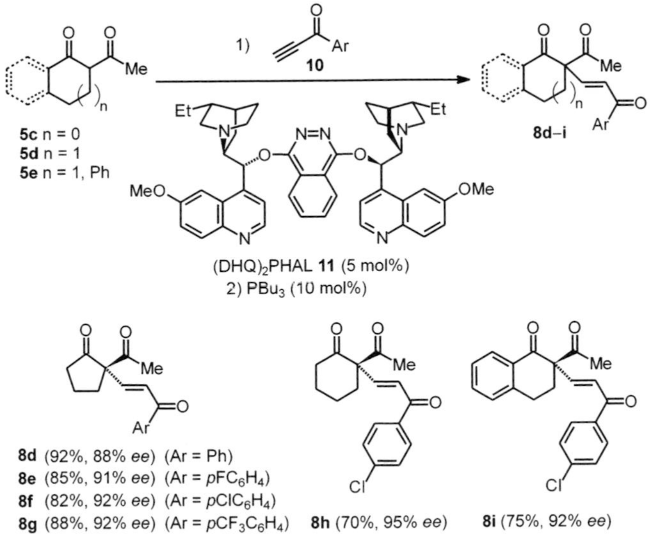

Scheme 4.5 Organocatalytic enantioselective conjugate addition to alkynones.

Closely related is the organocatalytic amination of α-substituted α-cyanoacetates **12** or β-dicarbonyl compounds **5** with di-*tert*-butyl azodicarboxylate **13**. This conjugate addition leads to valuable optically active building blocks **15a–i** (see Table 4.1) which can be further transformed to α-amino-functionalized derivatives by reductive N–N cleavage using, for example, SmI$_2$ [13]. The α-amination is catalyzed by the quinidine-derived alkaloid β-isocupreidine (β-ICD, **14**) (Scheme 4.6).

Table 4.1 Organocatalytic α-amination of compounds **12** to give products **15a–e.**

Product	Ar	Yield (%)	*ee* (%)
15a	Ph	99	>98
15b	2-F-C$_6$H$_4$	99	98
15c	4-O$_2$N-C$_6$H$_4$	99	91
15d	4-MeO-C$_6$H$_4$	95	89
15e	2-thienyl	99	97

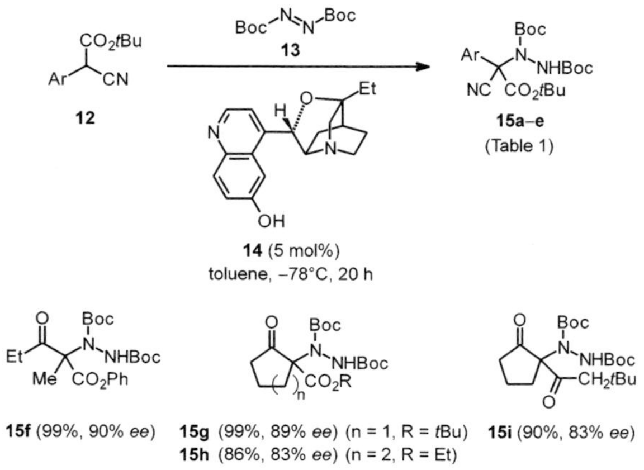

15f (99%, 90% *ee*)   **15g** (99%, 89% *ee*) (n = 1, R = *t*Bu)   **15i** (90%, 83% *ee*)
15h (86%, 83% *ee*) (n = 2, R = Et)

Scheme 4.6 Organocatalytic α-amination of α-cyanoacetates.

A modern concept for the synthesis of 1,4-dicarbonyl compounds utilizes the umpolung of aldehydes, e.g. by application of 1,3-dithiane derivatives [26] or the Stetter reagent [27]. The latter Michael–Stetter reaction is generally accepted to be an example of organocatalysis. Recently, Rovis and coworkers have reported on the application of aminoindanol-derived chiral triazolium salts like **17**, which are precursors for *N*-heterocyclic carbenes, in intramolecular asymmetric Michael–Stetter reactions (Scheme 4.7) [28]. Aldehydes **16** are converted to cyclic 1,4-dicarbonyl compounds **18** in the presence of 20 mol% **17**, with excellent yields and stereoselectivities.

4.2.4
Miscellaneous Examples

Macrocyclic polyethers were introduced by Pedersen in 1967 for the coordination of alkaline metal ions [29]. Cram and coworkers investigated Michael reactions of β-keto esters with methyl vinyl ketone (**6a**) in the presence of optically active 18-crown-6 derivatives with axially chiral binaphthalene moieties in the backbone [30]. Chiral crown derivative **19** complexed to KO*t*Bu as the base afforded product **8a** with excellent enantioselectivity (99%; Scheme 4.8). The potassium ion is proposed to be coordinated by ligand **19** as well as the diketonate derived from **5a**.

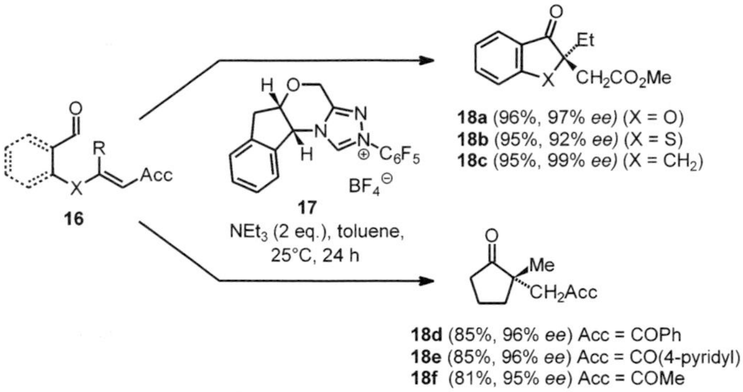
16
17
NEt_3 (2 eq.), toluene, 25°C, 24 h
18a (96%, 97% ee) (X = O)
18b (95%, 92% ee) (X = S)
18c (95%, 99% ee) (X = CH_2)
18d (85%, 96% ee) Acc = COPh
18e (85%, 96% ee) Acc = CO(4-pyridyl)
18f (81%, 95% ee) Acc = COMe

Scheme 4.7 The catalytic asymmetric Michael–Stetter reaction.

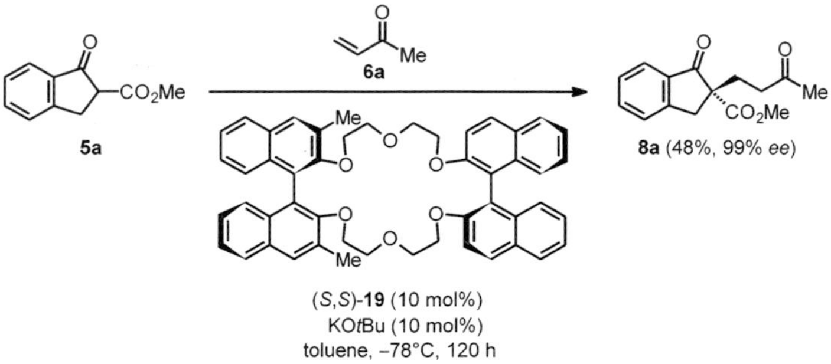
5a
6a
(S,S)-19 (10 mol%)
KOtBu (10 mol%)
toluene, –78°C, 120 h
8a (48%, 99% ee)

Scheme 4.8 Chiral crown ether–potassium complexes.

Tamai, Miyano, and coworkers utilized podand-type alkali metal 2′-substituted 1,1′-binaphthalen-2-oxides as ligands, as exemplified in Scheme 4.9 for compound **20**. With NaOH as base, three β-keto esters **5** were converted to give products **8** with selectivities of 51–64% *ee*. Interestingly, no ester saponification was observed under these reaction conditions, although an equimolar amount of NaOH was applied [31].

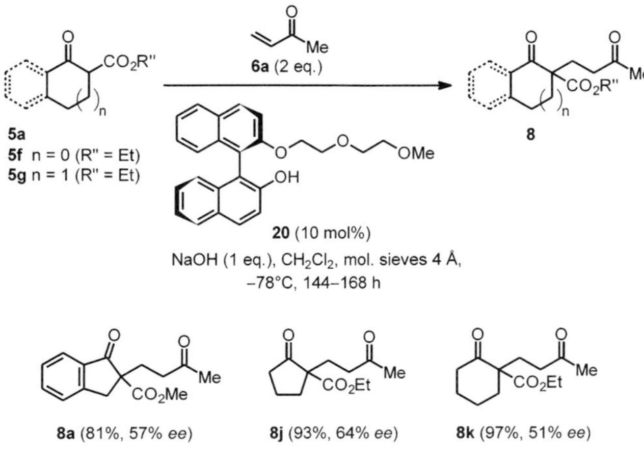
5a
5f n = 0 (R" = Et)
5g n = 1 (R" = Et)
6a (2 eq.)
20 (10 mol%)
NaOH (1 eq.), CH_2Cl_2, mol. sieves 4 Å, –78°C, 144–168 h
8
8a (81%, 57% ee)
8j (93%, 64% ee)
8k (97%, 51% ee)

Scheme 4.9 Chiral polyether-tethered binaphtholate ligands.

The lupine alkaloid (–)-sparteine **22** was proved to be a powerful chiral additive for lithium-carbanionic chemistry [32]. Beak et al. reported the conversion of benzylamines **21**, deprotonated by nBuLi/(–)-sparteine, with dinitriles **23** (Scheme 4.10) [33]. Products **24** with vicinal quaternary and tertiary stereocenters were isolated with both high diastereo- and enantioselectivity (Table 4.2). The respective conversions of allylamines gave lower diastereomeric ratios, although the enantiomeric ratios of the major isomer were excellent.

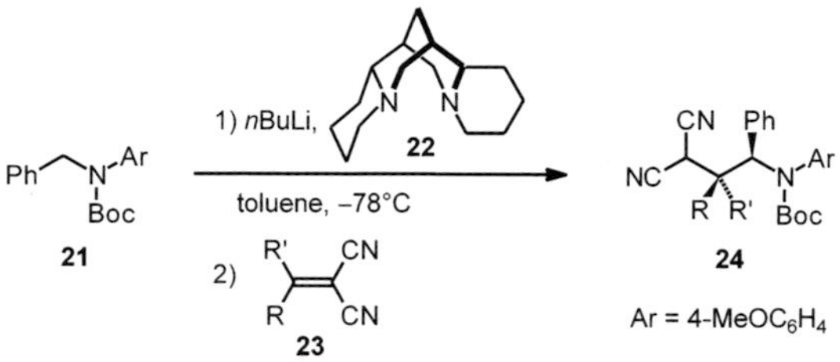

Scheme 4.10 Conjugate additions of chiral organolithium nucleophiles complexed to (–)-sparteine according to Beak.

4.3
Chiral Metal Complexes

4.3.1
Cobalt and Copper Catalysis

The first metal-catalyzed asymmetric Michael reaction was reported in 1984 by Brunner and Hammer [34]. A combination of Co(acac)$_2$ and optically active diamine **25** led to indanone derivative **8a** with 66% *ee* (Scheme 4.11). This study was continued and extended to other substrates, for example acrolein **6b** as acceptor. However, the best selectivity obtained for product **8l** was 39% *ee* together with a rather low chemical yield of 41% [35].

A comment on Michael addition product **8a** might be allowed here, since it seems to be a leading structure in all the investigations mentioned so far and also most of those mentioned later: In contrast to other products without an aromatic

Table 4.2 Reaction of chiral lithiated intermediates complexed to (–)-sparteine **22** with tetrasubstituted α,α-dinitrile activated olefins **23** giving products **24** with quaternary stereocenter.

23	R	R'	24	Yield (%)	*dr*	*er*
a	Ph	Me	a	91	92:8	95:5
b	Ph	Et	b	89	95:5	99:1
c	Me	nPr	c	94	87:13	98:2
d	Me	Et	d	60	78:22	98:2

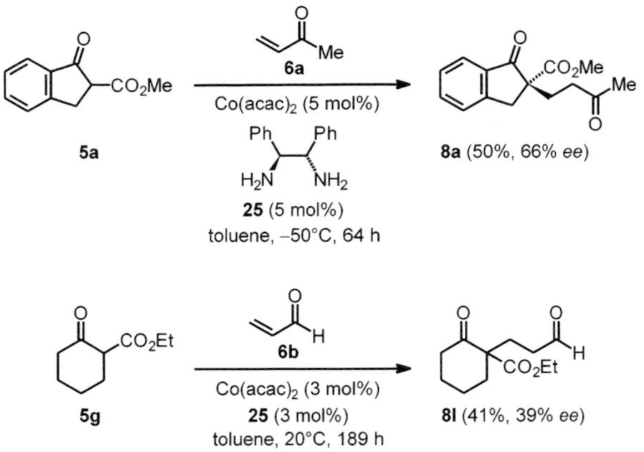

Scheme 4.11 Cobalt/diamine-catalyzed Michael addition according to Brunner.

ring, derivative **8a** is a solid material and therefore easy to purify by recrystallization, thus resulting in an enhancement of optical purity (but, of course, lowering the yield of isolated product). This might be one of the reasons why **5a** was an ubiquitous substrate in early studies of both metal- and organocatalysis.

Not surprisingly, the group led by Desimoni also focused their studies of optically active tetradentate Schiff base/copper(II) complexes **26** on substrate **5a**. In initial investigations product **8a** was obtained quantitatively with ca. 70% *ee* with dimeric complexes **26a** and **26b** whose ligands derive from salicylaldehyde and optically active aminodiols (Scheme 4.12) [36]. In further work a series of other amino alcohols was used for imine synthesis, and with complex **26c** the selectivity of **8a** could be enhanced to 75% *ee*, although a stoichiometric quantity of the catalyst was necessary in this case [37].

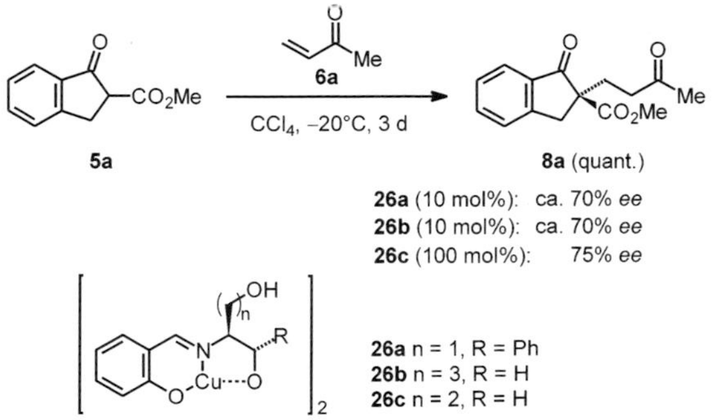

Scheme 4.12 Copper salicylimine complexes according to Desimoni.

4.3.2
Rhodium Catalysis

Ito and coworkers have developed an optically active diphosphanebiferrocene ligand **28** called PhTRAP with both planar and central chirality, which is a rare case of a diphosphane ligand that chelates Pt(II), Pd(II), and Rh(I) center metals in a *trans* manner [38]. First studies on asymmetric Michael addition of α-cyanopropionate **27a** to olefins **6a–d** with a Rh catalyst generated *in situ* from RhH(CO)(PPh$_3$)$_3$ and (*S,S*)-(*R,R*)-PhTRAP **28** gave products **29a–d** with a quaternary stereocenter in excellent yields and 85–87% *ee* selectivity (Scheme 4.13, Table 4.3) [39].

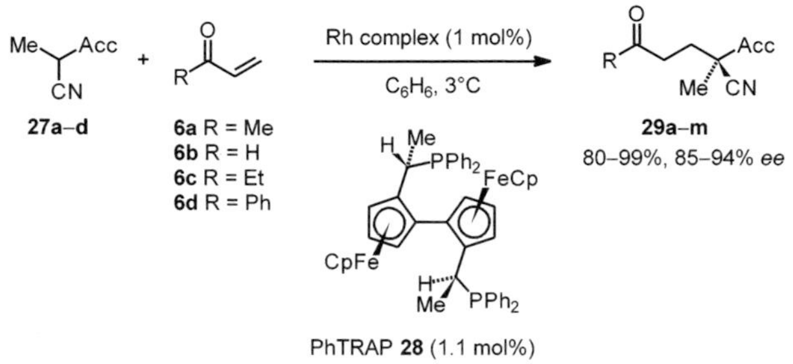

Scheme 4.13 Rhodium catalysis with optically active PhTRAP ligand **28** according to Ito.

After optimization, selectivities were improved up to 93% *ee* by application of the sterically demanding 2,4-dimethyl-3-pentyl ester **27b** [40]. These results could be forwarded on α-cyano Weinreb amides **27c** (Table 4.3) [41]. Moreover, the phosphine-free Rh complex Rh(acac)(CO)$_2$ turned out to be a better catalyst precursor than the compound used in the first studies. The *N*-methoxy-*N*-methylcarboxamide

Table 4.3 Rhodium-catalyzed Michael reaction of α-cyano substituted donors **27** and olefins **6** according to Ito's method shown in Scheme 4.13.

Donor	Acc	Olefin	R	Rh complex	Product	Yield (%)	*ee* (%)
27a	CO$_2$iPr	6a	Me	[RhH(CO)(PPh$_3$)$_3$]	29a	97	86
27a	CO$_2$iPr	6b	H	[RhH(CO)(PPh$_3$)$_3$]	29b	88	87
27a	CO$_2$iPr	6c	Et	[RhH(CO)(PPh$_3$)$_3$]	29c	98	85
27a	CO$_2$iPr	6d	Ph	[RhH(CO)(PPh$_3$)$_3$]	29d	95	83
27b	CO$_2$CHiPr$_2$	6a	Me	[RhH(CO)(PPh$_3$)$_3$]	29e	99	93
27b	CO$_2$CHiPr$_2$	6b	H	[RhH(CO)(PPh$_3$)$_3$]	29f	88	92
27b	CO$_2$CHiPr$_2$	6c	Et	[RhH(CO)(PPh$_3$)$_3$]	29g	99	91
27c	CON(OMe)Me	6a	Me	Rh(acac)(CO)$_2$	29h	98	93
27c	CON(OMe)Me	6b	H	Rh(acac)(CO)$_2$	29i	94	89
27c	CON(OMe)Me	6d	Ph	Rh(acac)(CO)$_2$	29j	99	94
27d	P(O)(OEt)$_2$	6a	Me	Rh(acac)(CO)$_2$	29k	96	93
27d	P(O)(OEt)$_2$	6b	H	Rh(acac)(CO)$_2$	29l	80	92
27d	P(O)(OEt)$_2$	6c	Et	Rh(acac)(CO)$_2$	29m	98	93

moiety (Weinreb amide) in products **29h–29j** is a versatile synthetic intermediate owing to its easy transformation into aldehydes, ketones, and carboxylic acid derivatives [42]. The scope of Rh catalysis was further proved by conversion of α-(cyanoethyl)phosphonate **27d** to give optically active products **29k** and **29l**, which can be converted into quaternary α-aminophosphonic acid derivatives [43].

The asymmetric reaction of α-cyanopropionates with acrolein **6b** was also investigated by the group led by Motoyama [44], which developed a chiral Rh catalyst generated *in situ* from [RhCl(coe)$_2$]$_2$ (coe = cyclooctene) and (Phebox)SnMe$_3$ **30** (Phebox = 2,6-bis(oxazolinyl)phenyl) as ligand precursor. Selectivities up to 86% *ee* were achieved as exemplified in Scheme 4.14.

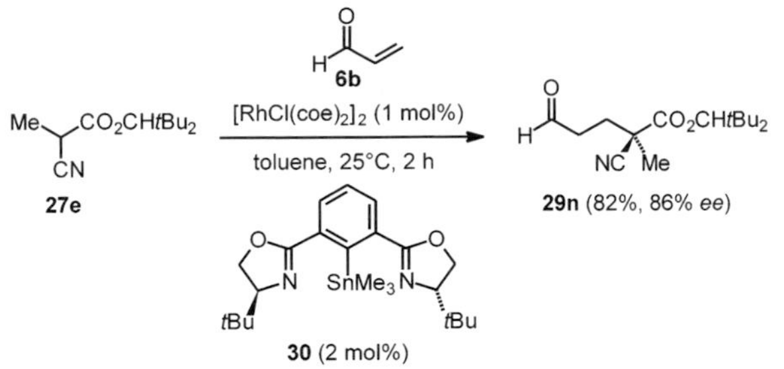

Scheme 4.14 Rhodium-catalyzed Michael reaction according to Motoyama.

Indanonecarboxylate **8a** was isolated quantitatively with a maximum enantiomeric excess of 75% by Suzuki et al. [45] using a catalyst prepared by conversion of [Cp*RhCl$_2$]$_2$ (Cp* = Me$_5$C$_5$) (1 eq.), sulfonamide **31** (2 eq.), and KOH (10 eq.) as depicted in Scheme 4.15.

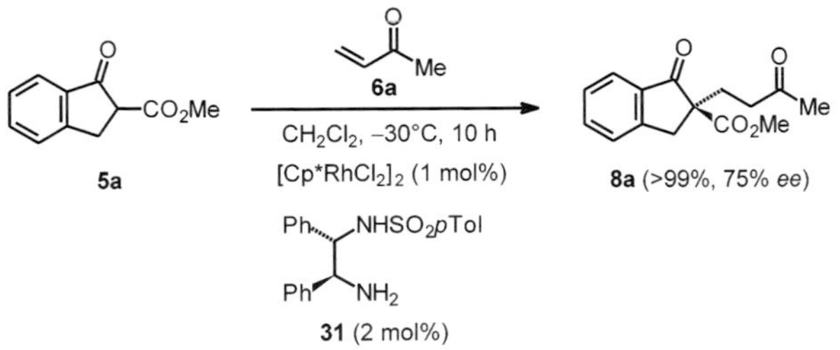

Scheme 4.15 Rhodium-catalyzed Michael addition according to Suzuki.

4.3.3
Heterobimetallic Catalysis

A significant contribution to the field of asymmetric catalysis was made by Shibasaki and his coworkers by the development of rare earth heterobimetallic BINOL complexes, which demonstrated their versatility in a number of organic transformations [46]. Asymmetric Michael reactions were reported to be promoted by either an alkali metal-free La-BINOL catalyst **32a** [47], lanthanum-sodium binaphtholate (LSB) **32b** [48], or aluminum-lithium binaphtholate (ALB) **32c** [49]. The results obtained for tertiary stereocenters are shown in Scheme 4.16 and collected in Table 4.4.

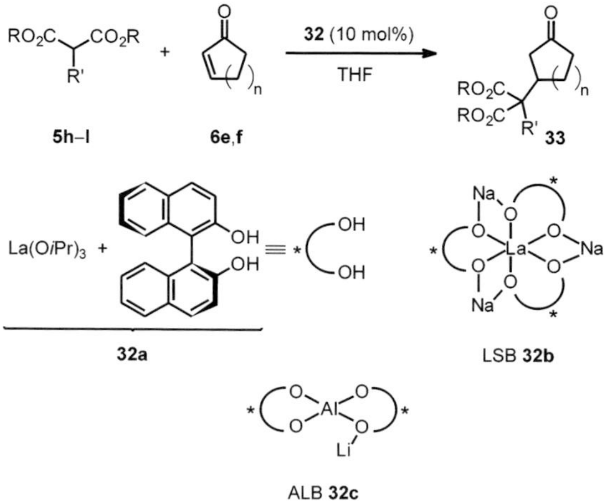

Scheme 4.16 Michael additions following the methodology developed by Shibasaki.

Obviously, ALB **32c** simply generated by *in situ* conversion of LiAlH$_4$ with BINOL (2 eq.) is the recommended catalyst since high selectivities are achieved even at ambient temperature, whereas **32a** and LSB **32b** result in lower enantiomeric excesses at 0 to –40°C. The ALB-catalyzed process has been applied to tandem Michael–aldol reactions, which can be utilized as a key step in prostaglandin synthesis [50], and, furthermore, is working reliably on a large scale [51].

Table 4.4 BINOL-catalyzed Michael additions according to Shibasaki.

Donor	R	R'	Acc	n	Catalyst	T (°C); time (h)	Product	Yield (%)	ee (%)
5h	Bn	Me	6e	1	32a	–20; 60	33a	97	95
5i	Bn	H	6e	1	32a	–20; 72	33b	96	92
5h	Bn	Me	6f	2	32a	0; 84	33c	83	87
5i	Bn	H	6f	2	32a	–10; 84	33d	94	92
5h	Bn	Me	6e	1	32b	–40; 36	33a	89	72
5h	Bn	Me	6f	2	32b	0; 24	33c	91	92
5i	Bn	H	6f	2	32b	0; 24	33d	97	88
5j	Et	Me	6e	1	32c	rt, 72	33e	84	91
5h	Bn	H	6e	1	32c	rt, 60	33b	93	91
5h	Bn	H	6f	2	32c	rt, 72	33d	88	99
5k	Me	H	6f	2	32c	rt, 72	33f	90	93
5l	Et	H	6f	2	32c	rt, 72	33g	87	95

The construction of quaternary stereocenters, however, requires rather a low temperature of –50°C (Scheme 4.17). In this case, LSB **32b** is the catalyst of choice, and, furthermore, CH_2Cl_2 rather than THF was found to be the optimal solvent [52]. The enantioselectivities of addition products **8j–m**, and **34a–c** with quaternary carbon center are generally lower than for the construction of tertiary stereocenters.

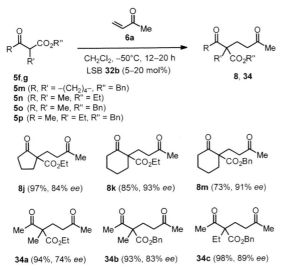

Scheme 4.17 Generation of quaternary stereocenters using Shibasaki's method.

Lanthanum-sodium binaphtholate **32b** is readily prepared from $LaCl_3$ and NaOtBu (3 eq.) together with binaphthol (BINOL). The structure of the bimetallic complexes has been elucidated and a mechanism has been established. In the LSB compound **32b** the La(III) center is in an octahedral environment of three chelating BINOL ligands. The three sodium cations each bridge a pair of BINOL ligands. The Lewis acidic lanthanum center can extend its coordination number to seven by binding to the carbonyl group of the acceptor, e.g. **6a**. The donor is deprotonated by one Brönstedt basic BINOLate, and the resulting dionate anion coordinates as a chelate to a sodium cation that still has contact with the coordination sphere of La via one BINOL oxygen, as outlined in Fig. 4.1. The key feature of the LSB catalyst is to combine both Lewis acidity and Brönstedt basicity. The deprotonated donor and the activated acceptor are held together in the coordination sphere of the bimetallic complex. The stereochemistry of the Michael reaction proceeds under the control of the chiral BINOL ligands.

Lanthanum-sodium binaphtholate can also be applied successfully to thia-Michael reactions giving tertiary sulfides in 56–94% yield with a range of 56–90% *ee* [53]. Feringa et al. reported ALB catalysis for Michael additions of α-nitroesters to α,β-unsaturated ketones, yielding quaternary stereocenters with the highest selectivity being 80% *ee* [54].

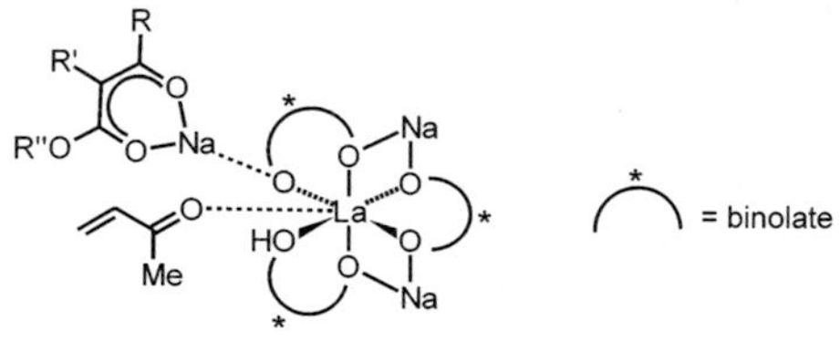

Fig. 4.1 Working model **35** of LSB **32b** catalysis shown for methyl vinyl ketone **6a**.

The development of efficient methods to facilitate the recovery and reuse of asymmetric catalysts remains an important goal in organic chemistry. To address this issue, Shibasaki and coworkers synthesized so-called linked-BINOL complexes with increased stability that meet these requirements [55]. One representative, generated from ligand **36** and La(OiPr)$_3$, is a complex that is air- and water-stable and storable over a long period. Results with the La-catalyst derived from **36** by asymmetric Michael reactions are satisfactory as far as the formation of tertiary stereocenters is concerned. However, as indicated in Scheme 4.18, although results with quaternary stereocenters are promising, they need further development [56]. Tertiary alcohols, for example compound **38**, were obtained with selectivities up to 96% from racemic α-hydroxyketone **37** and methyl vinyl ketone **6a** using a catalyst prepared *in situ* from Et$_2$Zn and linked BINOL **36** [57].

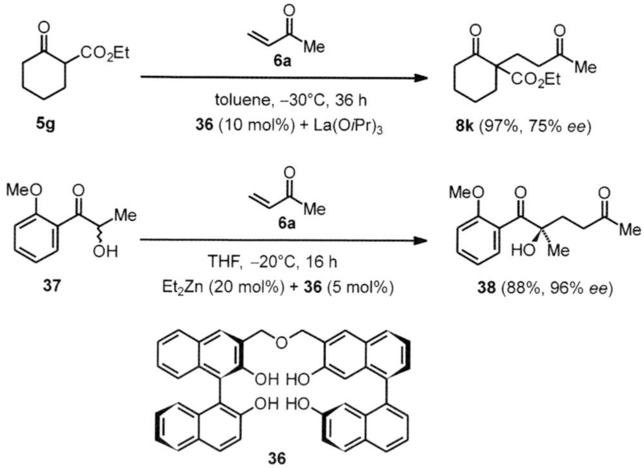

Scheme 4.18 Michael reactions with linked BINOL catalysts developed by Shibasaki.

4.3.4
Miscellaneous Examples

An important breakthrough in the field of enantioselective Michael reactions has been reported by Sodeoka and coworkers [58]. For the first time, quaternary stereocenters could be generated with >90% *ee* at the relatively high temperature of –20°C in a Michael reaction of β-dicarbonyl compounds **5** with enones **6** using palladium(II) as the catalytically active transition metal. The Pd(II)-diaqua-diphosphane complexes **39a** and **39b** are formed with chiral ligands (*R*)-Tol-BINAP and (*R*)-BINAP, respectively. Scheme 4.19 shows four typical reaction products **8n,o** and **34d,e** as examples. Both cyclic (**5q**) and acyclic (**5r**) β-keto esters react with methyl vinyl ketone **6a** at –20°C in a Michael reaction promoted by **39a** to give the corresponding products **8n** and **34d** in moderate to good yields with >90% *ee*. The use of the sterically demanding *tert*-butyl- and phenyl-β-keto esters is noteworthy. The reaction of β-diketone **5s** (X = CH₂Ar) in the presence of catalyst **39b** (10 mol%) at –10°C is an outstanding example of an asymmetric Michael addition to acyclic triketones such as **34e** with very high stereoselectivity, an achievement not previously reported. Furthermore, the reaction of **5q** with substituted enone **6g** catalyzed by **39b** at –20°C affords a mixture of two diastereoisomers (*dr* = 8:1), and an enantiomeric excess of 99% for the major diastereomer of **8o**.

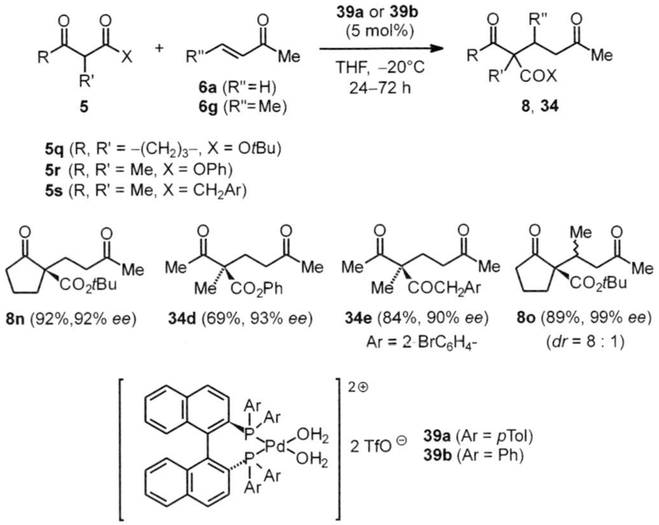

Scheme 4.19 Enantioselective Pd catalysis using the chiral ligands (*R*)-Tol-BINAP (**39a**) or (*R*)-BINAP (**39b**) according to Sodeoka.

Shibasaki's group was actually not the first to investigate rare earth metal complexes as catalysts for asymmetric Michael reactions. A rather curious example is the report of Scettri et al. on generation of a quaternary stereocenter with a maximum 36% *ee* by applying 30 mol% of the chiral NMR-shift reagent Eu(tfc)₃ (tfc = 3-(trifluoromethylhydroxymethylene)-*d*-camphorato) [59]. Kobayashi et al. published the first asymmetric Michael reactions in water, catalyzed by a Ag(I)-BINAP complex giving indanone derivative **8a** in 74% yield and 38% *ee* (Scheme 4.20) [60]. By applying optically active scandium-biquinolinedioxide complexes,

Nakajima and coworkers prepared indanone **8a** with 98% yield but low 39% *ee*, whereas the isobutyl ester **5t** gave the corresponding product **8p** with high 89% yield and 84% *ee* (Scheme 4.20) [61].

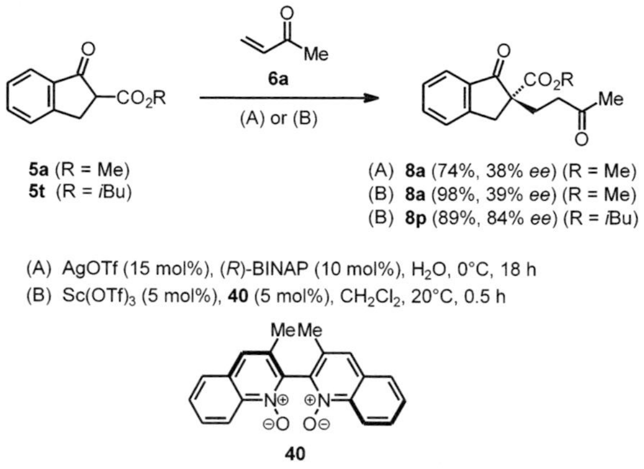

5a (R = Me)
5t (R = *i*Bu)

(A) **8a** (74%, 38% *ee*) (R = Me)
(B) **8a** (98%, 39% *ee*) (R = Me)
(B) **8p** (89%, 84% *ee*) (R = *i*Bu)

(A) AgOTf (15 mol%), (*R*)-BINAP (10 mol%), H₂O, 0°C, 18 h
(B) Sc(OTf)₃ (5 mol%), **40** (5 mol%), CH₂Cl₂, 20°C, 0.5 h

40

Scheme 4.20 Michael reactions in the presence of Ag(I)-BINAP and chiral biquinoline *N,N'*-dioxide Sc-complexes, respectively.

Chiral metal-salen complexes have been successfully introduced by Jacobsen for asymmetric epoxidations of unactivated olefins [62]. This extraordinary story of success was continued by extension to other organic transformations including conjugate additions to activated olefins as acceptors [63]. With respect to the formation of quaternary stereocenters, conjugate addition of α-cyanoacetates **41** to α,β-unsaturated imides **42** under catalysis of a dimeric μ-oxo-aluminum-salen complex (*S,S*)-[(salen)Al]₂O **43** has been reported [64]. The corresponding products, α-cyano-δ-keto esters **44**, were obtained in 76–98% yield (Scheme 4.21, Table 4.5). Two diastereoisomers are formed with *dr*s in the range of 14:1 to 35:1, while the enantiomeric excess of the major isomer ranges from 91% to 98%. In the case of α-(benzylamino)-α-cyanoacetate as donor, subsequent cyclization to γ-lactams **45** occurs. Again, high yields and *dr*s between 10:1 and 20:1 as well as enantioselectivities up to 97% *ee* are obtained (Scheme 4.21, Table 4.5).

4.4
Chiral Auxiliaries

4.4.1
α-Phenethylamine

Asymmetric catalysis is commonly regarded as superior to the use of optically active auxiliaries since the chiral information is required only in sub-stoichiometric amounts and, more importantly, the introduction and removal of the auxiliaries usually needs two additional synthetic steps. However, a number of reported

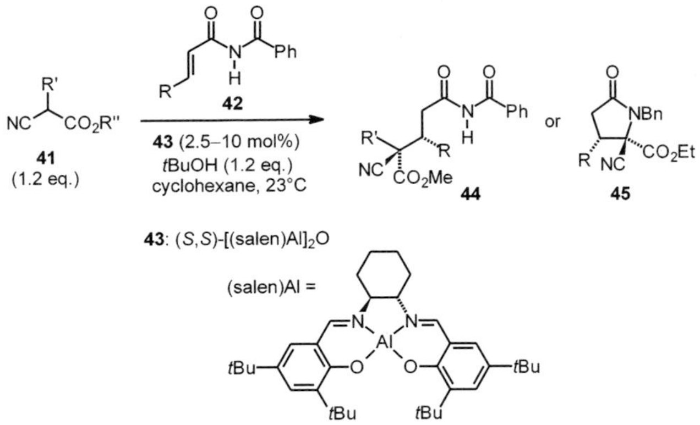

Scheme 4.21 Diastereo- and enantioselective conjugate addition of α-cyanoacetates according to Jacobsen.

auxiliary-mediated processes are successfully competing with chiral catalysts. A precondition is, of course, the ready availability of the chiral auxiliary at low cost, e.g. from the *chiral pool*. With regard to the formation of quaternary stereocenters by asymmetric Michael reactions, chiral amines often give yields and selectivities which are comparable or even superior to enantioselective catalysis. Therefore, some established methods will be discussed in this section. As mentioned, chiral primary and secondary amines are the dominating class of compounds, and these commonly form imines and enamines with ketones as Michael donors.

4.4.1.1 Monoketones as Donors

A pioneering method was developed in 1985 by the group led by d'Angelo with the use of α-phenethylamine **47** as auxiliary in the addition of cyclic α-alkylated ketones **46** to activated olefins **6** as acceptors (Scheme 4.22) [65]. The auxiliary is introduced by Brönstedt-acid-catalyzed imine formation. If water is removed, for

Table 4.5 Conjugate addition of α-cyanoacetates **41** to α,β-unsaturated N-benzoyl amides **42** according to Jacobsen.

Product	R	R'	Time (h)	Yield (%)	ee (%)	dr
44a	nPr	Ph	19	98	97	14:1
44b	nPr	3-CF$_3$C$_6$H$_4$	15	96	98	35:1
44c	nPr	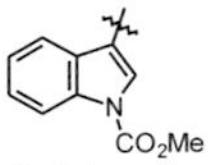 (indole-CO$_2$Me)	51	79	95	14:1
44d	nPr	2-thienyl	96	76	91	14:1
45a	Me	NHBn	26	74	92	10:1
45b	nPr	NHBn	23	92	91	14:1
45c	iBu	NHBn	60	89	97	13:1
45d	Ph	NHBn	54	68	96	20:1

example by azeotropic distillation, the yield of this step is generally quantitative. The obtained imines **48**, or in some cases their enamine tautomers **48′**, are directly converted with Michael acceptors **6** thus leading to intermediate products **49** with quaternary stereocenters, which can be hydrolyzed with dilute acetic acid to ketones **50**. The regioselectivity of this process is commonly high owing to the formation of the thermodynamically favored enamines **48′**. Table 4.6 summarizes some examples with various donors ranging from cyclic α-methyl (**46a,b**) [66], α-alkoxy (**46c**) [67], α-thiophenyl ketones (**46d**) [68], tetrahydrofuranone (**46f**) [69] to tetrahydrothiafuranone (**46g**) derivatives [70]. The yields given refer to the three-step sequence from **46** to **50**, which is often performed as a one-pot protocol.

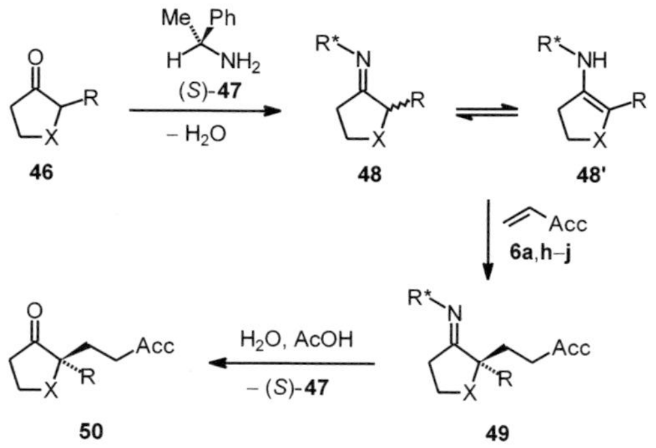

Scheme 4.22 Asymmetric Michael addition with the chiral auxiliary α-phenethylamine according to d'Angelo; R* = α-phenethyl.

Extensive theoretical and experimental investigations confirmed an aza-ene type mechanism of the reaction as depicted in Scheme 4.23 [71]. In order to

Table 4.6 Conversion of cyclic ketones **46** with (S)-phenethylamine **47** as auxiliary according to d'Angelo.

46	X	R	6	Acc	Solvent; T (°C); time (h)	50	Yield (%)	ee (%)
46a	CH$_2$	Me	6a	COMe	THF, 20; 72	50a	83	89
46a	CH$_2$	Me	6h	CO$_2$Me	THF, 20; 72	50b	79	90
46b	(CH$_2$)$_2$	Me	6a	COMe	THF, 20; 72	50c	88	91
46b	(CH$_2$)$_2$	Me	6h	CO$_2$Me	THF, 20; 72	50d	81	90
46c	(CH$_2$)$_2$	OMe	6h	CO$_2$Me	THF, 20; 72	50e	80	97
46d	(CH$_2$)$_2$	SPh	6h	CO$_2$Me	THF, 20; 24	50f	68	93
46d	(CH$_2$)$_2$	SPh	6a	COMe	THF, 0; 12	50g	60	95
46e	CH$_2$	SPh	6h	CO$_2$Me	THF, 20; 24	50h	25	92
46f	O	Me	6h	CO$_2$Me	none, 60; 72	50i	80	91
46g	S	Me	6h	CO$_2$Me	none, 45; 70	50j	63	≥95
46g	S	Me	6a	COMe	Et$_2$O, 20; 24	50k	25	–
46g	S	Me	6i	CN	none, 45; 72	50l	69	90
46g	S	Me	6j	SO$_2$Ph	none, 45; 72	50m	65	>95

avoid *syn*-pentane interaction, the phenethyl moiety adopts a conformation with hydrogen *syn* to the cyclohexene ring. In the case of the (*S*)-configured auxiliary, the phenyl group shields the back face and the acceptor attacks from the front giving the quaternary stereocenter in the configuration observed for products outlined in Scheme 4.22. The (*R*)-configured auxiliary affords the opposite product configuration.

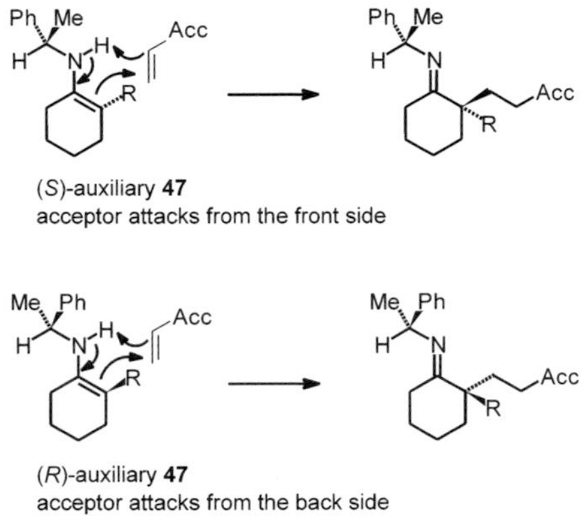

(*S*)-auxiliary **47**
acceptor attacks from the front side

(*R*)-auxiliary **47**
acceptor attacks from the back side

Scheme 4.23 Proposed mechanism of the phenethylamine-mediated Michael reaction.

α-Substituted acceptors such as acetoxy acrylonitrile **51a** give addition products with both a quaternary and a tertiary stereocenter. As shown in Scheme 4.24, diastereo- and enantioselectivity are quantitative and, moreover, absolute configurations are again in perfect accordance with an aza-ene-like mechanism [72]. Product **50n** in Scheme 4.24 can be further transformed to the optically active side chain of homoharringtonine, the naturally occurring derivative of antitumor-active cephalotaxine [73]. Acceptors with β-substituents, such as methyl crotonate **51b**, are almost inert to aza-ene reaction and reveal significant conversion only within several days at ambient temperature. Unfortunately, in these cases an increase of reaction temperature often results in unspecified decomposition of the starting materials [74]. As for other pericyclic reactions, a negative $\Delta V^{\neq}$ can be assumed, so that a pressure increase would accelerate the reaction [75]. Indeed, methyl crotonate **51b** reacts smoothly with imines **48a,b** at 1.2 GPa to products **50o** and **50p** with perfect diastereo- and enantioselectivity (Scheme 4.24) [76a].

An intramolecular addition of substrate **46i** gave spirocyclic compound **52** with 85% overall yield (Scheme 4.25). Despite elevated temperature (120°C in DMF) required for the aza-ene reaction, perfect stereoselectivities are obtained [76b].

A number of precursors for biologically active compounds have been prepared using aza-ene reactions as a key step. For example, a variety of tetralones **55** were transformed to hexahydrophenanthrones **56** with a quaternary stereocenter [77]. The yields in Scheme 4.26 relate to a four-step sequence involving imine formation, aza-ene reaction, cleavage of the auxiliary, and Robinson annulation.

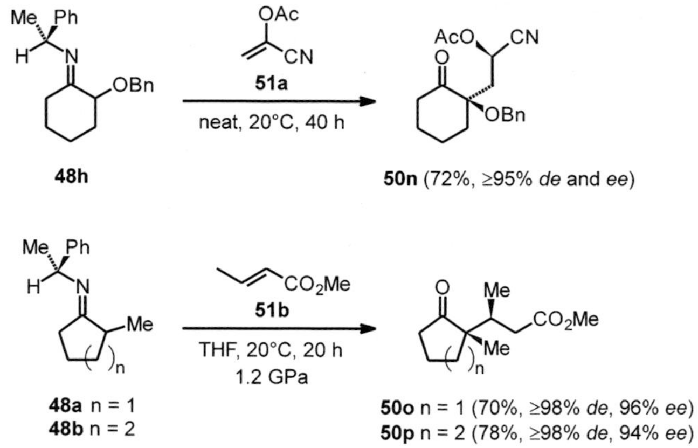

Scheme 4.24 Reaction of imines with acrylonitrile derivatives and β-substituted acrylates.

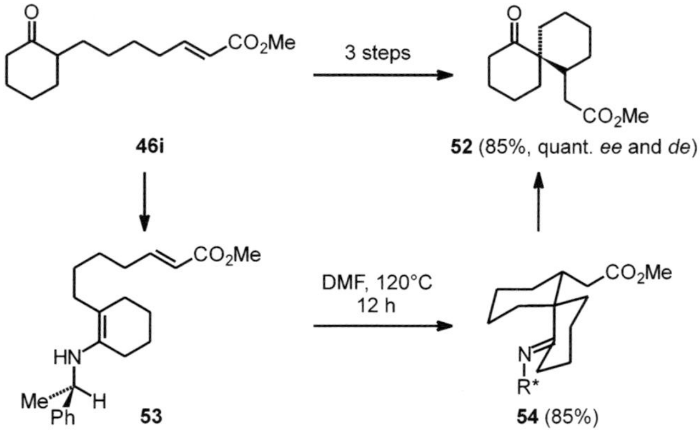

48h

50n (72%, ≥95% *de* and *ee*)

48a n = 1
48b n = 2

50o n = 1 (70%, ≥98% *de*, 96% *ee*)
50p n = 2 (78%, ≥98% *de*, 94% *ee*)

46i

52 (85%, quant. *ee* and *de*)

53

54 (85%)

Scheme 4.25 Intramolecular aza-ene reaction.

55

56

56a R,R" = H, R' = Me   (70%, 92% ee)
56b R = OMe, R" = H, R' = Me   (80%, 93% ee)
56c R = H, R" = OMe, R' = CH₂CO₂*t*Bu   (80%, ≥95% ee)

Scheme 4.26 Synthesis of hexahydrophenanthrone derivatives.

4.4.1.2 β-Dicarbonyl Compounds as Donors

Aza-ene type Michael reaction of β-dicarbonyl compounds **5** as donors was first investigated by Brunner and coworkers [78]. Conversion with (R)-phenethylamine **47** gave enamines without any imine tautomer detectable by NMR spectroscopy, which is obviously due to stabilization of the enamine by intramolecular hydrogen bonding (Scheme 4.27). Methylenemalonate **51c** reacted at ambient temperature to afford triester **50q** in satisfactory yield and excellent selectivity (≥95% *ee*). However, the addition to methyl vinyl ketone **6a** or methyl acrylate **6h** proceeded very slowly, and the obtained selectivities were in the range of 65–89% *ee* at elevated temperature. To overcome this disadvantage, Guingant studied reactions with Lewis-acidic additives or at high pressure. And indeed, conversion of β-keto ester **5g** was achieved even at −78°C with ZnCl$_2$ (1 equivalent), but again with relatively low enantiomeric excess of 79%. MgBr$_2$ (1 equivalent) or a pressure of 11 kbar turned out to have a beneficial effect on both reaction rate and selectivity in the addition reaction of acrylates **6** [79]. Consequently, d'Angelo et al. used MgBr$_2$ (1.2 equivalents) as additive for the reaction with methyl methacrylate **51d** as acceptor [80]. The formed product **50r** was isolated in 72% yield after 7 d at room temperature with quantitative diastereo- and enantioselectivity. Surprisingly, the reaction of β-keto ester **5f** with nitroolefin **51e** occurred rapidly at −15°C without additive or catalyst, yielding product **8u** (80%) with 90% *ee* [81]. Table 4.7 summarizes these results.

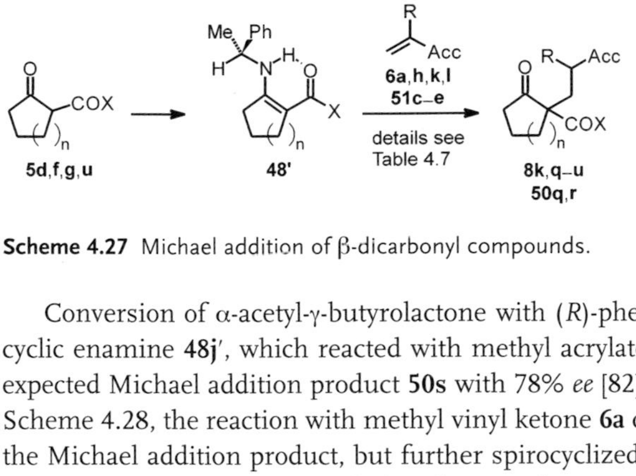

Scheme 4.27 Michael addition of β-dicarbonyl compounds.

Conversion of α-acetyl-γ-butyrolactone with (R)-phenethylamine **47** gave exocyclic enamine **48j′**, which reacted with methyl acrylate **6h** at 80°C to afford the expected Michael addition product **50s** with 78% *ee* [82]. In contrast, as shown in Scheme 4.28, the reaction with methyl vinyl ketone **6a** did not stop at the stage of the Michael addition product, but further spirocyclized to imine **57** in 83% yield and 86% *de*. The latter was hydrolyzed to spirolactone **58**.

Three groups applied the methodology to obtain optically active cyclohexenones **59** as important intermediates for natural product synthesis (Scheme 4.29) [83]. Michael reaction was promoted by ZnCl$_2$ (1 equivalent) and was followed by Robinson annulation under different basic conditions.

Stille and coworkers reported the preparation of δ-lactam derivatives **61** by treatment of chiral enamino esters originating from β-keto esters **60** with acryloyl chloride **6m** through formation of a C–C bond and *N*-acylation to generate the amide functionality and subsequent intramolecular enamine–Michael reaction

Table 4.7 Michael addition of various cyclic β-dicarbonyl compounds with acrylates.

Donor	*n*	X	Olefin	Acc	R	Solvent [cat./pressure]; *T* (°C); time (h)[a]	Prod.	Yield (%)	*ee* (%)
5g	2	OEt	51c	CO$_2t$Bu	CO$_2t$Bu	toluene; 20; 67	50q	67	≥95 (*R*)
5g	2	OEt	6a	COMe	H	toluene; 40; 306	8k	50	79 (*R*)
5g	2	OEt	6h	CO$_2$Me	H	toluene; 81; 92	8q	29	89 (*R*)
5d	2	Me	6a	COMe	H	toluene; 60; 70	8r	58	65 (*R*)
5g	2	OEt	6a	COMe	H	Et$_2$O [ZnCl$_2$]; −78; 1.5	8k	80	79 (*S*)
5g	2	OEt	6h	CO$_2$Me	H	Et$_2$O [MgBr$_2$]; 20; 2	8q	76	80 (*S*)
5g	2	OEt	6k	CO$_2t$Bu	H	Et$_2$O [MgBr$_2$]; 20; 3	8s	60	90 (*S*)
5g	2	OEt	6l	CO$_2$Et	H	THF [11 kbar]; 40; 65	8t	63	88 (*S*)
5u	2	OMe	51d	CO$_2$Me	Me	Et$_2$O; 20; 7 d	50r	72	≥95 ≥95 *de*
5f	1	OEt	51e	NO$_2$	H	THF; −15; 10 min	8u	80	90 (*S*)

a Unless otherwise stated.

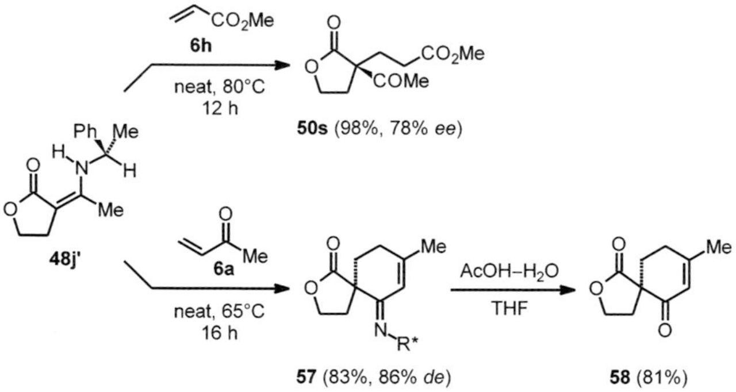

Scheme 4.28 Michael reaction of lactone-derived imine with various acceptors.

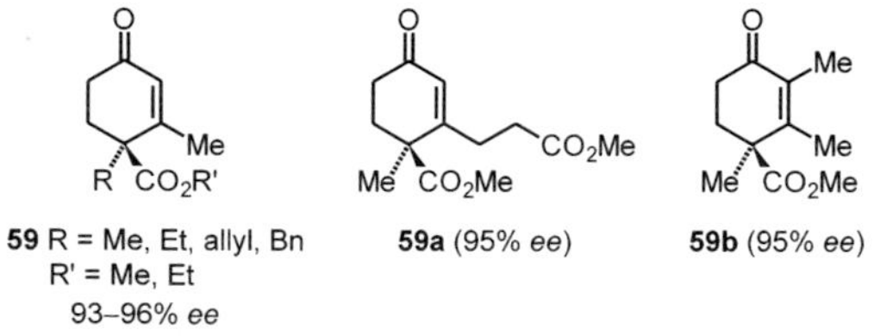

Scheme 4.29 Cyclohexenone derivatives as precursors for natural-product synthesis.

(Scheme 4.30) [84]. Phenethylamine is the chiral auxiliary as well as the lactam nitrogen source, and can be cleaved by catalytic hydrogenation in a subsequent synthetic step.

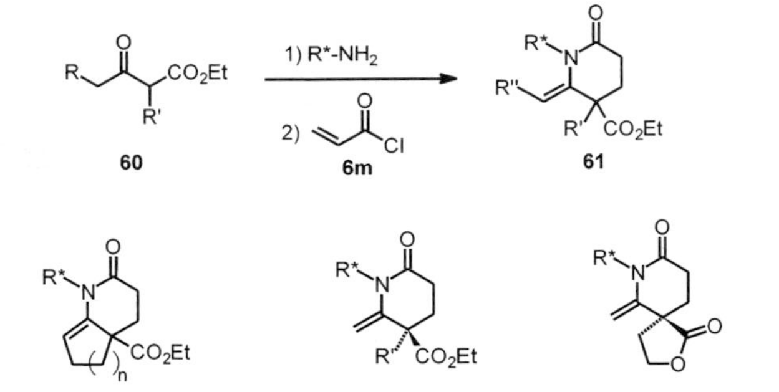

61a n = 1 (76%, *dr* 97 : 3)   **61c** R' = Me (92%, *dr* 97 : 3)   **61e** (80%, *dr* 94 : 6)
61b n = 2 (85%, *dr* >97 : 3)   **61d** R' = OBn (58%, *dr* 92 : 8)

Scheme 4.30 Synthesis of δ-lactam derivatives through aza-annulation according to Stille; R* = α-phenethyl.

The enantioselective synthesis of (+)-vincamine **65**, which was accomplished in 1997 by the group led by d'Angelo, should be mentioned as an example of a total synthesis. Vincamine **65**, as the major alkaloid encountered in periwinkle (*Vinca minor* L.), is reported to have cerebroprotective activity caused by a dilation of brain arteries. Enamine **63** is accessible from tryptamine **62**. Michael reaction with methyl acrylate **6h** results in diketo ester **64** with acceptable chemo- and stereoselectivity, and this is further transformed to the target structure **65**. The overall yield of (+)-vincamine is 1.2% over 15 steps starting from tryptamine **62** (Scheme 4.31) [85].

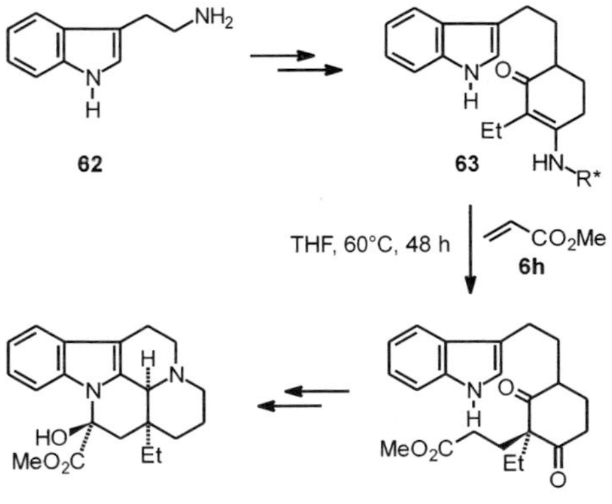

Scheme 4.31 Michael addition as the key step in the synthesis of (+)-vincamine **65** elaborated by d'Angelo; R* = α-phenethyl.

4.4.2
L-Valine Ester

The α-amino acid L-valine is a low-cost *chiral pool* material that has been successfully established by Evans as an auxiliary for asymmetric aldol reactions [86]. Koga and coworkers then reported the use of chiral enamines **66** prepared from β-keto esters **5g,n** and L-valine *tert*-butyl ester in the Michael reaction with methylene-malonate **51c**. Enaminoesters were first deprotonated with a stoichiometric amount of LDA and subsequently converted with the acceptor **51c** at low temperatures. After hydrolysis, enantiopure compounds **50q** and **50t** with a quaternary stereocenter were obtained. Interestingly, the absolute configuration of products **50** was found to be highly solvent dependent. With a mixture of toluene–HMPA (S)-configured products are isolated, whereas in THF the enantiomeric (R)-products are formed (Scheme 4.32, Table 4.8) [87].

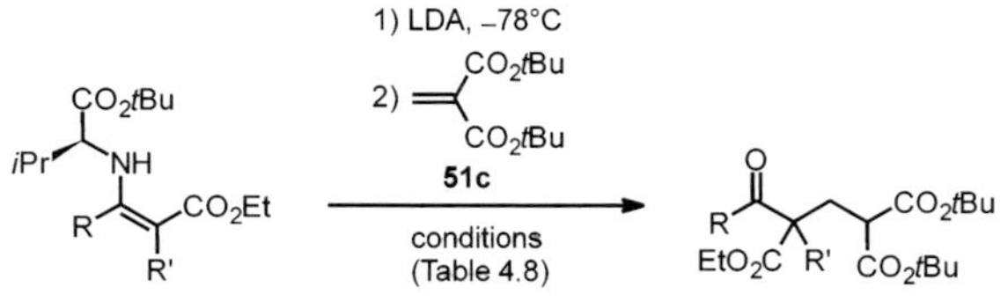

Scheme 4.32 L-Valine ester-promoted Michael reaction according to Koga.

A mechanistic rationale has been developed to explain this configuration-switch by solvent (Fig. 4.2), which is also observed for alkylation reactions with L-valine ester auxiliaries [88]. After deaggregation of the lithio-enamide, the strong ligand L = HMPA completely shields the bottom face of the complex by coordination to Li, resulting in top-side attack of the electrophile **51c**. On the other hand, the rather weakly coordinating ligand L = THF is readily displaced from Li by the Michael acceptor **51c** at the bottom face, which thus coordinates to Li and subsequently reacts from this face.

Following this method, contiguous quaternary and tertiary carbon centers could be constructed by application of an ethylidenemalonate as acceptor [89]. Apart from the need for a stoichiometric quantity of LDA and the rather low temperature (ca. $-100\,^{\circ}C$), the major disadvantage of Koga's method is the low reaction rate of simple acceptors like methyl vinyl ketone **6a**. In order to overcome this low reactivity, the group has investigated several Lewis-acidic additives for converting **6a** with lithioenamides. As a result, an over-stoichiometric amount of TMSCl (5 eq.) was found to be optimal (Scheme 4.33, Table 4.9) [90]. Again, either THF or toluene–HMPA acts as a configuration-switch. However, for addition

Table 4.8 Asymmetric Michael reaction of α-alkyl-β-keto esters via chiral enamines according to Koga and coworkers.

Donor	R	R'	Solvent; T (°C)	Product	Yield (%)	*ee* (%)
5g	-(CH₂)₄-		toluene–HMPA; −95	50q	73	92 (S)
5n	Me	Me	toluene–HMPA; −95	50t	82	87 (S)
5g	-(CH₂)₄-		THF; −105	50q	86	95 (R)
5n	Me	Me	THF; −105	50t	86	84 (R)

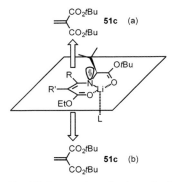

Fig. 4.2 Proposed mechanism of solvent-dependent diastereofacial selection according to Koga: (a) top-side attack of **51c** when L = HMPA; (b) bottom-side attack of **51c** when L = THF.

to methyl vinyl ketone **6a**, the yield and stereoselectivity are only satisfactory in THF [90].

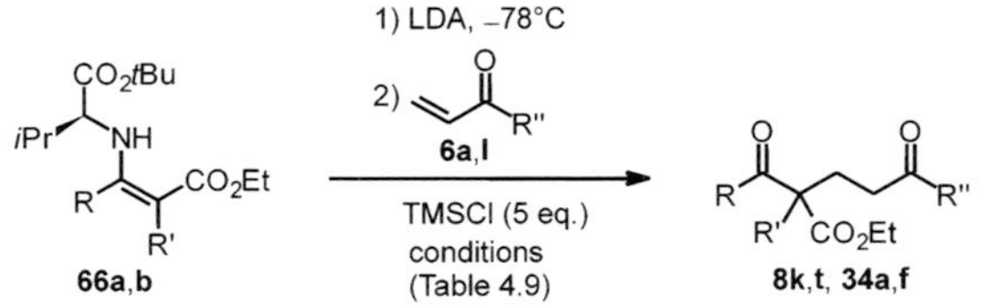

Scheme 4.33 Asymmetric Michael reaction with α,β-unsaturated acceptors in the presence of additives, according to Koga.

4.4.3
L-Valine Diethylamide

In 1999, Christoffers' group carried out an intensive screening program with several primary, chiral amines and transition metal salts as catalysts, which led to the development of a highly reliable process for the formation of quaternary

Table 4.9 Michael reaction with over-stoichiometric amounts of TMSCl according to Koga.

Donor	R	R′	Acc	R″	Solvent; T (°C)	Product	Yield (%)	ee (%)
5g	-(CH$_2$)$_4$-		6a	Me	THF; –100	8k	67	90 (R)
5n	Me	Me	6a	Me	THF; –100	34a	66	87 (R)
5g	-(CH$_2$)$_4$-		6l	OEt	THF; –100	8t	53	57 (R)
5n	Me	Me	6l	OEt	THF; –100	34f	43	79 (R)
5g	-(CH$_2$)$_4$-		6a	Me	toluene–HMPA; –95	8k	48	60 (S)
5n	Me	Me	6a	Me	toluene–HMPA; –95	34a	38	50 (S)
5g	-(CH$_2$)$_4$-		6l	OEt	toluene–HMPA; –95	8t	23	77 (S)
5n	Me	Me	6l	OEt	toluene–HMPA; –95	34f	16	41 (S)

stereocenters by Michael reaction [91]. L-Valine diethylamide **67** combined with catalytic quantities of $Cu(OAc)_2 \cdot H_2O$ turned out to be extraordinarily efficient for this purpose. Representative results are depicted in Scheme 4.34 [92].

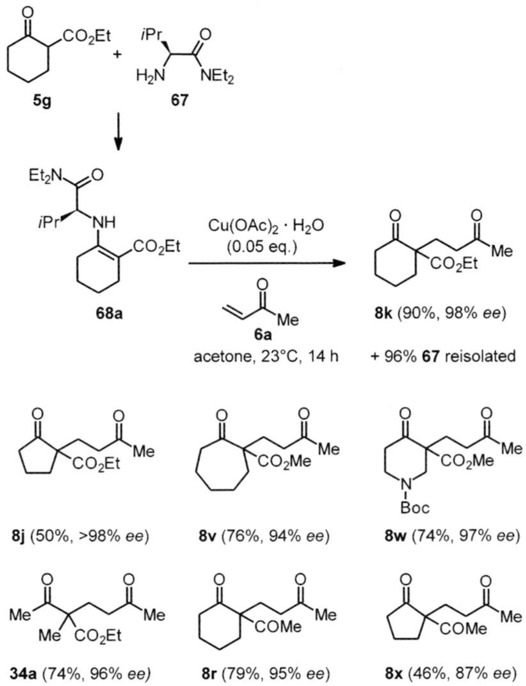

Scheme 4.34 Cu(II)-catalyzed asymmetric Michael reaction with L-valine diethylamide as auxiliary.

The developed procedure is of practical interest: conversion of enamines such as **68a** with **6a** in the presence of $Cu(OAc)_2 \cdot H_2O$ (1–5 mol%) proceeds at ambient temperature. Anhydrous or inert conditions are not required, and the solvent is simply acetone. After acidic workup, the products **8** and **34** were isolated in generally good yield, with selectivities up to 95–99% *ee*. The auxiliary could be separated from the reaction mixture by extraction and recovered almost quantitatively. The selectivities obtained for these products have, to date, not been exceeded by other methods. A special feature of the copper-catalyzed reaction is the compatibility with donor functions such as the carbamate moiety in product **8w** [92c]. Substrates of this type do not convert in reactions using Shibasaki's heterobimetallic catalysts.

The absolute configuration agrees perfectly with the working model outlined in Scheme 4.35 [93]. Enamines such as **68a** coordinate as tridentate ligands with one six-membered azadiketonate chelate and one five-membered ring to Cu(II). Because the isopropyl group shields the front face of the planar donor, the acceptor preferentially coordinates to and is activated from the back face of the complex. Thus, L-valine yields the (R)-configuration of the product **8k**. Scheme 4.35 also clarifies the successful role of amino acid amides as auxiliaries: The amide function coordinates through its carbonyl oxygen to copper, and the role of the amide nitrogen is just to provide electron density to the carbonyl moiety. Interestingly, the counterion of the catalyst is crucial and optimal with acetate, since the enamines need to be deprotonated prior to coordination to Cu(II).

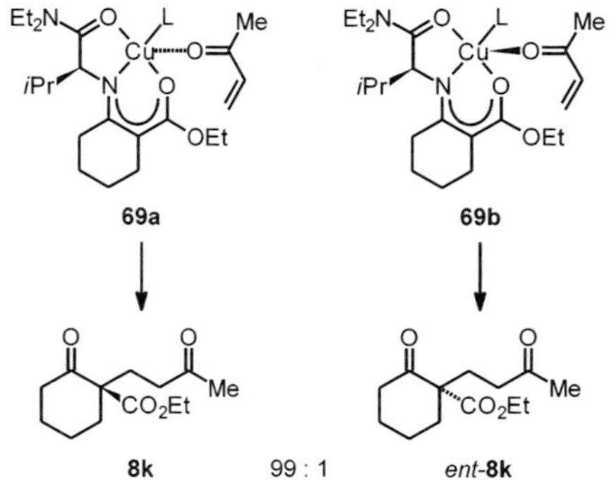

Scheme 4.35 The proposed origin of the stereoselectivity and absolute configuration in Cu(II)-catalyzed Michael reactions.

When α-acetyllactones or α-acetyllactams are converted with auxiliary **67**, the exocyclic enamines **68b** and **68c** are obtained (Scheme 4.36) [94]. In contrast to the endocyclic congeners such as **68a**, on Cu(II)-catalyzed conversion with ketone **6a** these exocyclic enamines give the spirocyclic products **69b** and **69c** in a sequence of Michael reaction and Robinson annulation. The imine moiety exhibits reasonable hydrolytic stability due to a neopentyl situation and is therefore retained in the products **69** (Scheme 4.36).

Acid-catalyzed hydrolysis finally yields the spiroketones **58** and **70c** [94]. When the exocyclic enamine **68d**, derived from α-acetylcyclohexanone **5d**, is converted along this sequence spirodiketone **70d** with (S)-configuration is obtained. Product **8r**, generated from the endocyclic enamine **68e** is spirocyclized to *ent*-**70d** with the opposite (R)-configuration. Evidently, exo- and endocyclic enamines give complementary stereochemical outcomes of the reaction although the same enantiomer of the auxiliary **67** is applied. This behavior is a direct consequence of the mechanistic picture shown in Scheme 4.35 and can be assumed to be additional evidence for the stereochemical model.

A precondition for this complementary stereochemistry is, of course, the control of the regioselectivity of enamine formation realized for compound **5d** as depicted in Scheme 4.37. The endocyclic enamine **68e** results as the thermodynamic product from acid-catalyzed conversion of donor **5d** with auxiliary **67**. The kinetic, exocyclic enamine **68d** is the product of the aminolysis of diketonato difluoroborate **71**.

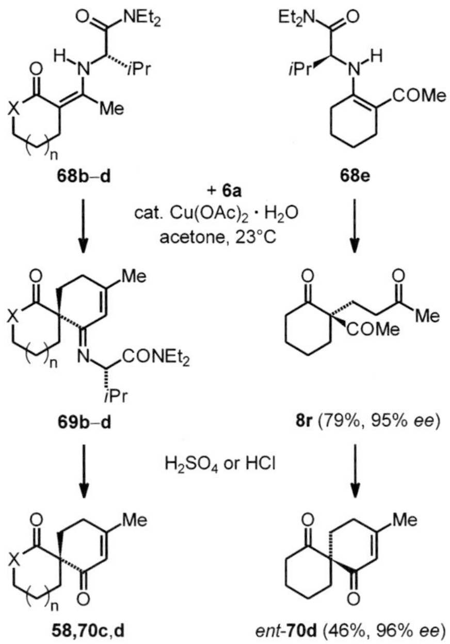

68	n	X	imine	yield	*de*	product	*ee*
b	0	O	69b	54%	>99%	58	>99%
c	1	NBn	69c	37%	>95%	70c	>95%
d	1	CH$_2$	69d	46%	86%	70d	87%

Scheme 4.36 Formation of (*R*) and (*S*) configured spiroketones from exo- and endocyclic enamines with auxiliary L-valine diethylamide.

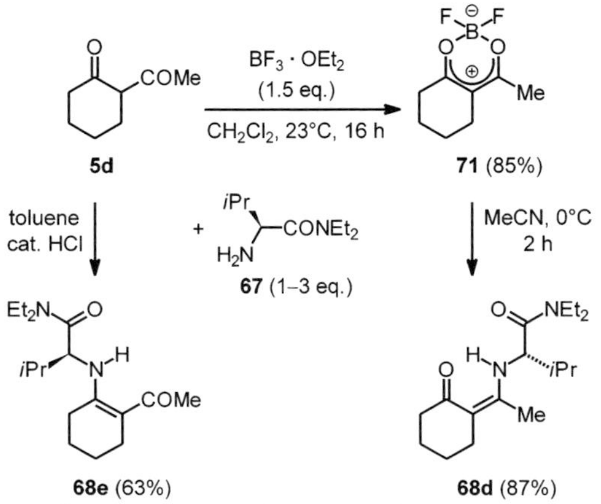

Scheme 4.37 Formation of endo- and exocyclic enamines.

4.4.4
Miscellaneous Examples

The L-proline-derived (*S*)-aminomethoxymethylpyrrolidine (SAMP) as well as its enantiomeric sibling have been successfully established by Enders as auxiliaries for various C–C bond-forming reactions including conjugate additions [95]. However, only a few examples lead to the formation of quaternary carbon centers. Lassaletta and coworkers, for example, have employed the formaldehyde SAMP-hydrazone **72** as a neutral formyl anion equivalent for conjugate additions to enones **73** promoted by dimethylthexylsilyltriflate (TDSOTf) [96]. Adducts can either be isolated as silyl enol ethers with up to 98% *de* or treated *in situ* with TBAF giving ketones **74** with a quaternary stereocenter in the β-position. The hydrazone moiety can be cleaved by ozonolysis to give aldehydes **75** or with Mg monoperoxyphthalate (MMPP) yielding nitriles **76**. Two examples are given in Scheme 4.38.

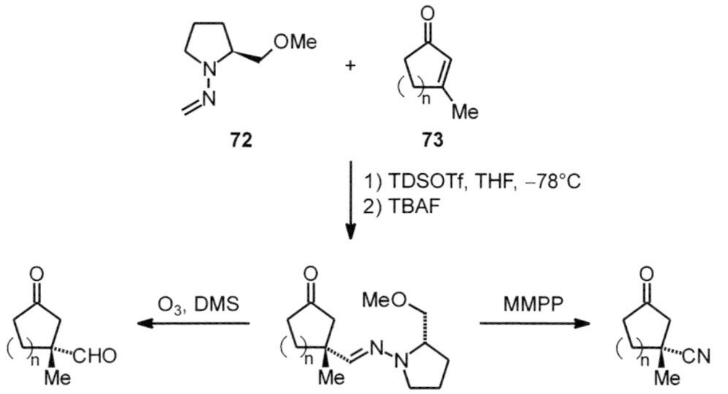

75a n = 1 (87%, ≥98% *ee*)
75b n = 2 (69%, 85% *ee*)
74 (60–68%, 85–98% *de*)
76a n = 1 (98%, ≥98% *ee*)
76b n = 2 (87%, 84% *ee*)

Scheme 4.38 Michael addition according to Lassaletta.

Palomo and his coworkers reported a highly diastereoselective intramolecular sulfur transfer giving tertiary thiols **78** from chiral *N*-enoyl oxazolidine-2-thiones **77**. Similarly to the application of the Evans auxiliary, the starting material was prepared by conversion of enoyl chlorides with 4-isopropyloxazolidine-2-thione. Subsequent treatment with an excess of Lewis acid $BF_3 \cdot OEt_2$ activated the enone moiety for intramolecular nucleophilic attack by the sulfur atom [97]. Upon aqueous workup thiols, now with an oxazolidinone residue, are obtained in good yield and high stereoselectivity, as exemplified in Scheme 4.39.

The authors first employed the starting material **77** as a single stereoisomer with $E/Z = 100{:}0$. However, if mixtures of isomers with variable E/Z composition ($E/Z = 30{:}70$ and 5:95) are converted, the reaction diastereoselectivity is essentially the same, regardless of the E/Z ratio of enoyl substrate **77** (Scheme 4.39). Obviously, BF_3 leads to rapid E/Z isomerization prior to nucleophilic addition, so only the thermodynamically favored *E*-isomers react. A mechanistic rationale is represented by structure **78**, which explains the observed product configuration by front-face attack on the C–C double bond (Scheme 4.39).

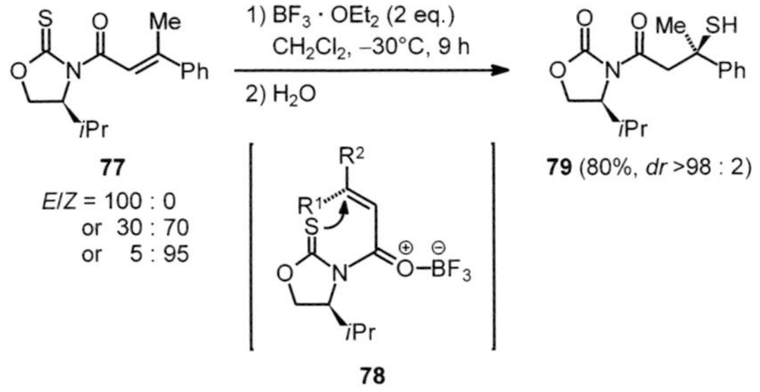

77

E/Z = 100 : 0
or   30 : 70
or   5 : 95

78

79 (80%, *dr* >98 : 2)

Scheme 4.39 Intramolecular sulfur transfer from *N*-enoyl oxazolidine-2-thione according to Palomo.

References

1 a) B. E. Rossiter, N. M. Swingle, *Chem. Rev.* **1992**, *92*, 771–806. b) J. Leonard, E. Díez-Barra, S. Merino, *Eur. J. Org. Chem.* **1998**, 2051–2061. c) M. P. Sibi, S. Manyem, *Tetrahedron* **2000**, *56*, 8033–8061. d) N. Krause, A. Hoffmann-Röder, *Synthesis* **2001**, 171–196. e) S. C. Jha, N. N. Joshi, *Arkivoc* **2002**, 167–196. f) J. Christoffers, in *Encyclopedia of Catalysis*, Vol. 5, ed. I. Horvath, J. Wiley & Sons, New York, **2003**, pp. 99–118.

2 a) R. G. Pearson, *Acc. Chem. Res.* **1993**, *26*, 250–255. b) R. G. Pearson, *Chemical Hardness*, Wiley-VCH, Weinheim, **1997**.

3 N. Krause, *Modern Organocopper Chemistry*, Wiley-VCH, Weinheim, **2002**.

4 a) B. L. Feringa, R. Naasz, R. Imbos, L. A. Arnold in *Modern Organocopper Chemistry*, ed. N. Krause, Wiley-VCH, Weinheim, **2002**, pp. 224–258. b) A. Alexakis, C. Benhaim, S. Rosset, M. Human, *J. Am. Chem. Soc.* **2002**, *124*, 5262–5263. c) M. T. Reetz, A. Gosberg, D. Moulin, *Tetrahedron Lett.* **2002**, *43*, 1189–1191.

5 a) T. Komnenos, *Liebigs Ann. Chem.* **1883**, *218*, 145–169. b) L. Claisen, *J. Prakt. Chem.* **1887**, *35*, 413–415.

6 a) A. Michael, *J. Prakt. Chem.* **1887**, *35*, 349–356. b) A. Michael, *J. Prakt. Chem.* **1887**, *36*, 113–114. c) A. Michael, O. Schulthess, *J. Prakt. Chem.* **1892**, *45*, 55–63. d) A. Michael, *Chem. Ber.* **1894**,

27, 2126–2130. e) A. Michael, *Chem. Ber.* **1900**, *33*, 3731–3769.

7 a) I. Denissova, L. Barriault, *Tetrahedron* **2003**, *59*, 10105–10146. b) C. J. Douglas, L. E. Overman, *Proc. Natl. Acad. Sci. USA* **2004**, *101*, 5363–5367. c) J. Christoffers, A. Baro, *Angew. Chem.* **2003**, *115*, 1726–1728; *Angew. Chem. Int. Ed.* **2003**, *42*, 1688–1690.

8 Review: a) P. I. Dalko, L. Moisan, *Angew. Chem.* **2004**, *116*, 5248–5286; *Angew. Chem. Int. Ed.* **2004**, *43*, 5138–5175. b) The thematic issue of *Account of Chemical Research* devoted to asymmetric organocatalysis, K. N. Houk, B. List (eds.), *Acc. Chem. Res.* **2004**, *37*, 487–631.

9 H. Wynberg, R. Helder, *Tetrahedron Lett.* **1975**, 4057–4060.

10 K. Hermann, H. Wynberg, *J. Org. Chem.* **1979**, *44*, 2238–2244.

11 S. Colonna, A. Re, H. Wynberg, *J. Chem. Soc., Perkin Trans. 1* **1981**, 547–552.

12 M. Bella, K. A. Jørgensen, *J. Am. Chem. Soc.* **2004**, *126*, 5672–5673.

13 a) S. Saaby, M. Bella, K. A. Jørgensen, *J. Am. Chem. Soc.* **2004**, *126*, 8120–8121. b) P. M. Pihko, A. Pohjakallio, *Synlett* **2004**, 2115–2118.

14 a) A. Kirschning, H. Monenschein, R. Wittenberg, *Angew. Chem.* **2001**, *113*, 670–701; *Angew. Chem. Int. Ed.* **2001**, *40*, 650–679. b) S. V. Ley, I. R. Baxendale, R. N. Bream, P. S. Jackson, A. G. Leach, D. A. Longbottom, M. Nesi, J. S. Scott, R. I.

Storer, S. J. Taylor, *J. Chem. Soc., Perkin Trans. 1* **2000**, 3815–4195.

15 K. Hermann, H. Wynberg, *Helv. Chim. Acta* **1977**, *60*, 2208–2212.

16 a) N. Kobayashi, K. Iwai, *J. Am. Chem. Soc.* **1978**, *100*, 7071–7072. b) N. Kobayashi, K. Iwai, *J. Polym. Sci., Polym. Chem.* **1980**, *18*, 923–932.

17 M. Inagaki, J. Hiratake, Y. Yamamoto, J. Oda, *Bull. Chem. Soc. Jpn.* **1987**, *60*, 4121–4126.

18 P. Hodge, E. Khoshdel, J. Waterhouse, *J. Chem. Soc., Perkin Trans. 1* **1983**, 2205–2209.

19 R. Alvarez, M.-A. Hourdin, C. Cavé, J. d'Angelo, P. Chaminade, *Tetrahedron Lett.* **1999**, *40*, 7091–7094.

20 a) U. Eder, G. Sauer, R. Wiechert, *Angew. Chem.* **1971**, *83*, 492–493; *Angew. Chem. Int. Ed. Engl.* **1971**, *10*, 496–497. b) Z. G. Hajos, D. R. Parrish, *J. Org. Chem.* **1974**, *39*, 1615–1621. c) J. Gutzwiller, P. Buchschacher, A. Fürst, *Synthesis* **1977**, 167–168. d) C. Agami, F. Meynier, C. Puchot, J. Guilhem, C. Pascard, *Tetrahedron* **1984**, *40*, 1031–1038. e) S. Terashima, S. Sato, K. Koga, *Tetrahedron Lett.* **1979**, 3469–3472.

21 a) P. Wieland, K. Miescher, *Helv. Chim. Acta* **1950**, *33*, 2215–2228. b) N. Cohen, *Acc. Chem. Res.* **1976**, *9*, 412–417.

22 a) M. Yamaguchi, T. Shiraishi, M. Hirama, *Angew. Chem.* **1993**, *105*, 1243–1245; *Angew. Chem. Int. Ed. Engl.* **1993**, *32*, 1176–1178. b) M. T. Hechavarria Fonseca, B. List, *Angew. Chem.* **2004**, *116*, 4048–4050; *Angew. Chem. Int. Ed.* **2004**, *43*, 3958–3960. c) B. List, *Acc. Chem. Res.* **2004**, *37*, 548–557.

23 D. Rajagopal, R. Narayanan, S. Swaminathan, *Tetrahedron Lett.* **2001**, *42*, 4887–4890.

24 a) N. Halland, P. S. Aburel, K. A. Jørgensen, *Angew. Chem.* **2004**, *116*, 1292–1297; *Angew. Chem. Int. Ed.* **2004**, *43*, 1272–1277. b) N. Halland, P. S. Aburel, K. A. Jørgensen, *Angew. Chem.* **2003**, *115*, 685–689; *Angew. Chem. Int. Ed.* **2003**, *42*, 661–665.

25 R. A. Johnson, K. B. Sharpless, in *Catalytic Asymmetric Synthesis*, ed. I. Ojima, 1st edn, VCH, Weinheim, **1993**, p. 227.

26 A. B. Smith III, C. M. Adams, *Acc. Chem. Res.* **2004**, *37*, 365–377.

27 D. Enders, T. Balensiefer, *Acc. Chem. Res.* **2004**, *37*, 534–541.

28 M. S. Kerr, T. Rovis, *J. Am. Chem. Soc.* **2004**, *126*, 8876–8877.

29 C. J. Pedersen, *Angew. Chem.* **1988**, *100*, 1053–1059; *Angew. Chem. Int. Ed. Engl.* **1988**, *27*, 1021–1027.

30 D. J. Cram, G. D. Y. Sogah, *J. Chem. Soc., Chem. Commun.* **1981**, 625–628.

31 Y. Tamai, A. Kamifuku, E. Koshiishi, S. Miyano, *Chem. Lett.* **1995**, 957–958.

32 a) D. Hoppe, H. Ahrens, W. Guarnieri, H. Helmke, S. Kolczewski, *Pure Appl. Chem.* **1996**, *68*, 613–618. b) D. Hoppe, T. Hense, *Angew. Chem.* **1997**, *109*, 2376–2410; *Angew. Chem. Int. Ed. Engl.* **1997**, *36*, 2282–2316.

33 D. O. Jang, D. D. Kim, D. K. Pyun, P. Beak, *Org. Lett.* **2003**, *5*, 4155–4157.

34 H. Brunner, B. Hammer, *Angew. Chem.* **1984**, *96*, 305–306; *Angew. Chem. Int. Ed. Engl.* **1984**, *23*, 312–313.

35 H. Brunner, J. Kraus, *J. Mol. Catal.* **1989**, *49*, 133–142.

36 G. Desimoni, P. Quadrelli, P. P. Righetti, *Tetrahedron* **1990**, *46*, 2927–2934.

37 a) G. Desimoni, G. Faita, G. Mellerio, P. P. Righetti, C. Zanelli, *Gazz. Chim. Ital.* **1992**, *122*, 269–273. b) G. Desimoni, G. Dusi, G. Faita, P. Quadrelli, P. P. Righetti, *Tetrahedron* **1995**, *51*, 4131–4144.

38 M. Sawamura, H. Hamashima, Y. Ito, *Tetrahedron: Asymmetry* **1991**, *2*, 593–596.

39 M. Sawamura, H. Hamashima, Y. Ito, *J. Am. Chem. Soc.* **1992**, *114*, 8295–8296.

40 M. Sawamura, H. Hamashima, Y. Ito, *Tetrahedron* **1994**, *50*, 4439–4454.

41 M. Sawamura, H. Hamashima, H. Shinoto, Y. Ito, *Tetrahedron Lett.* **1995**, *36*, 6479–6482.

42 J. Singh, N. Satyamurthi, I. S. Aidhen, *J. Prakt. Chem.* **2000**, *342*, 340–347.

43 M. Sawamura, H. Hamashima, Y. Ito, *Bull. Chem. Soc. Jpn.* **2000**, *73*, 2559–2562.

44 Y. Motoyama, Y. Koga, K. Kobayashi, K. Aoki, H. Nishiyama, *Chem. Eur. J.* **2002**, *8*, 2968–2975.

45 T. Suzuki, T. Torii, *Tetrahedron: Asymmetry* **2001**, *12*, 1077–1081.

46 Reviews: a) M. Shibasaki, H. Sasai, *Pure Appl. Chem.* **1996**, *68*, 523–530. b) M. Shibasaki, H. Sasai, T. Arai, *Angew. Chem.*

1997, *109*, 1290–1310; *Angew. Chem. Int. Ed. Engl.* **1997**, *36*, 1236–1256. c) M. Shibasaki, N. Yoshikawa, *Chem. Rev.* **2002**, *102*, 2187–2209. d) M. Shibasaki, M. Kanai, K. Funabashi, *Chem. Commun.* **2002**, 1989–1999. e) S. Matsunaga, T. Ohshima, M. Shibasaki, *Adv. Synth. Catal.* **2002**, *344*, 3–15. f) J.-A. Ma, D. Cahard, *Angew. Chem.* **2004**, *116*, 4666–4683; *Angew. Chem. Int. Ed.* **2004**, *43*, 4566–4583. g) S. Matsunaga, M. Shibasaki, in *Multimetallic Catalysts in Organic Synthesis*, eds. M. Shibasaki, Y. Yamamoto, Wiley-VCH, Weinheim, **2004**, pp. 121–142.

47 H. Sasai, T. Arai, M. Shibasaki, *J. Am. Chem. Soc.* **1994**, *116*, 1571–1572.

48 a) H. Sasai, T. Arai, Y. Satow, K. N. Houk, M. Shibasaki, *J. Am. Chem. Soc.* **1995**, *117*, 6194–6198. b) T. Arai, Y. M. A. Yamada, N. Yamamoto, H. Sasai, M. Shibasaki, *Chem. Eur. J.* **1996**, *2*, 1368–1372.

49 T. Arai, H. Sasai, K. Aoe, K. Okamura, T. Date, M. Shibasaki, *Angew. Chem.* **1996**, *108*, 103–105; *Angew. Chem. Int. Ed. Engl.* **1996**, *35*, 104–106.

50 K. I. Yamada, T. Arai, H. Sasai, M. Shibasaki, *J. Org. Chem.* **1998**, *63*, 3666–3672.

51 Y. Xu, K. Ohori, T. Ohshima, M. Shibasaki, *Tetrahedron* **2002**, *58*, 2585–2588.

52 H. Sasai, E. Emori, T. Arai, M. Shibasaki, *Tetrahedron Lett.* **1996**, *37*, 5561–5564.

53 E. Emori, T. Arai, H. Sasai, M. Shibasaki, *J. Am. Chem. Soc.* **1998**, *120*, 4043–4044.

54 E. Keller, N. Veldman, A. L. Spek, B. L. Feringa, *Tetrahedron: Asymmetry* **1997**, *8*, 3403–3413.

55 a) S. Matsunaga, T. Ohshima, M. Shibasaki, *Tetrahedron Lett.* **2000**, *41*, 8473–8478. b) R. Takita, T. Ohshima, M. Shibasaki, *Tetrahedron Lett.* **2002**, *43*, 4661–4665. c) K. Majima, R. Takita, A. Okada, T. Ohshima, M. Shibasaki, *J. Am. Chem. Soc.* **2003**, *125*, 15837–15845.

56 Y. S. Kim, S. Matsunaga, J. Das, A. Sekine, T. Ohshima, M. Shibasaki, *J. Am. Chem. Soc.* **2000**, *122*, 6506–6507.

57 S. Harada, N. Kumagai, T. Kinoshita, S. Matsunaga, M. Shibasaki, *J. Am. Chem. Soc.* **2003**, *125*, 2582–2590.

58 Y. Hamashima, D. Hotta, M. Sodeoka, *J. Am. Chem. Soc.* **2002**, *124*, 11240–11241.

59 F. Bonadies, A. Lattanzi, L. R. Orelli, S. Pesci, A. Scettri, *Tetrahedron Lett.* **1993**, *34*, 7649–7650.

60 S. Kobayashi, K. Kakumoto, Y. Mori, K. Manabe, *Israel J. Chem.* **2001**, *41*, 247–249.

61 a) M. Nakajima, Y. Yamaguchi, S. Hashimoto, *Chem. Commun.* **2001**, 1596–1597. b) M. Nakajima, S. Yamamoto, Y. Yamaguchi, S. Nakamura, S. Hashimoto, *Tetrahedron* **2003**, *59*, 7307–7313.

62 Reviews: a) E. N. Jacobsen, in *Catalytic Asymmetric Synthesis*, ed. I. Ojima, 1st edn, VCH, Weinheim, **1993**, pp. 159–202.
b) E. N. Jacobsen, *Acc. Chem. Res.* **2000**, *33*, 421–431.

63 a) J. K. Myers, E. N. Jacobsen, *J. Am. Chem. Soc.* **1999**, *121*, 8959–8960.
b) G. M. Sammis, E. N. Jacobsen, *J. Am. Chem. Soc.* **2003**, *125*, 4442–4443. c) G. M. Sammis, H. Danjo, E. N. Jacobsen, *J. Am. Chem. Soc.* **2004**, *126*, 9928–9929.

64 M. S. Taylor, E. N. Jacobsen, *J. Am. Chem. Soc.* **2003**, *125*, 11204–11205.

65 J. d'Angelo, D. Desmaële, F. Dumas, A. Guingant, *Tetrahedron: Asymmetry* **1992**, *3*, 459–505.

66 M. Pfau, G. Revial, A. Guingant, J. d'Angelo, *J. Am. Chem. Soc.* **1985**, *107*, 273–274.

67 D. Desmaële, J. d'Angelo, *Tetrahedron Lett.* **1989**, *30*, 345–348.

68 M. Nour, K. Tan, C. Cavé, D. Villeneuve, D. Desmaële, J. d'Angelo, C. Riche, *Tetrahedron: Asymmetry* **2000**, *11*, 995–1002.

69 D. Desmaële, J. d'Angelo, C. Bois, *Tetrahedron: Asymmetry* **1990**, *1*, 759–762.

70 D. Desmaële, S. Delarue-Cochin, C. Cavé, J. d'Angelo, G. Morgant, *Org. Lett.* **2004**, *6*, 2421–2424.

71 a) J. d'Angelo, A. Guingant, C. Riche, A. Chiaroni, *Tetrahedron Lett.* **1988**, *29*, 2667–2670. b) J. d'Angelo, G. Revial, A. Guingant, C. Riche, A. Chiaroni, *Tetrahedron Lett.* **1989**, *30*, 2645–2648. c) C. Cavé, D. Desmaële, J. d'Angelo, C. Riche, A. Chiaroni, *J. Org. Chem.* **1996**, *61*, 4361–4368.

72 L. Keller, C. Camara, A. Pinheiro, F. Dumas, J. d'Angelo, *Tetrahedron Lett.* **2001**, *42*, 381–383.

73 L. Keller, F. Dumas, J. d'Angelo, *Eur. J. Org. Chem.* **2003**, 2488–2497.

74 G. Revial, I. Jabin, M. Redolfi, M. Pfau, *Tetrahedron: Asymmetry* **2001**, *12*, 1683–1688.

75 a) M. Ciobanu, K. Matsumoto, *Liebigs Ann./Recueil* **1997**, 623–635. b) G. Jenner, *Tetrahedron* **1997**, *53*, 2669–2695.

76 a) C. Camara, D. Joseph, F. Dumas, J. d'Angelo, A. Chiaroni, *Tetrahedron Lett.* **2002**, *43*, 1445–1448. b) J. d'Angelo, C. Ferroud, C. Riche, A. Chiaroni, *Tetrahedron Lett.* **1989**, *30*, 6511–6514.

77 a) T. Volpe, G. Revial, M. Pfau, J. d'Angelo, *Tetrahedron Lett.* **1987**, *28*, 2367–2370. b) J. d'Angelo, G. Revial, T. Volpe, M. Pfau, *Tetrahedron Lett.* **1988**, *29*, 4427–4430. c) H. Sdassi, G. Revial, M. Pfau, J. d'Angelo, *Tetrahedron Lett.* **1990**, *31*, 875–878.

78 a) H. Brunner, J. Kraus, H.-J. Lautenschlager, *Monatsh. Chem.* **1988**, *119*, 1161–1167. b) K. D. Belfield, J. Seo, *Synth. Commun.* **1995**, *25*, 461–466.

79 a) A. Guingant, H. Hammami, *Tetrahedron: Asymmetry* **1991**, *2*, 411–414. b) A. Guingant, *Tetrahedron: Asymmetry* **1991**, *2*, 415–418.

80 C. Cavé, V. Daley, J. d'Angelo, A. Guingant, *Tetrahedron: Asymmetry* **1995**, *6*, 79–82.

81 J. d'Angelo, C. Cavé, D. Desmaële, A. Gassama, C. Thominiaux, C. Riche, *Heterocycles* **1998**, *47*, 725–747.

82 A. Felk, G. Revial, B. Viossat, P. Lemoine, M. Pfau, *Tetrahedron: Asymmetry* **1994**, *5*, 1459–1462.

83 a) A. Gassama, J. d'Angelo, C. Cavé, J. Mahuteau, C. Riche, *Eur. J. Org. Chem.* **2000**, 3165–3169. b) M. Nour, K. Tan, R. Jankowski, C. Cavé, *Tetrahedron: Asymmetry* **2001**, *12*, 765–769. c) R. K. Boeckman, Jr., M. d. R. Rico Ferreira, L. H. Mitchell, P. Shao, *J. Am. Chem. Soc.* **2002**, *124*, 190–191.

84 a) N. S. Barta, A. Brode, J. R. Stille, *J. Am. Chem. Soc.* **1994**, *116*, 6201–6206. b) P. Benovsky, G. A. Stephenson, J. R. Stille, *J. Am. Chem. Soc.* **1998**, *120*, 2493–2500.

85 D. Desmaële, K. Mekouar, J. d'Angelo, *J. Org. Chem.* **1997**, *62*, 3890–3901.

86 D. J. Ager, I. Prakash, D. R. Schaad, *Aldrichimica Acta* **1997**, *30*, 3–12.

87 a) K. Tomioka, K. Ando, K. Yasuda, K. Koga, *Tetrahedron Lett.* **1986**, *27*, 715–716. b) K. Ando, K. Yasuda, K. Tomioka, K. Koga, *J. Chem. Soc., Perkin Trans. 1* **1994**, 277–282.

88 K. Ando, Y. Takemasa, K. Tomioka, K. Koga, *Tetrahedron* **1993**, *49*, 1579–1588.

89 K. Tomioka, K. Yasuda, K. Koga, *J. Chem. Soc., Chem. Commun.* **1987**, 1345–1346.

90 a) K. Tomioka, W. Seo, K. Ando, K. Koga, *Tetrahedron Lett.* **1987**, *28*, 6637–6640. b) K. Ando, W. Seo, K. Tomioka, K. Koga, *Tetrahedron* **1994**, *50*, 13081–13088.

91 Review: J. Christoffers, *Chem. Eur. J.* **2003**, *9*, 4862–4867.

92 a) J. Christoffers, A. Mann, *Angew. Chem.* **2000**, *112*, 2871–2874; *Angew. Chem. Int. Ed.* **2000**, *39*, 2752–2754. b) J. Christoffers, A. Mann, *Chem. Eur. J.* **2001**, *7*, 1014–1027. c) J. Christoffers, H. Scharl, *Eur. J. Org. Chem.* **2002**, 1505–1508. d) J. Christoffers, K. Schuster, *Chirality* **2003**, *15*, 777–782. e) J. Christoffers, H. Scharl, W. Frey, A. Baro, *Eur. J. Org. Chem.* **2004**, 2701–2706. f) J. Christoffers, H. Scharl, W. Frey, A. Baro, *Org. Lett.* **2004**, *6*, 1171–1173.

93 J. Christoffers, W. Frey, H. Scharl, A. Baro, *Z. Naturforsch.* **2004**, *59b*, 375–379.

94 a) J. Christoffers, B. Kreidler, H. Oertling, S. Unger, W. Frey, *Synlett* **2003**, 493–496. b) J. Christoffers, B. Kreidler, S. Unger, W. Frey, *Eur. J. Org. Chem.* **2003**, 2845–2853.

95 A. Job, C. F. Janeck, W. Bettray, R. Peters, D. Enders, *Tetrahedron* **2002**, *58*, 2253–2329.

96 J.-M. Lassaletta, R. Fernández, E. Martín-Zamora, E. Díez, *J. Am. Chem. Soc.* **1996**, *118*, 7002–7003.

97 C. Palomo, M. Oiarbide, F. Dias, R. López, A. Linden, *Angew. Chem.* **2004**, *116*, 3369–3372; *Angew. Chem. Int. Ed.* **2004**, *43*, 3307–3310.

5
Rearrangement Reactions

Annett Pollex and Martin Hiersemann

5.1
Introduction

A rearrangement is a migration of an atom or a group of atoms from one atom to another in the same molecule [1]. Various interesting applications in natural-product synthesis prove the synthetic power of rearrangements as tools for stereoselective carbon–carbon bond formation. Since the 1980s the scope of rearrangement reactions has expanded substantially. The number of applications that cover the more challenging formation of quaternary instead of the more easily accessible tertiary chiral carbon atoms has increased remarkably. (We are aware that the unambiguous semantic of stereochemistry does not accept the term "quaternary chiral carbon atom". However, we prefer to use this catchy term instead of the more precise description "non-hydrogen-substituted stereogenic chirotopic carbon atom".) It is certainly undesirable on this occasion to provide a comprehensive summary of all synthetic methods based on rearrangements that have been developed for the stereoselective synthesis of quaternary chiral carbon atoms. Because every synthetic method is only as good as its applications, we will present rearrangements that have a proven track record in target-oriented natural-product synthesis. Natural products in whose total syntheses the problem of constructing such quaternary chiral carbon atoms was solved by way of rearrangement strategies are depicted in Schemes 5.1–5.4. The quaternary chiral carbon atoms that were created by a rearrangement reaction are labeled by a '*'. This chapter includes both simple molecular reorganizations without change in total number of carbon–carbon bonds (e.g. [1,2]-shifts) and pericyclic (e.g. [2,3]- or [3,3]-sigmatropic) rearrangements in which an additional carbon–carbon bond is formed. Space limitations have required us to confine ourselves to only 26 syntheses of secondary natural products (e.g. terpenes, alkaloids, glycosides, amino acids, citric acid-derived products). We would like to apologize to all those who have planned and realized equally elegant and important rearrangements in the context of natural-product synthesis which could not be considered here.

Reactions that at the same time build up carbon frameworks and chirality centers with a high degree of diastereo- and enantioselectivity are of pivotal

Quaternary Stereocenters: Challenges and Solutions for Organic Synthesis. Edited by Jens Christoffers, Angelika Baro
Copyright © 2005 WILEY-VCH Verlag GmbH & Co. KGaA, Weinheim
ISBN: 3-527-31107-6

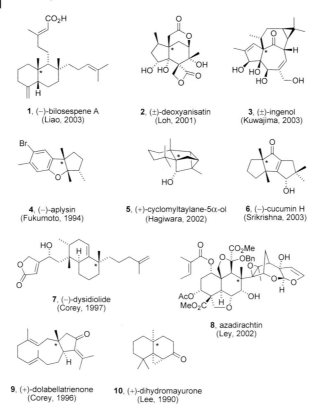

1, (−)-bilosespene A
(Liao, 2003)

2, (±)-deoxyanisatin
(Loh, 2001)

3, (±)-ingenol
(Kuwajima, 2003)

4, (−)-aplysin
(Fukumoto, 1994)

5, (+)-cyclomyltaylane-5α-ol
(Hagiwara, 2002)

6, (−)-cucumin H
(Srikrishna, 2003)

7, (−)-dysidiolide
(Corey, 1997)

8, azadirachtin
(Ley, 2002)

9, (+)-dolabellatrienone
(Corey, 1996)

10, (+)-dihydromayurone
(Lee, 1990)

Scheme 5.1 Terpenoids as targets in natural product total synthesis.

importance for most natural-product syntheses. Rearrangements are particularly interesting in this respect because if one carefully chooses the rearrangement type, they proceed with a reliable and predictable stereochemical result based on a concerted mechanism (no intermediate, bond breaking and bond making are proceeding at the same time but are not necessarily equally advanced in the transition state). Although the simple diastereoselectivity of a rearrangement may be predictable based on its mechanism, the required induced diastereoselectivity can often be very capricious and it will be very instructive to notice how the induced diastereoselectivity has been accomplished in the following syntheses.

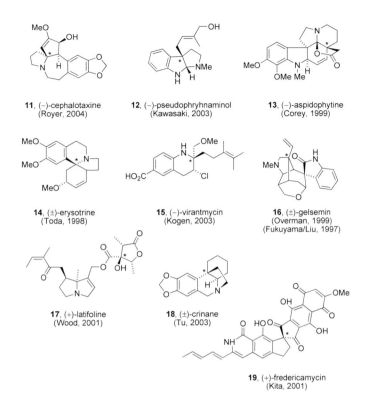

11, (–)-cephalotaxine
(Royer, 2004)

12, (–)-pseudophrynaminol
(Kawasaki, 2003)

13, (–)-aspidophytine
(Corey, 1999)

14, (±)-erysotrine
(Toda, 1998)

15, (–)-virantmycin
(Kogen, 2003)

16, (±)-gelsemin
(Overman, 1999)
(Fukuyama/Liu, 1997)

17, (+)-latifoline
(Wood, 2001)

18, (±)-crinane
(Tu, 2003)

19, (+)-fredericamycin
(Kita, 2001)

Scheme 5.2 Alkaloids as targets in natural product total synthesis.

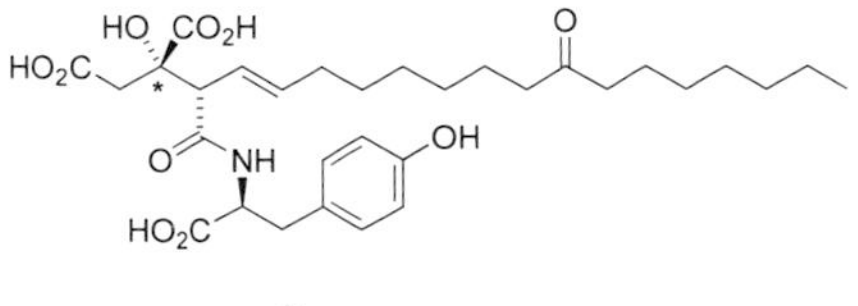

20, (–)-viridiofungin A
(Hiersemann, 2004)

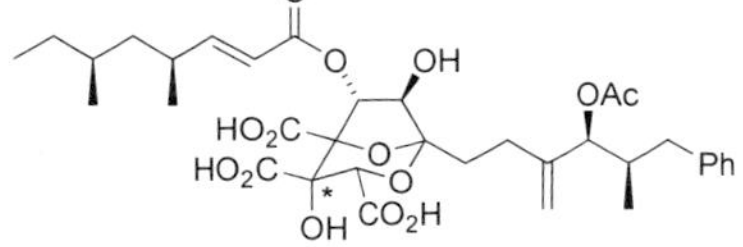

21, (+)-squalestatin S1
(Tomooka, 2000)

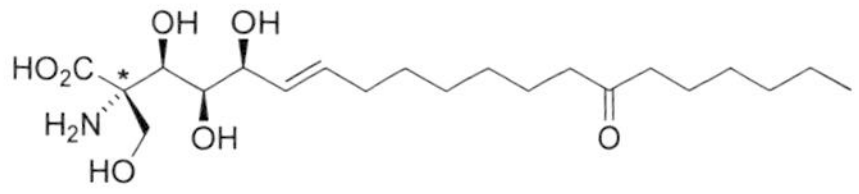

22, (–)-sphingofungin E
(Chida, 2002)

Scheme 5.3 Citric acid-derived targets in natural-product synthesis.

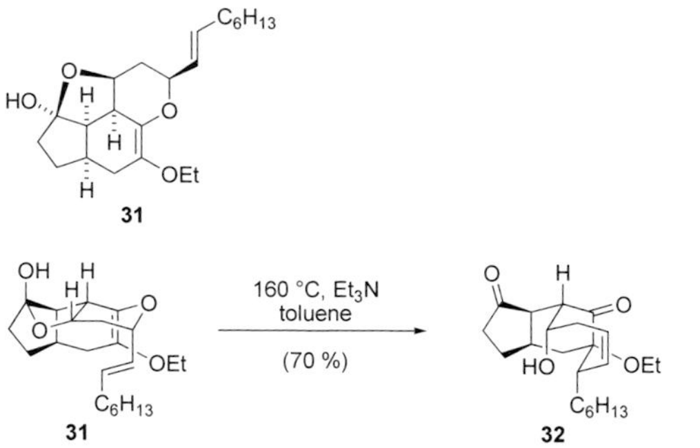

(1)

Equation (1) represents a high-end application of a thermal Claisen rearrangement in natural product synthesis. The configuration of the tetrasubstituted vinyl ether double bond, the simple- and the substrate-induced diastereoselectivity as well as the Z-configuration of the stereogenic double bond generated by the rearrangement are all predetermined by the substrate structure.

One of the very few examples in which the induced diastereoselectivity of the Claisen rearrangement can be controlled by a chiral reagent was developed by Corey. In an application of this method, Corey et al. used a chiral boron reagent to create **34** with excellent diastereo- and enantioselectivity (Scheme 5.7). Using the combination of boron Lewis acid and the amine base afforded a macrocyclic allyl vinyl ether containing an *E*-configured vinyl ether double bond. The stereochemical course of the enolate Claisen rearrangement may be explained by assuming the formation of a chiral enol borinate and the ability of the chiral ligand on the boron to differentiate between the two possible diastereomeric chair-like transition states. The rearrangement product (**34**) was further transformed to (+)-dolabellatrienone **9** [10].

Scheme 5.7 Application of Corey's asymmetric Claisen rearrangement in natural-product synthesis. Following the formation of an *E*-configured boron enolate, the chiral ligand on boron determines the reagent-induced diastereoselectivity.

The Claisen rearrangement of a vinyl propargyl ether affords an allene. The allene may be part of a target molecule or may be used for further transformations [Eq. (2)].

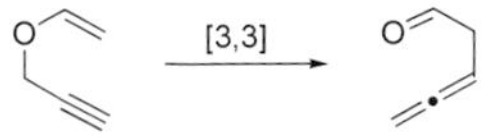

$$(2)$$

Ley et al. utilized a Claisen rearrangement of a propargyl vinyl ether as part of a sequence leading to the synthesis of azadirachtin **8** [11]. This rearrangement resulted in the formation of the allene moiety in **36** that was later used for an intramolecular radical cyclization (Scheme 5.8). The tetrasubstituted stereogenic vinyl ether double bond was part of a polycyclic ring system that also provided the substrate-induced diastereoselectivity by efficient blocking of the potential *Re*-approach of the propargylic ether fragment of the propargyl vinyl ether.

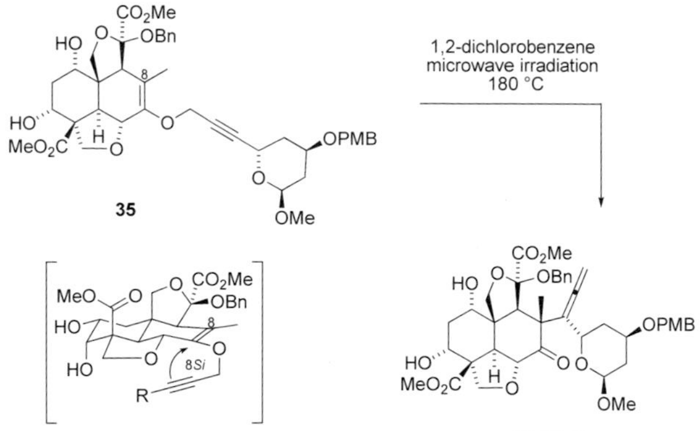

Scheme 5.8 Microwave-initiated Claisen rearrangement of the vinyl propargyl ether **35**. Substrate-induced diastereoselectivity is based on the rigid polycyclic structure of the substrate for the rearrangement (PMB = p-MeO–C_6H_4).

The generation of a consistently configured stereogenic vinyl ether double bond is critical for the utilization of the full synthetic power of the Claisen rearrangement. An interesting strategy to obtain a configurationally defined tetra-substituted vinyl ether double bond was developed by Wood [12]. In his approach the reactive *Z*-configured enol intermediate **40** was formed during the Rh^{II}-catalyzed coupling of the allylic alcohol **38** with the α-diazo-β-ketoester **37** (Scheme 5.9). The highly selective formation of the *Z*-configured enol ether is apparently a result of an intramolecular proton transfer [12c]. The sequential [3,3]/[1,2]-rearrangement leads to the formation of **39** as an intermediate towards the synthesis of (+)-latifoline **17** [13].

The 1,3-chirality transfer during [3,3]-sigmatropic rearrangements is the most powerful and most frequently utilized strategy to achieve an excellent substrate-induced diastereoselectivity. The absolute configuration of the oxygen-substituted carbon atom C-1′ is immolated to generate a new all-carbon-substituted chirality center with complete stereocontrol [14]. This strategy was successfully implemented in the synthesis of **39**.

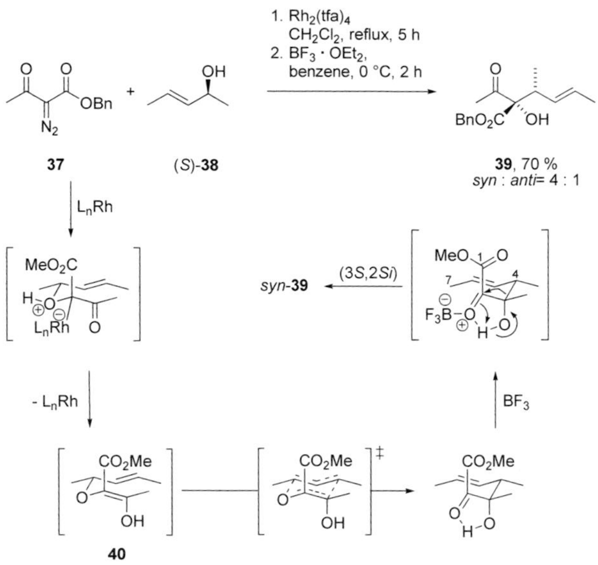

Scheme 5.9 figures and reaction schemes:

1. Rh$_2$(tfa)$_4$
   CH$_2$Cl$_2$, reflux, 5 h
2. BF$_3$ · OEt$_2$,
   benzene, 0 °C, 2 h

37   (*S*)-**38**

39, 70 %
syn : anti= 4 : 1

L$_n$Rh

syn-**39**   (3*S*,2*Si*)

- L$_n$Rh

BF$_3$

40

Scheme 5.9 Transition metal-catalyzed formation of a tetra-substituted stereogenic vinyl ether double bond followed *in situ* by a thermal Claisen rearrangement to provide an α-keto ester with a quaternary chirality center in the β-position.

The Caroll rearrangement of a β-keto allylic ester is usually terminated by a decarboxylation to provide a γ,δ-unsaturated carbonyl compound [Eq. (3)] [3].

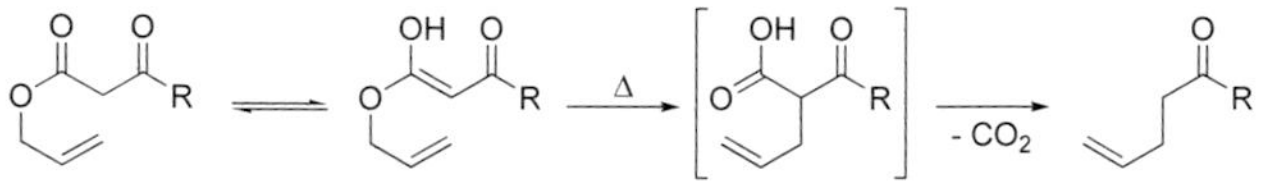

$$\text{(3)}$$

For the total synthesis of (−)-malyngolide **26**, Enders et al. took advantage of an auxiliary-induced diastereo- and enantioselective Caroll rearrangement (Scheme 5.10) [15, 16]. It is apparent that the RAMP auxiliary in concert with the lithium cation induced a rigid geometry in the dienolate **42**. That forces the allyl ether portion of the allyl vinyl ether to approach the vinyl ether moiety from the less hindered *Si*-face. The carboxylic acid moiety of the rearrangement product **43** was immediately reduced to the corresponding alcohol, affording **44** with high diastereoselectivity and in reasonable yield.

Johnson et al. developed a variation of the Claisen rearrangement wherein a 2-alkoxycarbonyl-substituted allyl vinyl ether is formed *in situ* by the reaction of an allylic alcohol with an ortho ester in the presence of an acid catalyst [Eq. (4)] [5].

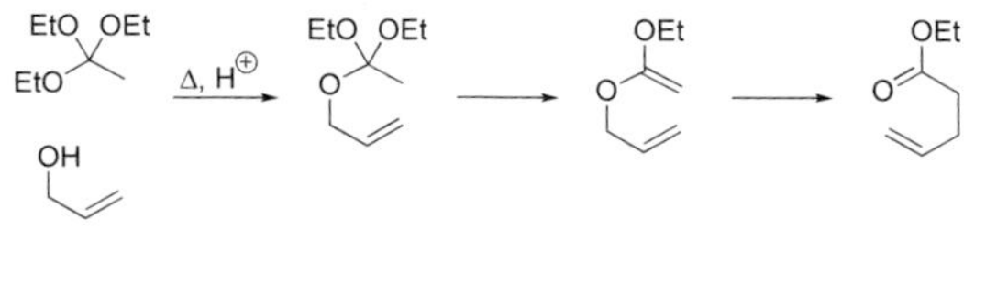

$$\text{(4)}$$

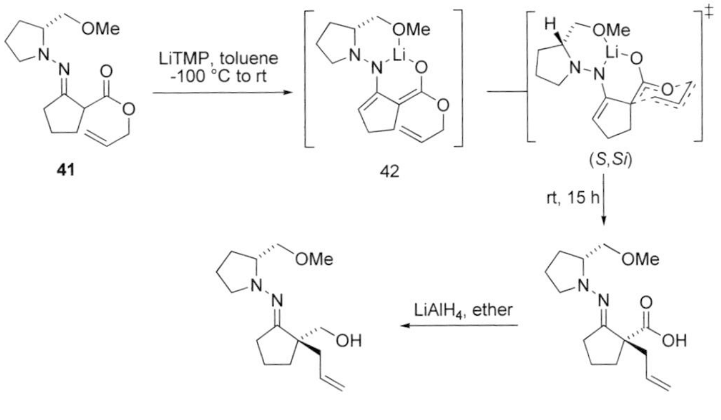

Scheme 5.10 Asymmetric Caroll rearrangement by auxiliary-induction. Formation of a chelated enolate is a powerful concept for the control of the configuration of the tetrasubstituted vinyl ether double bond (LiTMP = lithium 2,2,6,6-tetramethylpiperidine).

Several applications of the Johnson–Claisen rearrangement in natural-product synthesis have been published [17]. An example for an enantioselective Johnson–Claisen rearrangement can be found in the total synthesis of (–)-cucumin H **6** (Scheme 5.11) [18]. This application is a nice example of how a chiral carbon atom that is not directly located at the allyl vinyl ether framework can control the stereochemical course of the rearrangement. In the event, the *Si*-face of the allylic ether double bond of **45** was effectively shielded against the approach of the isopropenyl substituent.

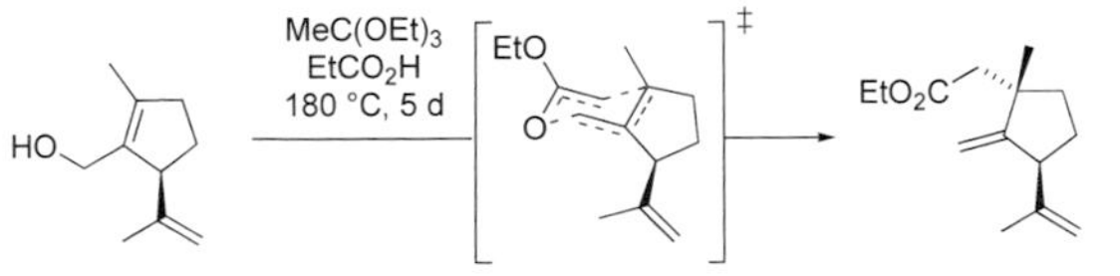

Scheme 5.11 Substrate-induced diastereoselectivity by 1,3-asymmetric induction – an alternative to the 1,3-chirality transfer, which would require a chiral secondary allylic alcohol and would afford a stereogenic double bond in the rearrangement product.

Three advantages make the Ireland–Claisen rearrangement [Eq. (5)] one of the most successful [3,3]-sigmatropic rearrangements [6]. First, synthesis of the allyl vinyl ether is straightforward by deprotonation of an allyl ester and subsequent silyl ketene acetal formation at low temperature. Second, the otherwise difficult-to-set configuration of a stereogenic vinyl ether double bond can be conveniently controlled by the deprotonation conditions. Third, the rearrangement usually proceeds at low temperature during the warming of the reaction mixture.

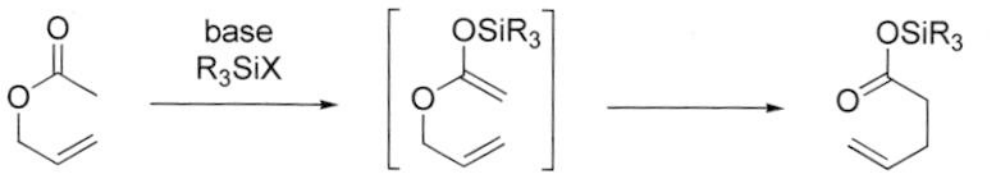

$$(5)$$

Mulzer et al. applied an Ireland–Claisen rearrangement wherein the vinyl ether double bond configuration was controlled by the formation of the chelated ester enolate *E*-**49** prior to the formation of the silyl ketene acetal *Z*-**50** (Scheme 5.12) [19]. This conversion allows the rearrangement to proceed at the low temperature that is typical for the Ireland–Claisen rearrangement. Decomposition of the ester enolate at higher temperatures prevents direct Claisen rearrangement of the ester enolate. A 1,3-chirality transfer served efficiently to provide the substrate-induced diastereoselectivity for the synthesis of the bis(tetrahydrofuran) fragment of (+)-asteltoxin **25**.

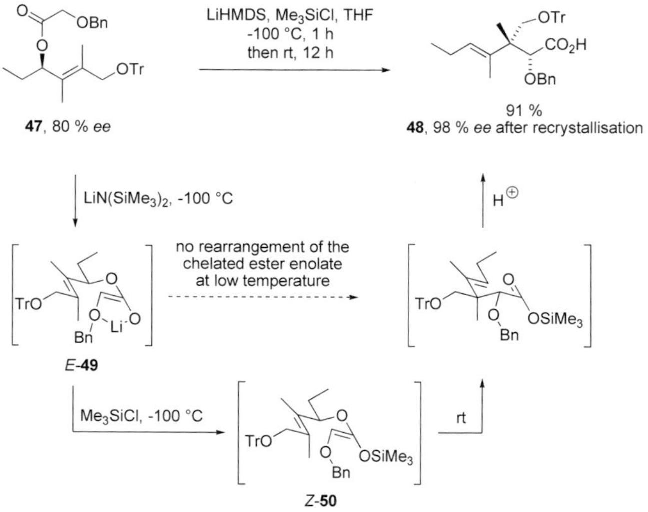

Scheme 5.12 The formation of a chelated enolate prior to the formation of the silyl ketene acetal was utilized to generate the vinyl ether double bond with the required configuration.

Corey et al. used the Ireland–Claisen rearrangement and the concept of 1,3-chirality transfer during their total synthesis of (–)-aspidophytine **13** (Scheme 5.13) [20]. Starting from the chiral allyl acetate **51** the single quaternary stereocenter in **52** was constructed enantioselectively. The rearrangement proceeded through a strained bicyclo[3.2.1]octane-type transition state what demanded an increased reaction temperature.

The Eschenmoser–Claisen rearrangement involves the condensation of an allylic alcohol with an amide acetal (usually *N,N*-dimethylacetamide dimethyl acetal) followed by elimination and *in situ* formation of a mixed *N,O*-ketene acetal that undergoes the [3,3]-sigmatropic rearrangement under the increased reaction temperature required for its formation [Eq. (6)] [4]. The usefulness of the rearrangement product as a building block can be hampered by

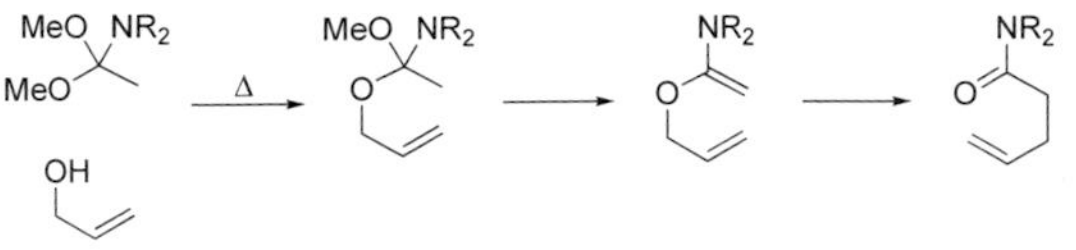

(6)

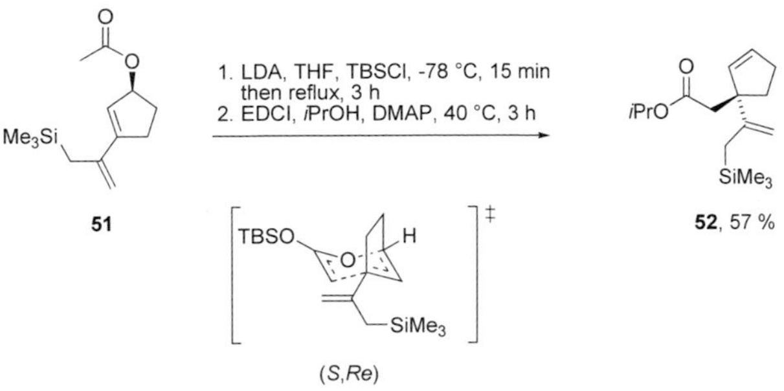

Scheme 5.13 The stereochemical course of an Ireland–Claisen rearrangement is efficiently controlled by 1,3-chirality transfer (TBS = tert-BuMe₂Si).

the presence of the amide function which may be difficult to convert to other functionalities.

An instructive example of the power of the concept of 1,3-chirality transfer during an Eschenmoser–Claisen rearrangement is depicted in Scheme 5.14. The conformationally restricted bicyclic structure of the allyl vinyl ether and the suprafaciality of the rearrangement governed the stereochemical course of the rearrangement, even though the vinyl ether double bond had to approach the allylic ether double bond from the *Re*-face that was shielded by the methyl group. Subsequent double-bond isomerization into conjugation under the reaction conditions afforded racemic **54**, which was used as an intermediate for the total synthesis of (±)-deoxanisatin **2** [21].

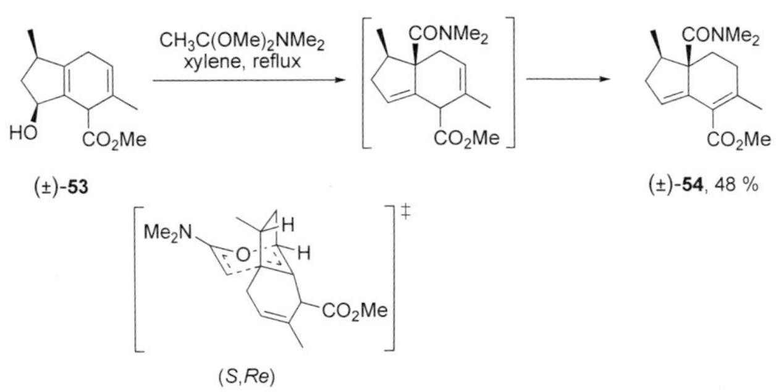

Scheme 5.14 Example of an Eschenmoser–Claisen rearrangement.

A heteroaromatic Claisen rearrangement in which the allyl vinyl ether was generated in situ was used during the total synthesis of (−)-pseudophrynaminol **12** (Scheme 5.15) [22]. This example represents another neat application of the concept of 1,3-chirality transfer. The stereogenic tetrasubstituted vinyl ether double bond is part of a ring system and, therefore, its configurational integrity was not of concern. The phenyl substituent preferentially adopts an equatorial position on the six-membered chair-like transition state. Thereby, the E-configuration of the newly formed stereogenic double bond is unambiguously generated.

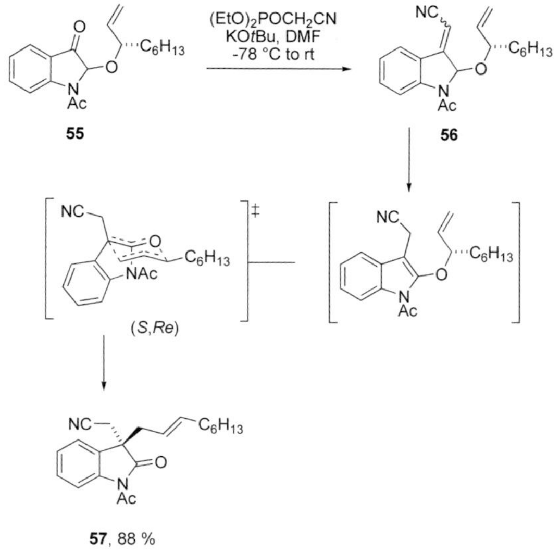

Scheme 5.15 A heteroaromatic Claisen rearrangement that complies with the concept of 1,3-chirality transfer.

5.2.2
The Overman Rearrangement

The [3,3]-sigmatropic rearrangement of trichloroacetimidates of allylic alcohols affords allylic trichloroacetamides [Eq. (7)]. This so-called Overman rearrangement is a heteroatom variation of the Claisen rearrangement so that no new carbon–carbon bond will be formed. Consequently, in our context, the Overman rearrangement provides access to nitrogen-substituted quaternary chiral carbon atoms [23].

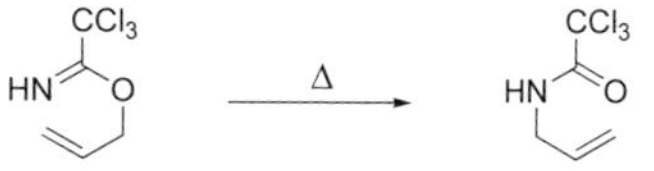

(7)

Substrate-induced diastereoselectivity was utilized to control the stereochemical course of the allylic imidate Claisen rearrangement of **58** to afford the nitrogen-substituted quaternary chiral carbon atom of the building block **59**, an intermediate in the total synthesis of (–)-sphingofungine E **22** [24]. The substrate-induced diastereoselectivity may be explained by application of the allylic strain concept [25]. Assuming that the 1,2-allylic strain (*synperiplanar* arrangement of the PMBOCH$_2$-substituent of the double bond and the hydrogen atom on the chiral carbon atom next to the allylic ether double bond) determines the preference for the conformer **60**, the benzyloxy moiety could effectively block the *Si*-face which would lead to the depicted preference for the *Re*-approach. The formation of the minor diastereomer may be explained by the availability of a second transition state **61** whose conformation is determined by the 1,3-allylic strain (*synperiplanar* arrangement of the hydrogen atom on the allylic ether double bond and the hydrogen on the chiral carbon atom next to the allylic ether double bond). Consequently, the benzyloxymethyl moiety does now block the *Re*-face of the double bond and forces the imidate to approach the double bond from the *Si*-face (see Scheme 5.16).

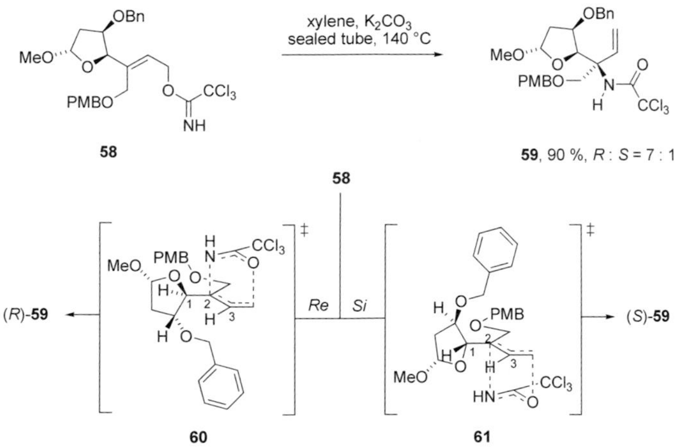

Scheme 5.16 Application of allylic-strain considerations to explain the substrate-induced diastereoselectivity of the Overman rearrangement (PMB = *para*-methoxybenzyl).

5.2.3
The Cope Rearrangement

The [3,3]-sigmatropic rearrangement of 1,5-hexadienes is known as the Cope rearrangement [26]. Since this reaction is inherently reversible, variations of the substrate structure are required to make this rearrangement synthetically useful [27]. The most successful variation attaches a hydroxyl function at the 3-position of the 1,5-hexadiene. The corresponding 3-oxy-Cope rearrangement affords an enol that tautomerizes to a carbonyl function, which usually renders the rearrangement irreversible [Eq. (8)]. Tremendous rate acceleration can be achieved if the hydroxyl group is converted to a negatively charged oxy-substituent by deprotonation prior to the then so-called anionic oxy-Cope rearrangement [28].

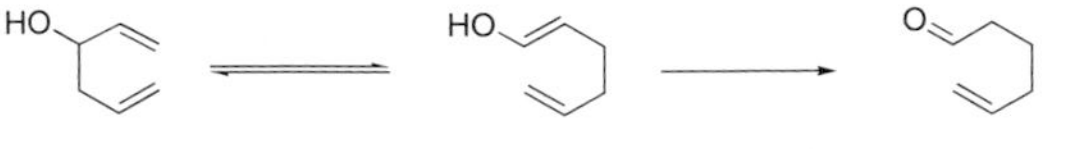

$$(8)$$

Liao et al. successfully converted the bicyclo[2.2.2]octane skeleton into the *cis* decalin framework of biolespene A **1** using the anionic version of the oxy-Cope rearrangement [29]. The absolute configuration of the newly generated quaternary chiral carbon atom is a consequence of the rigid tricyclic structure of the substrate **62** (Scheme 5.17).

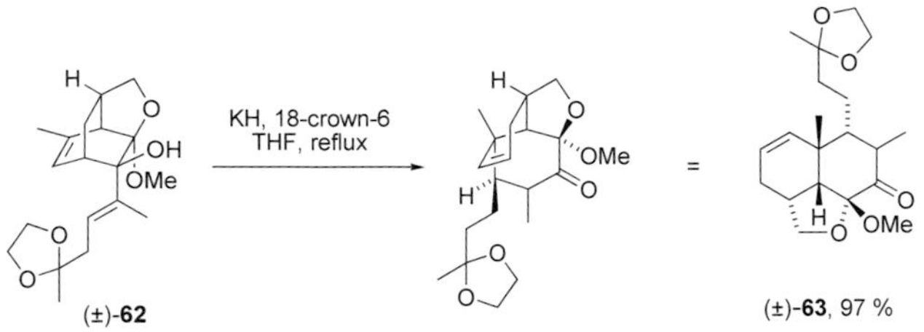

Scheme 5.17 The oxy-Cope rearrangement is frequently applied to convert bicylic carbon skeletons into annulated ring systems.

As expected, the 1,3-chirality transfer is also a legitimate concept for stereo-control in Cope rearrangement chemistry. For instance, anionic oxy-Cope rearrangement of the enantiomerically enriched 3-hydroxy-1,5-hexadiene **64** afforded the aldehyde **65** which was utilized in the total synthesis of (+)-dihydro-mayurone **10** (Scheme 5.18) [30].

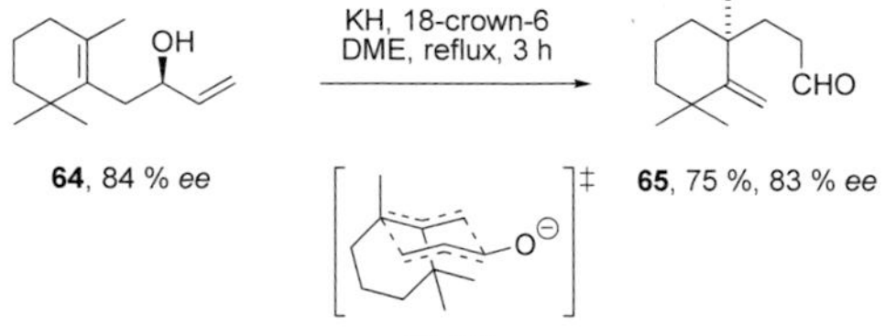

Scheme 5.18 Asymmetric induction based on 1,3-chirality transfer successfully applied to an anionic oxy-Cope rearrangement (DME = dimethoxyethane).

An anionic 2-aza-4-oxy-Cope rearrangement was employed during the total synthesis of (±)-gelsemin **16** [31]. In the event, deprotonation of the hydroxyl function and the generation of an imine double bond, which served as one of the two required double bonds of the rearranging system, was realized by treatment of the amino alcohol (±)-**66** with potassium hydride. *In situ* methoxycarbonylation of the intermediate imino enolate **69** afforded carbonate **70**, which was cleaved by treatment with a base to give the bicyclic *N*-acyl enamine **68** as a single diastere-omer (Scheme 5.19). The stereochemical course of the rearrangement was a consequence of the bicyclo[2.2.2]octane skeleton in which the rearranging system was embedded.

Release of ring strain is another common strategy to provide the required driving force that makes the Cope rearrangement an irreversible and, therefore, synthetically useful process. The *cis*-divinyl cyclopropane to cycloheptadiene rearrangement profits from the release of ring strain and is a well-known

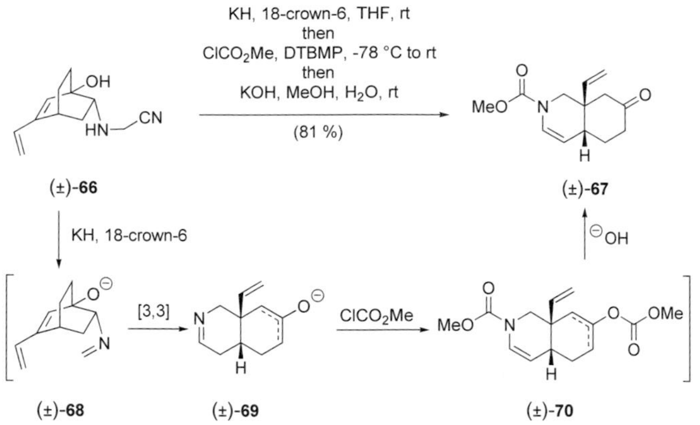

Scheme 5.19 One-pot generation of the rearrangement prone 2-aza-4-oxy-1,5-hexadiene **66**, oxyanionic 3-oxy-Cope rearrangement, bis(methoxycarbonylation) and carbonate saponification to provide the substituted hexahydro-1 *H*-isoquinoline **67** (DTBMP = 2,6-di-*tert*-butyl-4-methylpyridine).

variation of the Cope rearrangement [32]. An application in the total synthesis of (±)-gelsemine **16** is depicted in Scheme 5.20 [33]. Notice that both double bonds of the substrate are part of conjugated systems, which should significantly increase the activation energy. However, the release of ring strain and the formation of one α,β-unsaturated ester more than compensates for this loss of conjugation. The stereochemical course of the rearrangement is dictated by the absolute configuration of the three stereogenic carbon atoms that build up the cyclopropane ring and by the necessity of a boat-like transition state for the *cis*-divinyl cyclopropane to cycloheptadiene rearrangement.

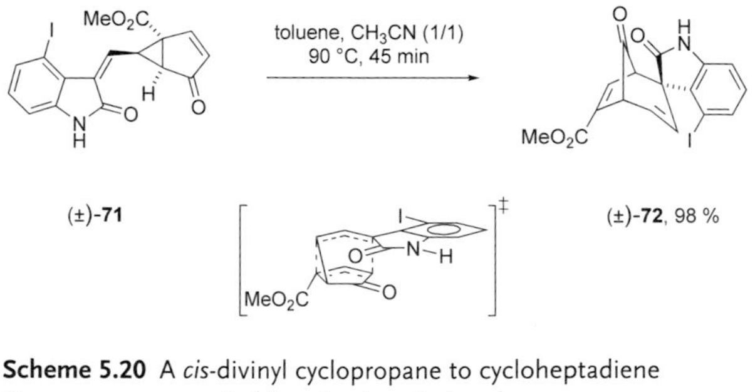

Scheme 5.20 A *cis*-divinyl cyclopropane to cycloheptadiene Cope rearrangement, driven by ring-strain release.

5.2.4
The Wittig Rearrangement

The [2,3]-sigmatropic rearrangement of an allyloxy-substituted carbanion is called a [2,3]-Wittig rearrangement [Eq. (9)] [34]. The rearrangement proceeds through a five-membered cyclic transition state, and transposition of the negative charge from a carbon to an oxygen atom provides the driving force for the rearrangement (G = electron withdrawing group).

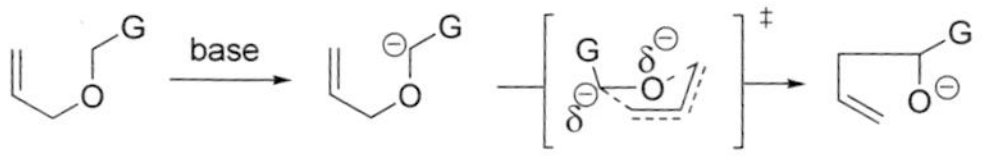

(9)

Although applications of the [2,3]-Wittig rearrangement are numerous [35], examples of the generation of quaternary chiral carbon atoms are very limited. Two strategies are possible; the first utilizes a substrate containing a stereogenic trisubstituted allylic ether double bond. A good example of this procedure was used during the synthesis of the D ring of cobyric acid **23** [36]. Initial formation of the carbanion was achieved by transmetallation of the stannane **73** with *n*-butyllithium, a method originally developed by Still [37]. The concept of 1,3-chirality transfer is also valid for the [2,3]-Wittig rearrangement because the rearrangement proceeds by a concerted suprafacial mechanism. Besides the major product **74**, which was formed enantioselectively based on the 1,3-chirality transfer, 10% of the so-called [1,2]-Wittig rearrangement product **75** were observed. As depicted in Scheme 5.21, the [1,2]-Wittig rearrangement proceeds by a stepwise mechanism featuring radicals as intermediates. Without the concerted nature of the transition state of the rearrangement, no chirality transfer would be expected. This expectation is apparently validated by the result of the example discussed above.

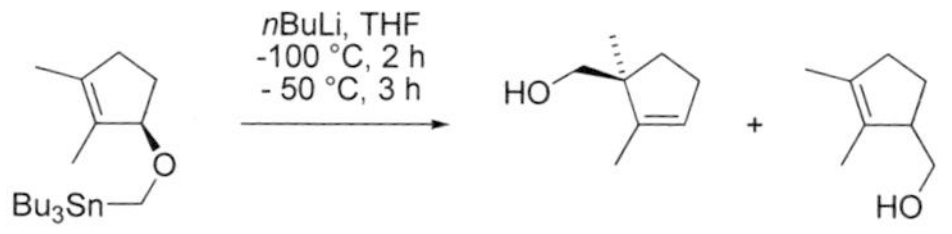

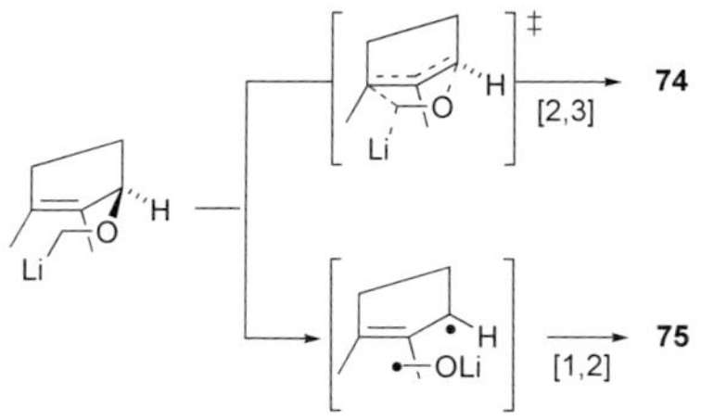

Scheme 5.21 A [2,3]-Wittig–Still rearrangement used to create a quaternary chiral carbon atom. Based on the concerted and suprafacial nature of the [2,3]-Wittig rearrangement, the concept of 1,3-chirality transfer is effective.

The second strategy to form a quaternary chiral carbon atom by a [2,3]-Wittig rearrangement is to employ a non-hydrogen-substituted carbanion. The ester dienolate [2,3]-Wittig rearrangement represents a recently developed example of this strategy [38]. It was demonstrated that the ester dienolate generated from the α,β-unsaturated ester **76** was prone to undergo a [2,3]-Wittig rearrangement to afford the highly substituted 1,5-hexadiene **77** (Scheme 5.22). The rearrangement product **77** was generated with a high diastereoselectivity and later used in the total synthesis of the alkyl citrate viridiofungin A **20** [39].

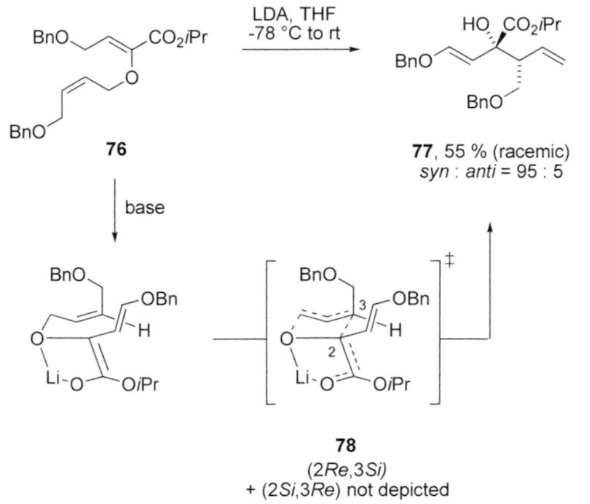

Scheme 5.22 The simple diastereoselectivity of the ester dienolate [2,3]-Wittig rearrangement may be explained by the preference for a chelated bicyclic transition state in which the benzyloxymethyl-substituent is directed toward the *convex* face of the bicyclic transition state (LDA = LiN*i*Pr$_2$).

The stereochemical course of this rearrangement may be explained by assuming that the lithium cation is chelated between the ester enolate and the etheroxygen atom. The resulting transition state **78** would resemble a bicyclo[3.3.0]octane with a *convex* and a *concave* face. It is reasonable to assume that the benzyloxymethyl-substituent at C-3 is preferentially directed toward the *convex* face of the bicyclic transition state and this interpretation nicely explains the experimentally observed relative configuration of the rearrangement product **77**.

We indicated above that the [1,2]-Wittig rearrangement proceeds by a stepwise mechanism with radicals as intermediates. One could be immediately suspicious about the synthetic usefulness of such a rearrangement particularly with respect to stereocontrol. However, the following example instructively demonstrates that a radical mechanism does not necessarily preclude stereochemical control. It proceeds through a radical dissociation–recombination mechanism preferentially with retention of the configuration at the migration center, and was employed in the total synthesis of (+)-squalestatin S1 **21** (Scheme 5.23) [40],[41]

The stereochemical course of the rearrangement is apparently dictated by the configurational stability of the anomeric radical and the formation of a radical pair **80** held together by lithium coordination. Recombination of the two radicals then proceeds with substrate-induced diastereoselectivity. The dialkinyl-substituted radical preferentially approaches the anomeric radical in such a way that the more bulky TPS-substituent is directed away from the highly substituted tetrahydrofurane derivative **81**.

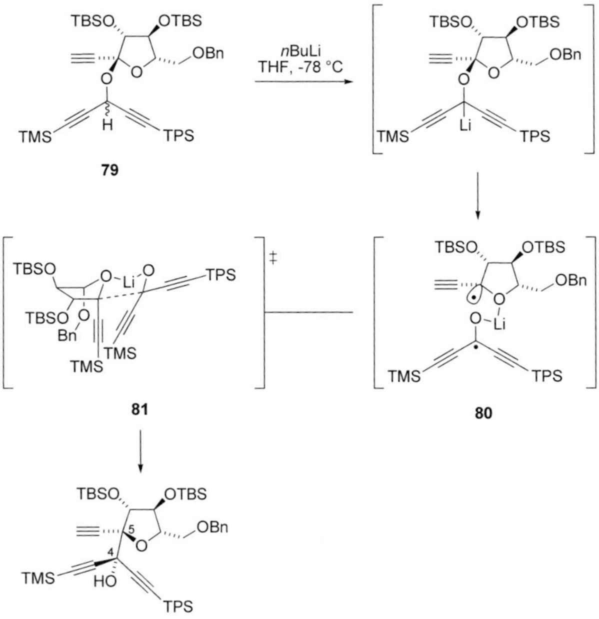

Scheme 5.23 A diastereoselective [1,2]-Wittig rearrangement that proceed by a stepwise mechanism with radical intermediates (TMS = Me$_3$Si, TPS = *tert*-BuPh$_2$Si).

5.2.5
Semipinacol Rearrangements

Intramolecular 1,2-chirality transfer during the 1,2-shift of a carbon–carbon bond can be exploited to create chiral quaternary carbon atoms [14]. In semipinacol-type rearrangements, a carbocation or a positive partial charge is formed in the α-position to an oxygenated carbon atom. The positive charge can be stabilized by the 1,2-shift of a group R that results in the formation of a carbonyl function at the origin of the migration [Eq. (10)].

(10)

During the synthesis of (–)-cephalotaxin **11**, the pinacol rearrangement of the *N*-acyl iminium ion **86** proceeded with good substrate-induced diastereoselectivity to afford the azaspirocyclic building block **87** (Scheme 5.24) [42]. The *N*-acyl iminium ion was generated under acidic conditions from the substituted 1,5 dihydro-pyrrol-2-one **84**, possibly by a 1,2-hydrogen-shift of the protonated intermediate **85** [43].

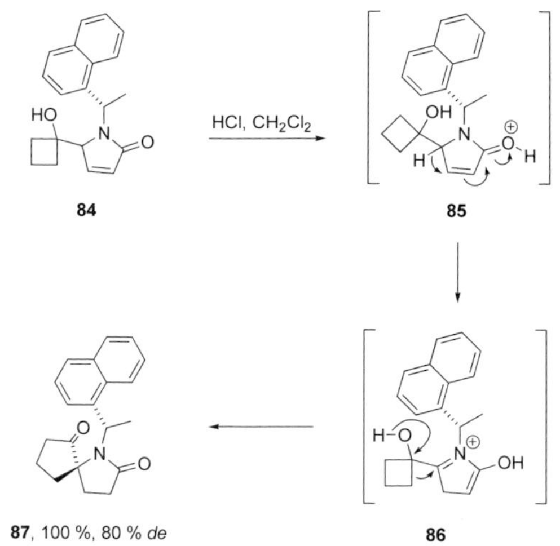

Scheme 5.24 Pinacol rearrangement of a *N*-acyl iminium ion. The stereochemical course of the rearrangement is dictated by the presence of the chiral 1-(α-naphthyl)ethyl moiety which had to be removed reductively. It is not a chiral auxiliary because it cannot be recycled after its removal.

A more attractable alternative to the pinacol rearrangement via a positively charged intermediate is the so-called epoxide rearrangement. Using a suitable Lewis acid that coordinates to the epoxide oxygen atom, it is possible to polarize the epoxide oxygen–carbon bond to an extent that is sufficient to trigger a pinacol-type rearrangement. The different variations of this theme can be classified according to their precise mechanism, the epoxide substituents, the migrating group and the Lewis acid that promotes the rearrangement.

The Suzuki–Tsuchihashi rearrangement utilizes nonracemic 2,3-epoxy alcohols, which are easily accessible [44], for instance by a Katsuki–Sharpless epoxidation [45]. The migrating group is a vinyl or aryl group [Eq. (11)].

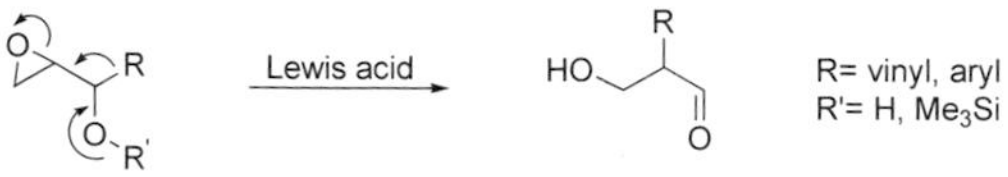

$$R= \text{vinyl, aryl} \qquad (11)$$
$$R'= H, Me_3Si$$

An example of a Suzuki–Tsuchihashi rearrangement can be found in the total synthesis of (+)-asteltoxin **25** (Scheme 5.25) [46]. The stereochemical course of the rearrangement can be explained assuming a concerted mechanism with a rear-side attack of the migrating group on the polarized C–O bond that is cleaved during the process.

A related strategy was used to construct the tetracyclic carbon skeleton of (±)-ingenol **3** starting from trans-decaline **90** (Scheme 5.26) [47]. The rearrangement proceeded with high chemo- and diastereoselectivity because only one substituent in the constrained tetracyclic ring system was able to attack the oxiran ring in an *antiperiplanar* fashion.

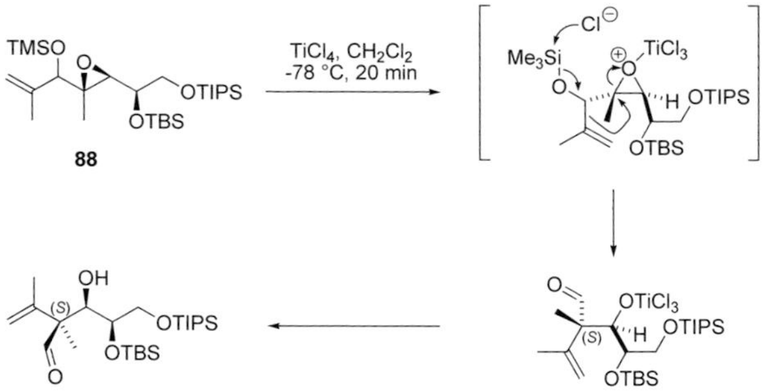

Scheme 5.25 A 1,2-chirality transfer based on a concerted mechanism with an S$_N$-2-type attack of the migrating group (TIPS = *i*Pr$_3$Si).

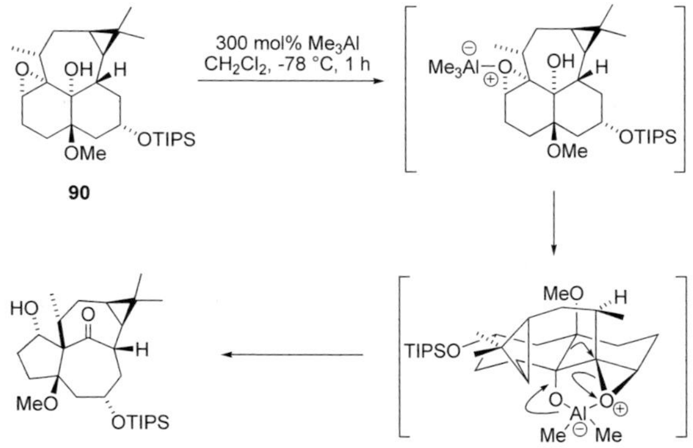

Scheme 5.26 Using a pinacol-type rearrangement for the construction of the unusual *trans*-bridged hydroazulene AB ring system of ingenol, an ingenane diterpene from plants of the genus *Euphorbia*.

The Jung rearrangement utilizes simple vinyl- or aryl-substituted epoxides as substrates for a pinacol-type rearrangement promoted by a Lewis acid [Eq. (12)] [48]. The vinyl or aryl substituent apparently influences the degree of carbon–oxygen-bond polarization (as represented by the different mesomeric structures in the VB formalism) and/or allows additional transition-state stabilization. As a result, the regioselectivity of the rearrangement is predefined by the presence of the vinyl substituent.

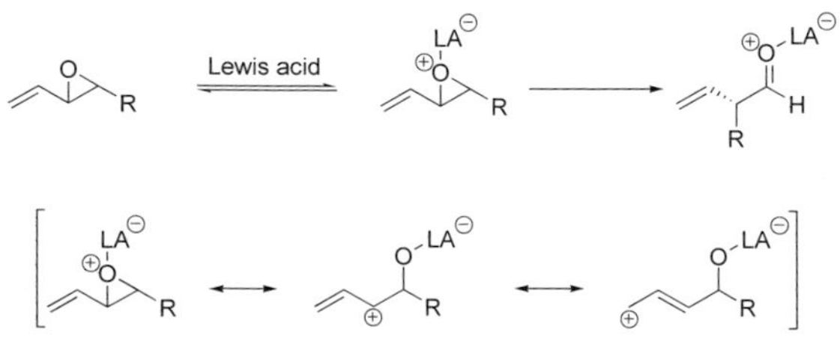

(12)

A rearrangement of this type was utilized in a synthetic approach toward (+)-fredericamycin **19** [49]. The epoxide **92** was subjected to BF_3OEt_2 and rearranged to **93** in excellent yield and diastereoselectivity (Scheme 5.27).

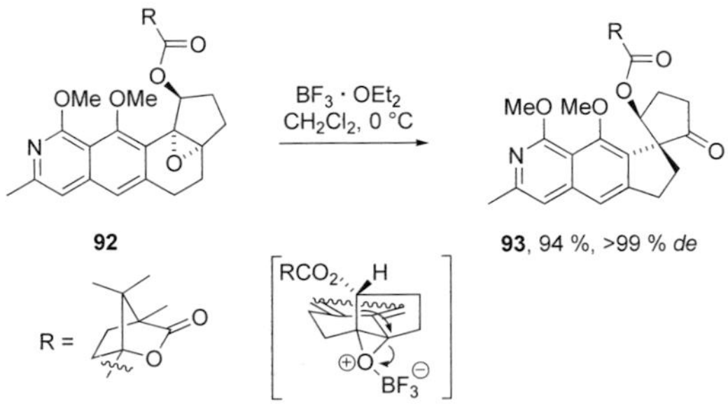

Scheme 5.27 Preference for an *antiperiplanar* attack of the migrating group at the benzylic position determines the regio- and diastereoselectivity of the process.

The Fukumoto variant of an epoxide rearrangement takes advantage of the ring strain of an oxa-spiro[2.2]pentane [50]. If 2-cyclopropylidene-ethanol is subjected to a Katsuki–Sharpless epoxidation, the spirocyclic oxirane spontaneously rearranges to a cyclobutanone [(Eq. (13)].

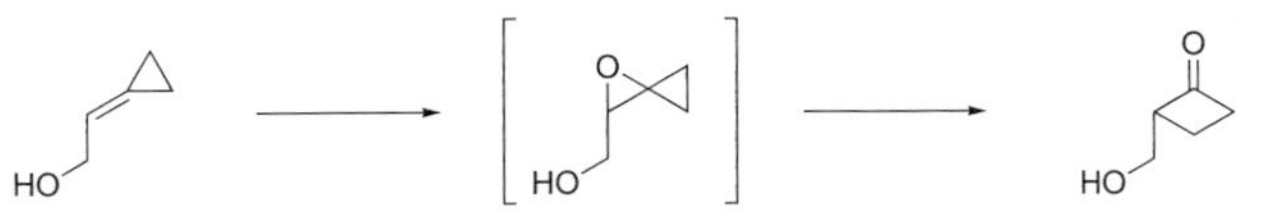

$$(13)$$

An example of this epoxide rearrangement can be found in the total synthesis of (−)-aplysin **4** [51]. Following the enantioselective epoxidation of **94**, the unstable epoxidation product rearranges to afford cyclobutanone **95** as a single enantiomer (Scheme 5.28).

5.2.6
Miscellaneous

In this section we summarize some uncommon rearrangements that we could not classify as one of the rearrangement types already discussed.

A tributylphosphine/CCl_4-mediated rearrangement was employed for the total synthesis of (−)-virantmycin **15** [52, 53]. The rearrangement of α,α-disubstituted indoline-2-methanol **96** involved the formation of the phosphorylated intermediate **97** that underwent an intramolecular S_N2-substitution to the aziridine, which was in turn ring-opened by the S_N2-attack of a chloride ion at the more easily accessible carbon atom of the aziridine moiety (Scheme 5.29).

A reductive rearrangement of the azaspirocyclic enone (±)-**100** afforded the azaspirocyclic ketone (±)-**101**, which is a key intermediate for the total synthesis of (±)-cephalotaxine **11** [54]. The 1,2-scrambling of the nitrogen atom proceeds via the aziridinium ion (±)-**102** and this determines the configuration of the newly generated chiral carbon atom. Reduction of the

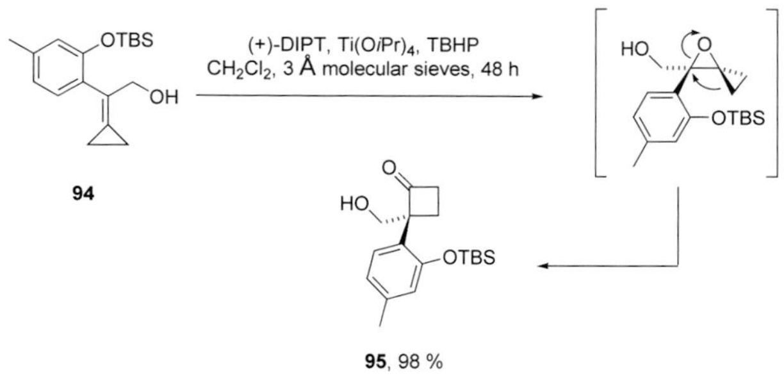

Scheme 5.28 An epoxide rearrangement according to
Fukumoto. The S_N2-type process ensures the stereochemical
integrity of the reaction [(+)-DIPT = (+)-diisopropyl tartrate,
TBHP = *tert*-butylhydroperoxide].

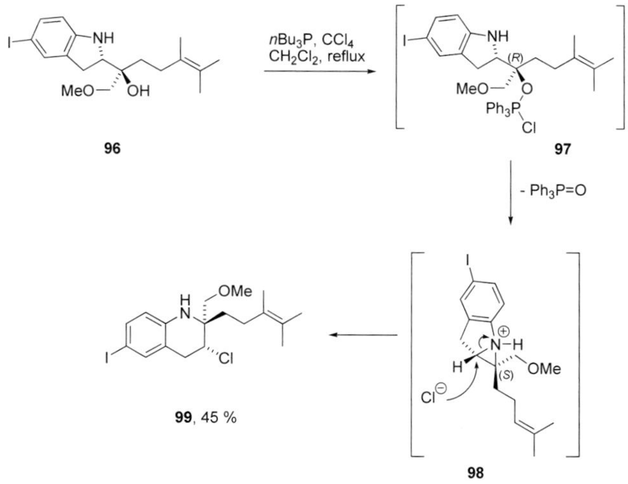

Scheme 5.29 Exploiting stereospecific reaction mechanisms
for the generation of quaternary chiral carbon atoms. A tertiary
alcohol was converted to a tertiary amine with inversion of
configuration based on the Bu_3P–CCl_4–alcohol reaction.

intermediate 4-(methylene-cyclohexa-2,5-dienylidene)-oxonium ion (±)-**103** renders the rearrangement irreversible (Scheme 5.30).

A Lewis-acid-initiated Wagner–Meerwein rearrangement was used in the synthesis of (–)-dysidiolide **7** (Scheme 5.31) [55]. Ionization by BF_3 followed by a suprafacial 1,2-methyl-shift generates the carbenium ion **108**, which is stabilized by the β-effect of the silicon atom [56]. Subsequent elimination of trimethylsilanol affords the product **105** of the ionization/Wagner–Meerwein rearrangement/elimination sequence.

An aza-analogue to the aforementioned epoxide rearrangement was applied during the synthesis of (±)-crinane **18** (Scheme 5.32) [57]. Zinc(II) was used as a Lewis-acid catalyst and a concerted mechanism would account for the observed diastereoselectivity, which was predefined by the aziridine configuration.

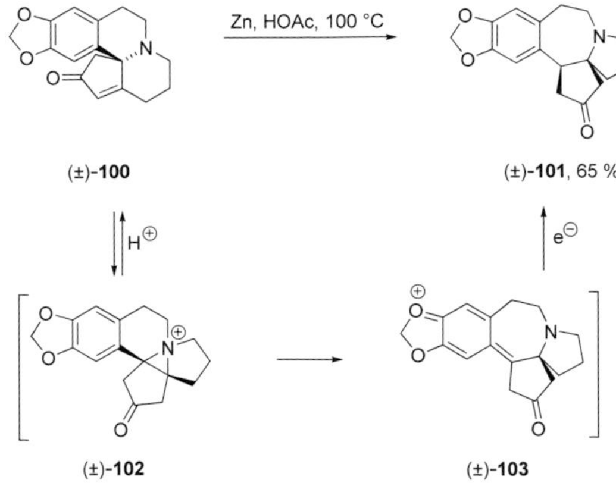

Scheme 5.30 A reductive rearrangement that proceeds by a stepwise mechanism transforms one azaspirocycle into another.

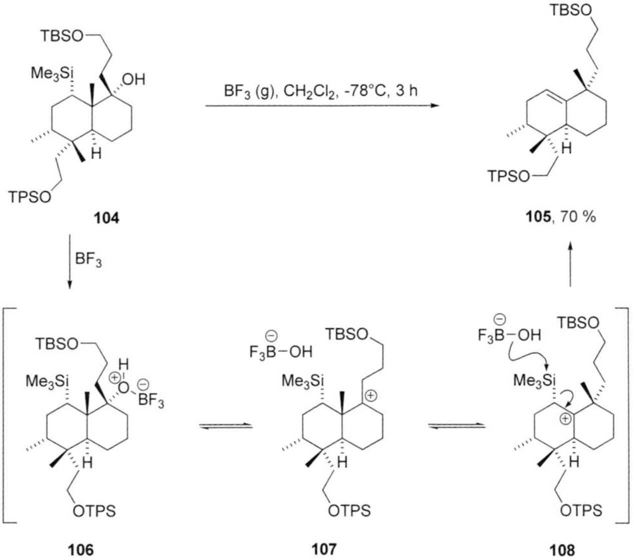

Scheme 5.31 Use of a "biomimetic" Wagner–Meerwein shift for the diastereoselective generation of a quaternary chiral carbon atom as part of the total synthesis of a sesterpene.

5.3
Summary

As indicated in our introduction, we have proven that rearrangements that proceed by concerted mechanisms are very powerful tools for the construction of quaternary chiral carbon atoms. The concerted nature of those rearrangements allowed us to suggest simple explanations for the most important criterion for the

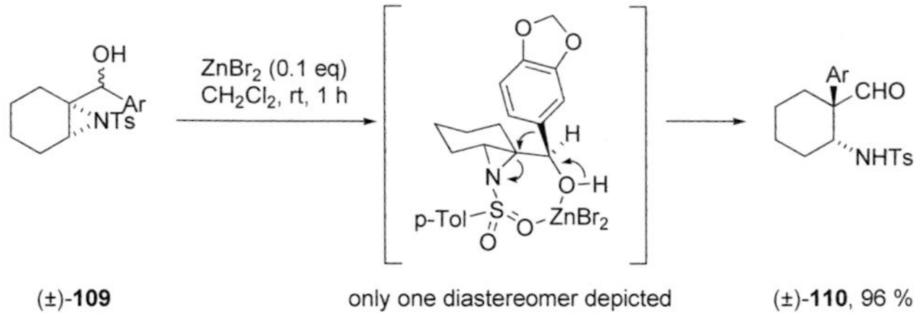

(±)-**109**   only one diastereomer depicted   (±)-**110**, 96 %

Scheme 5.32 Using an aziridine rearrangement, a racemic
mixture of two diastereomers afforded a single racemic
diastereomer because one stereogenic center was destroyed
during the rearrangement (Ts = *p*-Me-C₆H₄SO₂).

success of any transformation: selectivity. If it is easy and straightforward to suggest explanations, then it should be possible to make successful predictions about the outcome of a rearrangement. And it is exactly the predictability of selectivity expected from a concerted rearrangement that motivates chemists to apply rearrangements in complex target-oriented syntheses.

Acknowledgment

Financial support by the German Research Foundation (DFG) and the Fund of the Chemical Industry (FCI) is gratefully acknowledged. M.H. thanks Paul A. Grieco for his hospitality during a two-month stay as a visiting scientist at the Montana State University in Bozeman, Montana.

References

1 J. March, *Advanced organic chemistry: reactions, mechanisms, and structure*, 4th edn, J. Wiley & Sons, New York, **1992**, pp. 1051–1157.

2 L. Claisen, *Chem. Ber.* **1912**, *45*, 3157–3167.

3 (a) M. F. Caroll, *J. Chem. Soc.* **1940**, 704–706. (b) M. F. Caroll, *J. Chem. Soc.* **1940**, 1266–1268. (c) M. F. Caroll, *J. Chem. Soc.* **1941**, 507–510.

4 (a) A. E. Wick, D. Felix, K. Steen, A. Eschenmoser, *Helv. Chim. Acta* **1964**, *47*, 2425–2429. (b) A. E. Wick, D. Felix, K. Gschwend-Steen, A. Eschenmoser, *Helv. Chim. Acta* **1969**, *52*, 1030–1042.

5 W. S. Johnson, L. Werthemann, W. R. Bartlett, T. J. Brocksom, T. T. Li, D. J. Faulkner, M. R. Peterson, *J. Am. Chem. Soc.* **1970**, *92*, 741–743.

6 (a) R. E. Ireland, R. H. Mueller, *J. Am. Chem. Soc.* **1972**, *94*, 5897. (b) R. E.

Ireland, A. K. Willard, *Tetrahedron Lett.* **1975**, 3975–3978. (c) R. E. Ireland, R. H. Mueller, A. K. Willard, *J. Am. Chem. Soc.* **1976**, *98*, 2868–2877.

7 D. Enders, M. Knopp, R. Schiffers, *Tetrahedron: Asymmetry* **1996**, *7*, 1847–1882.

8 H. Hagiwara, T. Uchiyama, Sakai, Y. Ito, N. Morita, T. Hoshi, T. Suzuki, M. Ando, *J. Chem Soc., Perkin Trans. 1* **2002**, 583–591.

9 L. Barriault, P. J. A. Ang, R. M. A. Lavigne, *Org. Lett.* **2004**, *6*, 1317–1319.

10 E. J. Corey, R. S. Kania, *J. Am. Chem. Soc.* **1996**, *118*, 1229–1230.

11 T. Durand-Reville, L. B. Gobbi, B. L. Gray, S. V. Ley, J. S. Scott, *Org. Lett.* **2002**, *4*, 3847–3850.

12 (a) J. L. Wood, A. A. Holubec, B. M. Stoltz, M. M. Weiss, J. A. Dixon, B. D.

Doan, M. F. Shamji, J. M. Chen, T. P. Heffron, *J. Am. Chem. Soc.* **1999**, *121*, 6326–6327. (b) J. L. Wood, G. A. Moniz, D. A. Pflum, B. M. Stoltz, A. A. Holubec, H. J. Dietrich, *J. Am. Chem. Soc.* **1999**, *121*, 1748–1749. (c) J. L. Wood, G. A. Moniz, *Org. Lett.* **1999**, *1*, 371–374. (d) J. L. Wood, B. M. Stoltz, H.-J. Dietrich, D. A. Pflum, D. T. Petsch, *J. Am. Chem. Soc.* **1997**, *119*, 9641–6951.

13 I. Drutu, E. S. Krygowski, J. L. Wood, *J. Org. Chem.* **2001**, *66*, 7025–7029.

14 K. Fuji, *Chem. Rev.* **1993**, *93*, 2037–2066.

15 (a) D. Enders, M. Knopp, J. Runsink, G. Raabe, *Angew. Chem.* **1995**, *107*, 2442–2445. (b) D. Enders, M. Knopp, J. Runsink, G. Raabe, *Liebigs Ann.* **1996**, 1095–1116.

16 D. Enders, M. Knopp, *Tetrahedron* **1996**, *52*, 5805–5818.

17 (a) A. Srikrishna, S. Nagaraju, *J. Chem Soc., Perkin Trans. 1* **1991**, 657–658. (b) A. Srikrishna, K. Krishnan, *Tetrahedron* **1992**, *48*, 3429–3436. (c) A. Srikrishna, K. Krishnan, *J. Chem Soc., Perkin Trans. 1* **1993**, 667–673. (d) T. Schlama, R. Baati, V. Gouverneur, A. Valleix, J. R. Falck, C. Mioskowski, *Angew. Chem.* **1998**, *110*, 226–228. (e) A. Srikrishna, D. Vijaykuma, *J. Chem Soc., Perkin Trans. 1* **1999**, 1265–1271. (f) A. Srikrishna, C. V. Yelamaggad, P. P. Kumar, *J. Chem Soc., Perkin Trans. 1* **1999**, 2877–2881. (g) A. Srikrishna, M. Srinivasa Rao, *Tetrahedron Lett.* **2002**, *43*, 151–154. (h) A. Srikrishna, G. Satyanarayana, *Tetrahedron Lett.* **2003**, *44*, 1027–1030.

18 A. Srikrishna, D. H. Dethe, *Org. Lett.* **2003**, *5*, 2295–2298.

19 J. Mulzer, J.-T. Mohr, *J. Org. Chem.* **1994**, *59*, 1160–1165.

20 F. He, Y. Bo, J. D. Altom, E. J. Corey, *J. Am. Chem. Soc.* **1999**, *121*, 6771–6772.

21 T.-P. Loh, Q.-Y. Hu, *Org. Lett.* **2001**, *3*, 279–281.

22 T. Kawasaki, A. Ogawa, Y. Takashima, M. Sakamoto, *Tetrahedron Lett.* **2003**, *44*, 1591–1593.

23 (a) L. E. Overman, *J. Am. Chem. Soc.* **1974**, *96*, 597–599. (b) L. E. Overman, *J. Am. Chem. Soc.* **1976**, *98*, 2901–2910. (c) L. E. Overman, *Acc. Chem. Res.* **1980**,

13, 218–224. (d) T. Nishikawa, M. Asai, N. Ohyabu, M. Isobe, *J. Org. Chem.* **1998**, *63*, 188–192.

24 T. Oishi, K. Ando, K. Inomiya, H. Sato, M. Iida, N. Chida, *Bull. Chem. Soc. Jpn.* **2002**, *75*, 1927–1947.

25 (a) F. Johnson, S. K. Malhotra, *J. Am. Chem. Soc.* **1965**, *87*, 5492–5493. (b) S. K. Malhotra, F. Johnson, *J. Am. Chem. Soc.* **1965**, *87*, 5493–5495. (c) F. Johnson, *Chem. Rev.* **1968**, *68*, 375–413. (c) R. W. Hoffmann, *Chem. Rev.* **1989**, *89*, 1841–1860.

26 (a) A. C. Cope, E. M. Hardy, *J. Am. Chem. Soc.* **1940**, *62*, 441–444. (b) A. C. Cope, K. E. Hoyle, D. Heyl, *J. Am. Chem. Soc.* **1941**, *63*, 1843–1852. (c) A. C. Cope, C. M. Hofmann, E. M. Hardy, *J. Am. Chem. Soc.* **1941**, *63*, 1852–1857. (d) H. Levy, A. C. Cope, *J. Am. Chem. Soc.* **1944**, *66*, 1684–1688.

27 J. A. Berson, M. Jones, *J. Am. Chem. Soc.* **1964**, *86*, 5019–5020.

28 D. A. Evans, A. M. Golob, *J. Am. Chem. Soc.* **1975**, *97*, 4765–4766.

29 D.-S. Hsu, C.-C. Liao, *Org. Lett.* **2003**, *5*, 4741–4743.

30 E. Lee, I. J. Shin, T. S. Kim, *J. Am. Chem. Soc.* **1990**, *112*, 260–264.

31 A. Madin, C. J. O'Donnell, T. Oh, D. W. Old, L. E. Overman, M. J. Sharpe, *Angew. Chem.* **1999**, *38*, 3110–1112, *Angew. Chem. Int. Ed.* **1999**, *38*, 2934–2936.

32 J. E. Baldwin, C. Ullenius, *J. Am. Chem. Soc.* **1974**, *96*, 1542–1547.

33 T. Fukuyama, G. Liu, *Pure Appl. Chem.* **1997**, *69*, 501–505.

34 (a) G. Wittig, H. Döser, I. Lorenz, *Liebigs Ann. Chem.* **1949**, *562*, 192–205. (b) U. Schöllkopf, K. Fellenberger, *Liebigs Ann.* **1966**, *698*, 80–85. (c) Y. Makisumi, S. Notzumoto, *Tetrahedron Lett.* **1966**, *7*, 6393–6397.

35 (a) T. Nakai, K. Mikami, *Chem. Rev.* **1986**, *86*, 885–902. (b) M. A. Marshall, in *Comprehensive Organic Synthesis*, Vol. 3, ed. B. M. Trost, I. Fleming, Pergamon, Oxford, **1991**; pp 975–1014. (c) M. A. Marshall, in *Comprehensive Organic Synthesis*, Vol. 6, ed. B. M. Trost, I. Fleming, Pergamon, Oxford, **1991**, pp. 873–903. (d) T. Nakai, K. Mikami, *Org. React.* **1994**, *46*, 105–209. (e) J. Kallmerten, in *Houben Weyl, Stereoselective*

Synthesis, Vol. E 21d, ed. G. Helmchen, R. W. Hoffmann, J. Mulzer, E. Schaumann, Thieme, Stuttgart, **1995**, pp. 3757–3809.

36 J. Mulzer, D. Riether, *Org. Lett.* **2000**, *2*, 3139–3141.

37 W. C. Still, A. Mitra, *J. Am. Chem. Soc.* **1978**, *100*, 1927–1928.

38 (a) M. Hiersemann, *Tetrahedron* **1999**, *55*, 2625–2638. (b) M. Hiersemann, C. Lauterbach, A. Pollex, *Eur. J. Org. Chem.* **1999**, 2713–2724. (c) M. Hiersemann, L. Abraham, A. Pollex, *Synlett* **2003**, 1088–1095.

39 A. Pollex, L. Abraham, J. Müller, M. Hiersemann, *Tetrahedron Lett.* **2004**, *45*, 6915–6918.

40 (a) K. Tomooka, T. Igarashi, T. Nakai, *Tetrahedron Lett.* **1993**, *34*, 8139–8142. (b) K. Tomooka, H. Yamamoto, T. Nakai, *Liebigs Ann./Recueil* **1997**, 1275–1281. (c) K. Tomooka, H. Yamamoto, T. Nakai, *Angew. Chem.* **2000**, *112*, 4674–4676.

41 K. Tomooka, M. Kikuchi, K. Igawa, M. Suzuki, P.-H. Keong, T. Nakai, *Angew. Chem.* **2000**, *112*, 4676–4679; *Angew. Chem. Int. Ed.* **2000**, *39*, 4502–4505.

42 L. Planas, J. Pérard-Viret, J. Royer, *J. Org. Chem.* **2004**, *69*, 3087–3092.

43 L. Planas, J. Pérard-Viret, J. Royer, M. Selkti, A. Thomas, *Synlett* **2002**, 1629–1632.

44 M. Shimazaki, H. Hara, K. Suzuki, G. I. Tsuchihashi, *Tetrahedron Lett.* **1987**, *28*, 5891–5894.

45 T. Katsuki, K. B. Sharpless *J. Am. Chem. Soc.* **1980**, *102*, 5974–5976.

46 J. V. Raman, H. K. Lee, R. Vleggaar, J. K. Cha, *Tetrahedron Lett.* **1995**, *36*, 3095–3098.

47 K. Tanino, K. Onuki, K. Asano, M. Miyashita, T. Nakamura, Y. Takahashi, I. Kuwajima, *J. Am. Chem. Soc.* **2003**, *125*, 1498–1500.

48 M. E. Jung, D. C. D'Amico, *J. Am. Chem. Soc.* **1995**, *117*, 7379–7388.

49 Y. Kita, Y. Higuchi, Y. Yoshida, K. Iio, S. Kitagaki, S. Ueda, S. Akai, H. Fujioka, *J. Am. Chem. Soc.* **2001**, *123*, 3214–3222.

50 H. Nemoto, H. Ishibashi, M. Nagamochi, K. Fukumoto, *J. Org. Chem.* **1992**, *57*, 1707–1712.

51 H. Nemoto, M. Nagamochi, H. Ishibashi, K. Fukumoto, *J. Org. Chem.* **1994**, *59*, 74–79.

52 M. Ori, N. Toda, K. Takami, K. Tago, H. Kogen, *Angew. Chem.* **2003**, *115*, 2644–2647; *Angew. Chem. Int. Ed.* **2003**, *42*, 2540–2543.

53 For a discussion of the mechanism of the R_3P/CCl_4-mediated S_N2-process, see: L. A. Jones, C. E. Sumner, B. Franzus, T. T. S. Huang, E. I. Snyder *J. Org. Chem.* **1978**, *43*, 2821–2827.

54 W.-D. Z. Li, Y.-Q. Wang, *Org. Lett.* **2003**, *5*, 2931–2934.

55 E. J. Corey, B. E. Roberts, *J. Am. Chem. Soc.* **1997**, *119*, 12425–12431.

56 J. B. Lambert, *Tetrahedron* **1990**, *46*, 2677–2689.

57 Z. L. Song, B. M. Wang, Y. Q. Tu, C. A. Fan, S. Y. Zhang, *Org. Lett.* **2003**, *5*, 2319–2321.

6
Cycloaddition Reactions
Giovanni Desimoni and Giuseppe Faita

6.1
Introduction

The stereoselective construction of molecules containing at least one quaternary carbon stereocenter, i.e. a carbon atom with four different non-hydrogen substituents, has attracted a great deal of interest, and the art of organic synthesis [1] made impressive progress as a result of this research. Among the different methods allowing successful achievement of this important target, cycloaddition reactions are probably the most powerful processes for their flexibility and predictability, the latter property being the enormous heritage left to organic chemists by Woodward and Hoffmann, as well as for their intrinsic ability to control the relative stereochemistry of the newly created stereocenters.

Methods for the enantioselective formation of quaternary carbon stereocenters through cycloadditions may involve two different approaches:

1) the use of a reactant incorporating a stoichiometric amount of a chiral auxiliary that must be removed and (possibly) recycled at the end of the process;
2) the use of a chiral catalyst that, like a *molecular robot* [2], assembles the achiral cycloaddends, limiting the use of enantiopure material with a favorable application of the atom economy principle.

Completeness conflicts with the limits assigned to this chapter, several excellent results cannot be discussed, and the many examples reviewed reflect personal choices by the authors. The following sections will discuss the different type of cycloadditions that have been reported to produce at least one quaternary stereocenter through asymmetric synthesis. Although the use of optically pure starting materials, which incorporate the chiral information in the reaction product, will not be usually discussed, a few examples will be illustrated for the particular elegance of the synthesis.

Quaternary Stereocenters: Challenges and Solutions for Organic Synthesis. Edited by Jens Christoffers, Angelika Baro
Copyright © 2005 WILEY-VCH Verlag GmbH & Co. KGaA, Weinheim
ISBN: 3-527-31107-6

6.2
[2+1] Cycloaddition Reactions

The thermal decomposition of diazocompounds, catalyzed by a chiral salicylaldimine copper complex, in the presence of olefins to give *asymmetric cyclopropanation reactions* was first reported by Nozaki and his coworkers in 1966 [3]. The development of ligand **1** by Aratani [4] was usefully applied in the cyclopropanation between isobutene **2** and ethyl diazocarboxylate **3** and led to the factory-scale production of ethyl (*1S*)-2,2-dimethylcyclopropanecarboxylate **4**, the key intermediate of Cilastatine **5**, an enzyme inhibitor for dehydropeptidase (Scheme 6.1).

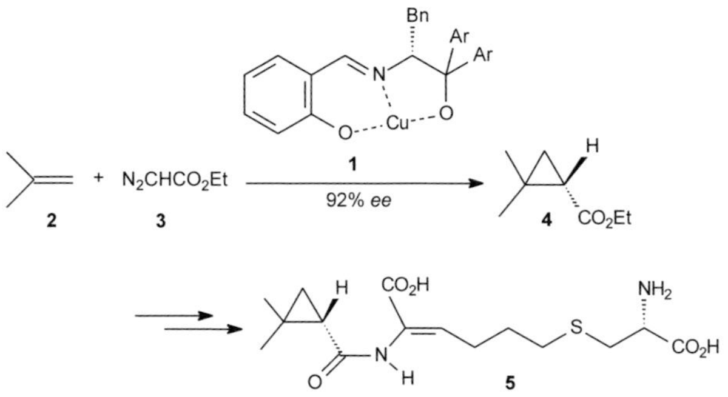

Scheme 6.1

Since then, enormous efforts have been spent in the design of new chiral transition metal catalysts and a review [5] illustrates the developments achieved in the field. For reasons of space, the relevant contribution of several groups (e.g. Brunner, Evans, Katsuki, Masamune, Nishiyama, Pfaltz) must be considered as reported in that review, and only a selected choice of inter- and intramolecular cyclopropanation reactions, as well as some examples of applications of the protocol to the syntheses of biologically relevant compounds, will be discussed in this section.

In addition to Cu, Rh(II) and sometimes Ru(II) are cations involved in the formation of the chiral catalysts; the ligands are 2,6-bis(2-oxazolinyl)pyridines (pybox), 2,2′-methylenebis(2-oxazolines) (box), and other amino acid-based derivatives.

Among *intermolecular cyclopropanation reactions*, the decomposition of vinyldiazomethanes **6** in the presence of arylbenzenes **7**, catalyzed by Rh(II) N-(arylsulfonyl)azacycloalkyl carboxylates **8**, is an excellent illustrative example [6] of the influence of reagents and catalysts on the enantioselectivity observed in the formation of **9** (Scheme 6.2 and Table 6.1).

In Table 6.1, a comparison of entries 1 and 7 demonstrates that pentane is a better solvent than dichloromethane. The enantioselectivity clearly decreases as the steric hindrance of the ester group increases (entries 1–4), while it is not influenced by the electronic character of the substituent in **7** (entries 9–11). Even

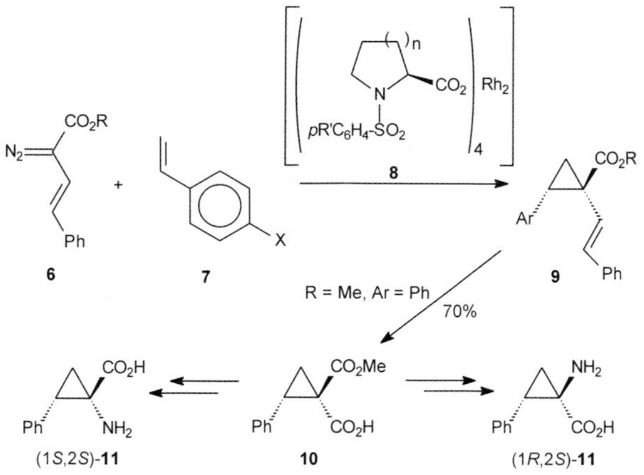

Scheme 6.2

though a cyclic amino acid derivative is required as ligand, a flexibility in terms of ring size can be easily tolerated (entries 1, 12 and 13), and different *N*-sulfonyl functions (entries 5–8) give comparable levels of asymmetric induction.

Under the conditions reported in entry 9, methyl (1*S*,2*S*)-2-phenyl-1-[2-(*Z*)-styryl]cyclopropane-1-carboxylate (**9**: R = Me, Ar = Ph) was obtained in a highly diastereoselective and enantioselective mode, and through the common intermediate **10**, it was stereodivergently converted into (1*S*,2*S*)- and (1*R*,2*S*)-2-phenylcyclopropan-1-amino acids **11**.

The topic illustrated in Scheme 6.2 is still up to date: the cyclopropanation of styrene with (silanyloxyvinyl)diazoacetates, catalyzed by Rh(II) complexes of

Table 6.1 Effect of substituents on reagents and catalyst on the asymmetric cyclopropanation of **6** and **7** [6].

Entry	Solvent	T (°C)	6 (R)	7 (X)	8 (R′, n)	Yield (%)	(1S,2S)-9 ee(%)
1	pentane	25	Me	H	tBu, 1	79	90
2	pentane	25	Et	H	tBu, 1	–	84
3	pentane	25	iPr	H	tBu, 1	–	76
4	pentane	25	tBu	H	tBu, 1	–	50
5	CH_2Cl_2	–	Me	H	MeO, 1	–	76
6	CH_2Cl_2	–	Me	H	NO_2, 1	–	83
7	CH_2Cl_2	–	Me	H	tBu, 1	–	74
8	CH_2Cl_2	–	Me	H	$C_{12}H_{25}$, 1	–	79
9	pentane	–78	Me	H	$C_{12}H_{25}$, 1	68	98
10	pentane	–78	Me	Cl	$C_{12}H_{25}$, 1	70	>97
11	pentane	–78	Me	OMe	$C_{12}H_{25}$, 1	41	90
12	pentane	–	Me	H	tBu, 0	–	81
13	pentane	–	Me	H	tBu, 2	–	81

1,8-naphthoyl-protected amino acids, has been studied [7] and excellent results in terms of diastereo- and enantioselectivity were obtained. The same catalyst was also the most effective within a Rh(II) series in the asymmetric cyclopropanation [8] of 1,1-diphenylethylene and 1-hexyne with 3,3,3-trifluoro-2-diazopropionate, the alkyne giving ethyl 2-butyl-1-trifluoromethylcycloprop-2-ene-1-carboxylate (51%; 24% *ee*) with the quaternary stereocenter on a cyclopropene ring.

Several *intramolecular cyclopropanation reactions* are known to involve the decomposition of allylic and homoallylic diazoesters **12** to give condensed cyclopropyllactones **13** in good yields and sometimes excellent enantioselectivity [9–12] (Scheme 6.3 and Table 6.2).

The reactions can be catalyzed either by a Rh(II)-based complex of methyl (5 S)-2-pyrrolidone-5-carboxylate **14a** (Table 6.2, entries 1–4), sometimes with heteroatoms in position 3 (**14b,c**, entries 5–11), or by Cu(I) complexes of box ligands **15a–d** and **16**. Among the effects of the different substituents on **12**, it can be emphasised that *cis* double bonds give better *ee* than *trans* ones (entries 6,7 versus 8,9), while with hindered diazoacetoacetates good results are obtained only with a box **16**-based catalyst (entry 18).

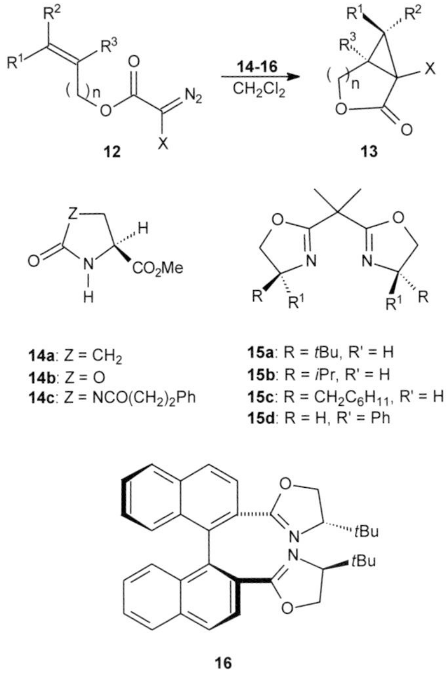

14a: Z = CH$_2$
14b: Z = O
14c: Z = NCO(CH$_2$)$_2$Ph

15a: R = *t*Bu, R' = H
15b: R = *i*Pr, R' = H
15c: R = CH$_2$C$_6$H$_{11}$, R' = H
15d: R = H, R' = Ph

16

Scheme 6.3

The construction of the optically active phorbol CD-ring skeleton **18** [13] (phorbol derivatives control intracellular signal transduction through protein kinase) by Cu(I)triflate/box-catalyzed intramolecular cyclopropanation of enol silyl ether **17** (Scheme 6.4) can be considered as a bridge between intramolecular cyclopropanation and the use of this reaction as a route to the synthesis of

Table 6.2 Effect of substituents on reagent and catalyst on the asymmetric cyclopropanation of **12** to **13**.

Entry	Ligand	Cation	R^1	R^2	R^3	X	n	Yield (%)	13 ee (%) (config.)	Ref.
1	14a	Rh(II)	H	H	Me	H	1	72	7 (1R,5S)	9
2	14a	Rh(II)	(a)	Me	H	H	1	79	93 (1S,5R)	9
3	14a	Rh(II)	Me	(a)	H	H	1	88	95 (1S,5R)	9
4	14a	Rh(II)	H	H	Me	H	2	76	83 (1R,6S)	9
5	14b	Rh(II)	Me	Me	H	Me	1	81	71	10
6	14b	Rh(II)	nPr	H	H	Me	1	62	85	10
7	14b	Rh(II)	Ph	H	H	Me	1	65	78	10
8	14b	Rh(II)	H	nPr	H	Me	1	46	52	10
9	14b	Rh(II)	H	Ph	H	Me	1	70	43	10
10	14c	Rh(II)	H	H	Me	H	1	75	89 (1S,5R)	11
11	14c	Rh(II)	H	H	nBu	H	1	82	93 (1S,5R)	11
12	15a^b	Cu(I)	H	H	Me	H	1	58	87 (1S,5R)	11
13	15a^b	Cu(I)	H	H	nBu	H	1	73	82 (1S,5R)	11
14	15a^c	Cu(I)	H	H	H	CO$_2$tBu	1	53	40	12
15	15b^c	Cu(I)	H	H	H	CO$_2$tBu	1	73	28	12
16	15c^c	Cu(I)	H	H	H	CO$_2$tBu	1	81	35	12
17	15d^c	Cu(I)	H	H	H	CO$_2$tBu	1	78	12	12
18	16	Cu(I)	H	H	H	CO$_2$tBu	1	24	83	12

a Me$_2$C=CH(CH$_2$)$_2$. b Complex from Cu(I)(PF$_6$)(MeCN)$_4$. c Complex from Cu(II)(OTf)$_2$.

biologically relevant compounds. The enantioselectivity of the reaction is strongly influenced by the structure of the box **15** and the results, reported in Table 6.3, show that only *tert*-butyl- and (2-trimethylsilyloxy)isopropyl-substituted ligands (**15a,f**) give good or excellent *ee*.

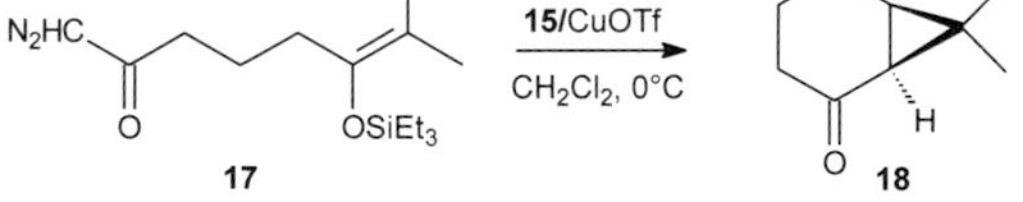

Scheme 6.4

Catalyzed intramolecular cyclopropanation was the key step in the synthesis of the natural (1R,2R,3R)-presqualene diphosphate **21**, an intermediate in the biosynthesis of squalene from farnesyl diphosphate. Diazoacetate **19** was decomposed in the presence of Rh/*ent*-**14a** catalyst and a single bicyclic lactone **20**

Table 6.3 Asymmetric intramolecular cyclopropanation of
18 to **17**, catalyzed by CuOTf/box complexes [13].

Entry	Ligand	Time (h)	Yield (%)	17 ee (%)
1	15a	19	82	78
2	*ent*-15d (R = Ph)	2	38	13
3	15e (R = CH₂CHMe₂)	12	55	36
4	15f (R = CMe₂OTMS)	19	70	92

was obtained [14] with excellent yield and enantioselectivity, with three (one
quaternary) stereocenters with the required configuration for its conversion to **21**
(Scheme 6.5).

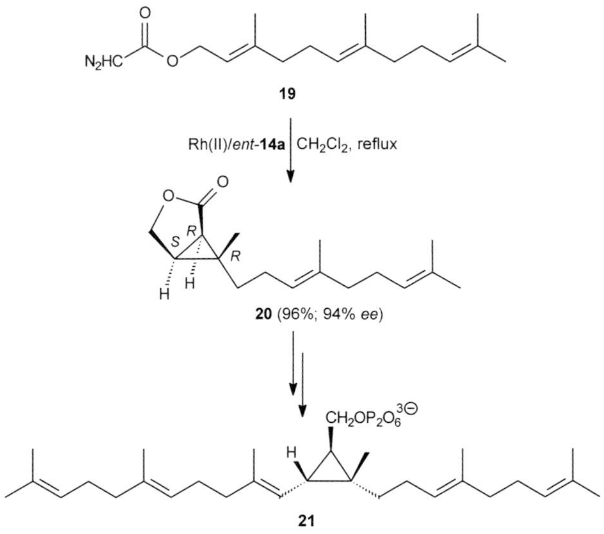

Scheme 6.5

The elegant and simple Corey enantioselective synthesis of sirenin **25** [15] is
the best example with which to close the cyclopropanation discussion. Diazoester
22 was stirred for 3 h in CH₂Cl₂ at 0 °C with Cu(I)/box catalyst **23**, and the bicyclic
intermediate **24** with three stereocenters (one quaternary) was cleanly obtained
and converted into **25** (53% yield), which was identical with authentic sirenin
(Scheme 6.6).

The formation of quaternary stereocenters via an *asymmetric aziridination
reaction* [16] is a process not yet fully explored and very few examples have been
reported in the literature. Even if the reaction of two chiral camphor *N*-enoylpyra-
zolidinones **26** (R = H or Me) with *N*-aminophthalimide (PhtN–NH₂) in the
presence of Pb(OAc)₄ [17] can be considered from a "formal" point of view as a
cycloaddition, the formation of **27**, with excellent *dr* when R is methyl and from

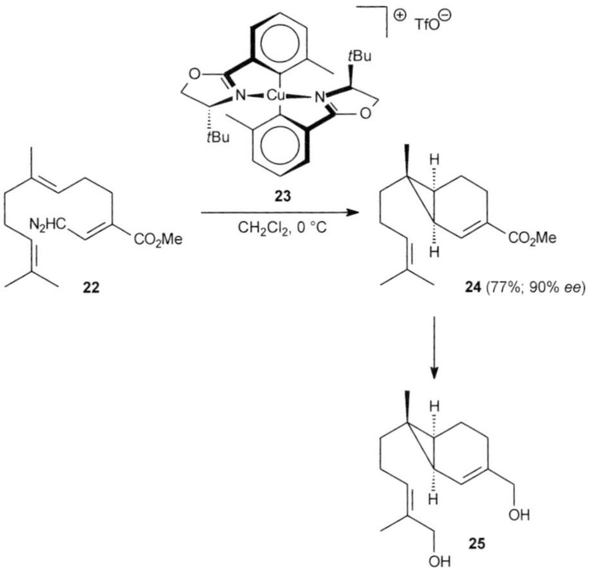

Scheme 6.6

which **28** is obtained with an easy recover of the auxiliary, could occur by an alternative mechanism (Scheme 6.7).

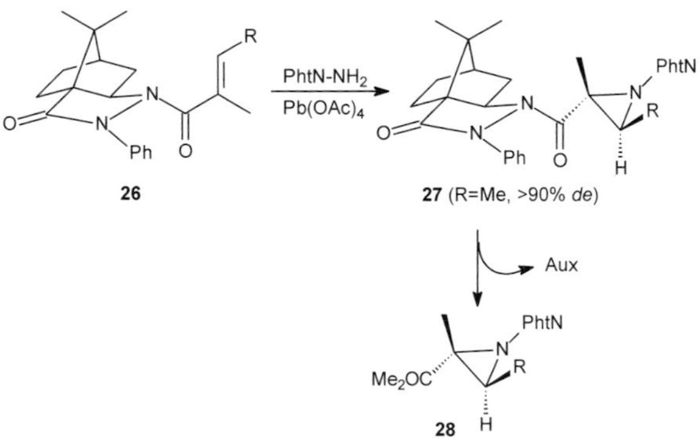

Scheme 6.7

6.3
[2+2] Cycloaddition Reactions

One of the few examples of asymmetric [2+2] cycloaddition is found in the enantioselective total synthesis of (+)-tricycloclavulone **33** [18], a marine prostanoid, where the cycloaddition represents the key step to induce enantioselectivity in the cyclobutane ring formation (Scheme 6.8). The reaction between

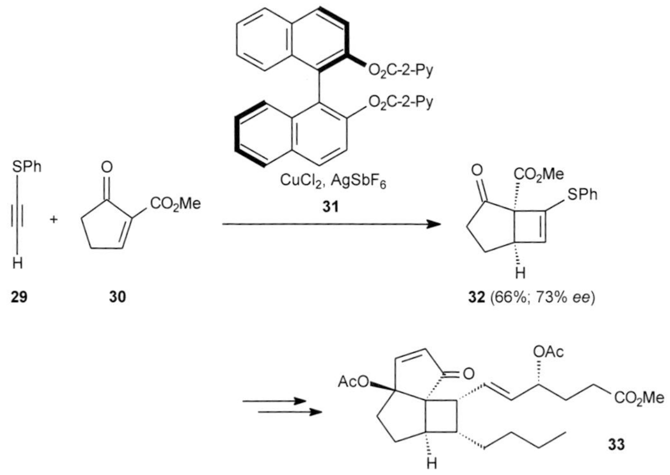

Scheme 6.8

phenylthioacetylene **29** and 2-methoxycarbonyl-2-cyclopenten-1-one **30** is catalyzed by the chiral Lewis acid **31**, and the bicyclic intermediate **32** is obtained in 66% yield with up to 73% *ee*.

6.4
1,3-Dipolar Cycloaddition Reactions

1,3-Dipolar cycloadditions represent one of the most useful synthetic routes to heterocyclic five-membered rings, and the asymmetric synthesis of adducts containing a quaternary stereocenter was usually achieved by using asymmetrically 1,1-disubstituted alkenes as dipolarophiles. Some examples falling within the limits of this chapter can be found in two reviews [19, 20].

6.4.1
Nitrone Cycloadditions

Nitrones provide easy access to enantiomerically pure isoxazolidines, which can be directly converted into 1,3-amino alcohols through reductive cleavage of the nitrogen–oxygen bond, the chiral amino alcohols being important synthons for the assembly of biologically active products. The enantioselective catalyzed reaction with α-substituted acroleins **34** allows both electronic and steric control of the cycloaddition to give isoxazoline-4- or -5-aldehydes (**36** and **37**). The effect of different catalysts on the cycloaddition with *N*-phenyl-*C*-arylnitrones **35** (Scheme 6.9), studied by three independent groups [21–23], are summarized in Table 6.4.

Some observations can be made on the data listed in the table:

• catalyst **38** is strongly endo-selective, but both regio- or enantioselectivity are unsatisfactory (entries 1–3);

- the DBFOX/Ph **39**-based complexes catalyze the 1,3-dipolar cycloaddition of α-bromoacrolein to produce the electronically controlled isoxazolidine-4-carbaldehydes **36** (entries 7–11), but only Zn(II)- and Co(II)-based catalysts give excellent ee;
- α-alkyl- or α-aryl-substituted acroleins react in the presence of Ni(II) complexes to furnish the sterically controlled isoxazoline-5-carbaldehydes **37** (entries 4, 5, 12–15), while Zn(II) gives a regioisomeric mixture (entry 6);
- the introduction of substituents R in the 5-position of **39** protects the metal center, suppresses the formation of aggregates, and usually increases the stereoselectivity of the reactions.

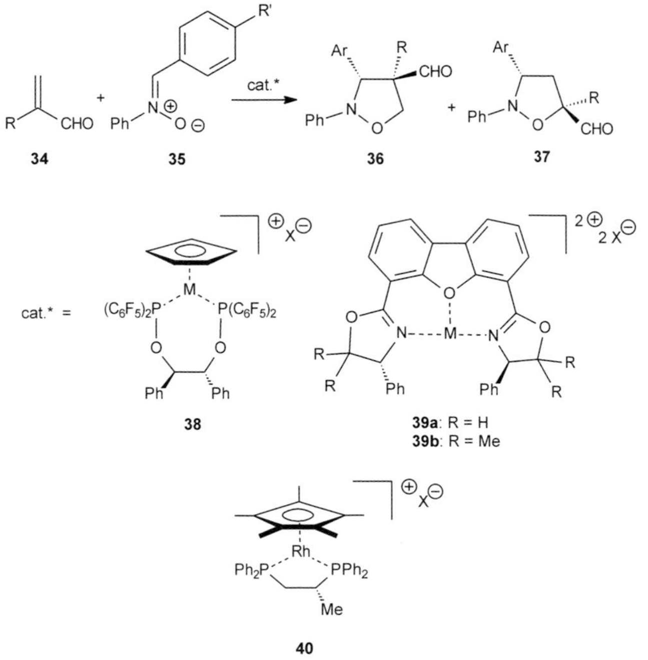

Scheme 6.9

The reacting complex **41** of methacrolein with **40** involved in the nitrone 1,3-dipolar cycloaddition was isolated and its crystal structure determined [24] (Fig. 6.1). The methacrolein molecule adopts an *s-trans*-conformation and, in this disposition, the upper *Re*-face is shielded by the methyl substituents on the apical ligand (C_5Me_5). The nitrone will approach methacrolein from the opposite *Si*-face, with the *exo* attack disfavored by the repulsion between the nitrone N–Ph group and the methyl substituent on the ligand. Through a careful inspection of the supramolecular structure involved in the catalytic cycle it is therefore possible to interpret the observed *endo* preference of the reaction (Table 6.4, entries 16–19) and to rationalize the stereochemical outcome. This approach may help to solve some of the unanswered questions relating to the relationship between the structure of the catalyst and the chirality transmitted to the product, and allow the design of tailor-made ligands for more stereoselective processes.

Table 6.4 Asymmetric 1,3-dipolar cycloadditions between α-substituted acroleins **34** and arylphenylnitrones **35**.

Entry	Catalyst	M	X⁻	R (34)	R′ (35)	Yield (%)	36:37	36 ee (%) (config.)	37 ee (%) (config.)	Ref.
1	38	Fe	SbF_6	Me	H	85	20:80	91	87	21
2	38	Ru	SbF_6	Me	H	92	40:60	94	76	21
3	38	Ru	SbF_6	Me	NO_2	80	0:100	–	66	21
4	39a	Ni	ClO_4	Me	H	73	0:100	–	96 (3R,5R)	22
5	39b	Ni	ClO_4	Me	H	98	0:100	–	99 (3R,5R)	23
6	39a	Zn	ClO_4	Me	H	98	45:55	83	95 (3R,5R)	22
7	39a	Zn	ClO_4	Br	H	85	100:0	98 (3R,4R)	–	22
8	39b	Ni	ClO_4	Br	H	59	94:6	18 (3R,4R)	–	23
9	39b	Zn	ClO_4	Br	H	96	100:0	99 (3R,4R)	–	23
10	39b	Co	BF_4	Br	H	92	100:0	98 (3R,4R)	–	23
11	39a	Zn	ClO_4	Br	NO_2	97	100:0	>99	–	22
12	39a	Ni	ClO_4	Et	H	98	2:98	55	77	22
13	39b	Ni	ClO_4	Et	H	100	0:100	–	92	23
14	39a	Ni	ClO_4	Ph	H	100	0:100	–	80	22
15	39b	Ni	ClO_4	Ph	H	67	0:100	–	84	23
16	40	Rh	SbF_6	Me	H	100	63:37	90 (3S,4S)	75 (3S,5S)	24
17	40	Rh	PF_6	Me	H	84	64:36	89 (3S,4S)	63 (3S,5S)	24
18	40	Rh	BF_4	Me	H	69	65:35	89 (3S,4S)	68 (3S,5S)	24
19	40	Rh	CF_3SO_3	Me	H	45	65:35	90 (3S,4S)	71 (3S,5S)	24

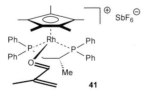

Fig. 6.1 Crystal structure of the reacting complex formed from **40** and methacrolein [24].

6.4.2
Other 1,3-Dipolar Cycloadditions

An example of asymmetric dipolar cycloaddition with diazoalkanes involves the commercially available (1S)-2,10-camphorsultam as chiral auxiliary bound to methacryloyl group **42**, which reacts with trimethylsilyl diazomethane to give, after workup of the adduct under acidic conditions, the "azaproline" derivative **43** (Scheme 6.10) [25]. The absolute configuration was established by X-ray analysis to be (S), and the observed facial selectivity was in accordance with the typical behavior of camphorsultam-derived systems.

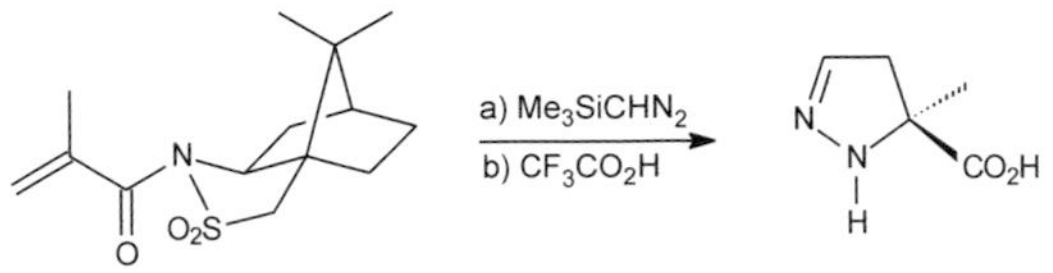

42 **43** (65%; 88% *ee*)

Scheme 6.10

Chiral asymmetric 1,3-dipolar cycloadditions of *azomethine ylides* have been achieved employing either chiral auxiliaries in the dipolarophile or chiral catalysts [26]. We report examples of the latter approach as the more promising, pointing out the importance of the Ag(I) cation to generate the dipole in the metal-complexed form in a stereospecific manner. In the pre-transition state, metal coordinates the aldiminoester **44**, through the nitrogen and carboxyl oxygen atoms, and as the chiral ligand **45** shields one face of the putative dipole, therefore the addition of electron-poor dipolarophiles gives the pyrrolidine adducts **46** (having a quaternary stereocenter) with good chemical yields and promising *ee* (Scheme 6.11).

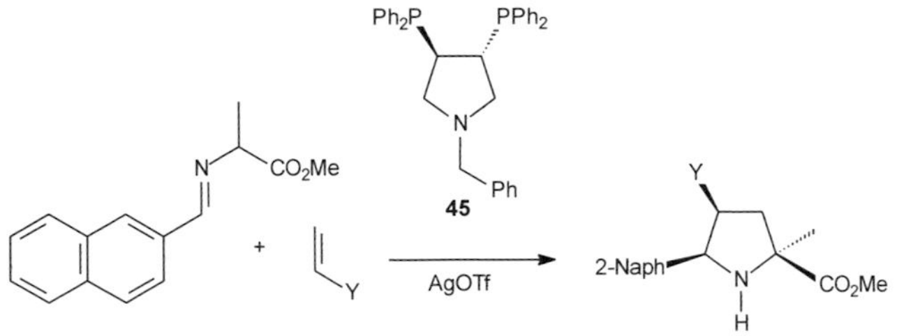

44

46a: Y = MeCO (83%; 70% *ee*)
46b: Y = SO_2Ph (64%; 70% *ee*)

Scheme 6.11

The intramolecular cyclization of diazocarbonyl compounds, catalyzed by rhodium(II), is a useful method for the generation of *carbonyl ylides* and their *in situ* reaction with dipolarophiles. If the cycloaddition is run in the presence of a chiral catalyst able to coordinate the dipolarophile, significant levels of enantioselectivity can be obtained. This "one-pot" approach has been followed with *o*-methoxycarbonyl-α-diazoacetophenone **47**, which gives carbonyl ylide **48** (see Scheme 6.12) with $Rh_2(OAc)_4$ [27]. The (*S*,*S*)-diisopropyl-2,6-bis(oxazolinyl)pyridine

[(*S,S*)-*i*Prpybox] **50a** and Sc(III) triflate give the chiral catalyst able to coordinate pyruvates **49a,b**, which undergo *exo*-selective cycloaddition (probably due to unfavorable dipolar interaction between the carbonyl group of the ylide and the ester in the *endo* approach) to give **51a,b**. Good results in terms of enantioselectivity (78–87% *ee*) are obtained only in the presence of pyruvic acid as additive, whose role, however, is still to be clarified.

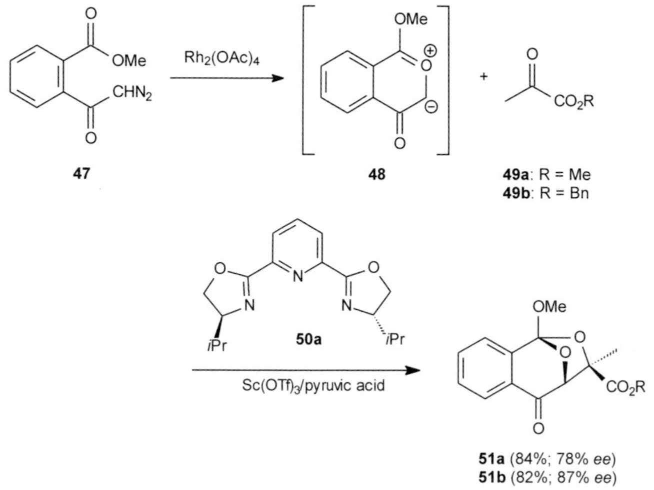

Scheme 6.12

6.5
Diels–Alder Reactions

The Diels–Alder (DA) reaction is one of the most powerful options in the art of organic synthesis for developing complex molecules carrying quaternary carbon stereocenters. The very first application of this concept was the very elegant synthesis of cortisone **55**, developed in 1952 by R. B. Woodward [28], through an early *endo* DA step reaction between quinone **52** and butadiene **53** to give **54**, which is reported at the beginning of this section as a due homage to the great scientist (Scheme 6.13).

Despite several remarkable syntheses of important and complex natural products developed in the subsequent twenty years and reported in a fascinating recent review by Nicolaou et al. [29], the control of the absolute stereochemistry of the DA reaction was not successfully achieved until more recent years.

Several applications of chiral auxiliaries (menthol, Evans's oxazolidinone, etc.) have been applied by several groups for asymmetric DA reactions, but these examples do not involve the generation of quaternary carbon centers. Nevertheless, before discussing the enantioselective catalytic approach, at least one example illustrating the potential power of the diastereoselective version of the DA reaction deserves a mention. As a personal choice, the elegant enantioselective total synthesis of (–)-chlorothricolide **60**, developed by Roush and Sciotti in 1994 [30], is

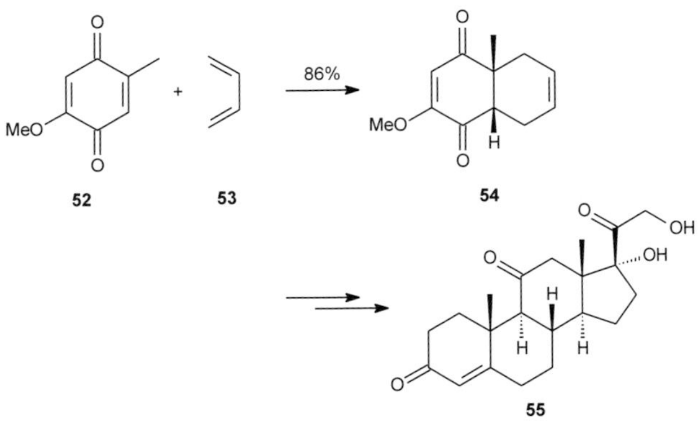

Scheme 6.13

illustrated in Scheme 6.14. The key step is the intermolecular *exo* DA of the poly-diene **56** with the (*R*)-dienophile **57**, to give the adduct **58**. Under the reaction conditions, this undergoes an intramolecular DA to give **59**, which in a further six

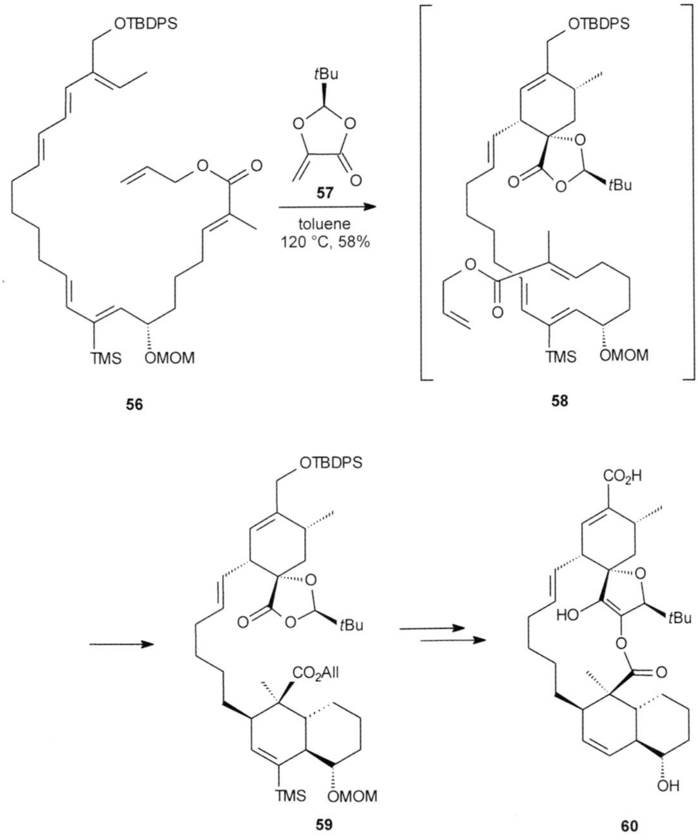

Scheme 6.14

steps is converted into the target. The regiochemical control was solved after several model studies: the diene system involved in the first step was activated, while the reactivity of the diene involved only in the intramolecular process was depressed by the strategically placed TMS group. One admirable feature of this amazing DA cascade process is that only two chiral centers on the starting reagents are able to induce seven new stereocenters, two of them being quaternary carbon atoms.

Chiral catalysts have been usefully applied to run enantioselective DA reactions, and excellent benchmarks to test the efficiency of different catalysts are the reactions involving 2-methyl- and 2-bromoacroleins (**34a,b**) as dienophiles (Scheme 6.15 reports the *endo*-**62** and *exo*-**63** adducts with cyclopentadiene **61**), not only because the reacting complexes in the catalytic cycles are good models for the study of the stereoselection mechanism, but also because the adducts derived from 2-haloacroleins are versatile synthons for natural products.

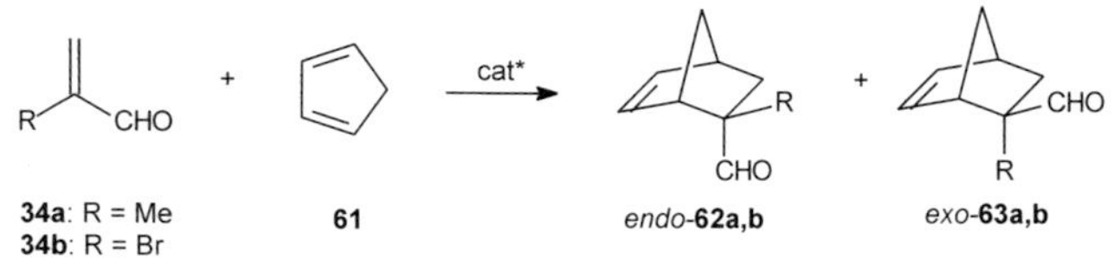

34a: R = Me   **61**   *endo*-**62a,b**   *exo*-**63a,b**
34b: R = Br

Scheme 6.15

The very first example of a catalyzed DA reaction between 2-methylacrolein and cyclopentadiene was reported by Koga et al. [31] and the catalysts were three chiral alkoxyaluminum dichlorides. The best result (*endo/exo* = 2:98; *ee* of *exo*-**63a** = 72%) was obtained with the catalyst derived from (1*R*,2*S*,5*R*)-neomenthol and this is the first of a long series of active catalysts, whose structures are reported in Fig. 6.2. These were tested over ten years by several groups [32–48], and the main results were illustrated in a review published in 2002 by Corey on the occasion of the centennial of the birth of Kurt Alder [49]. The diastereo- and enantioselectivity of these catalysts with three dienes [cyclopentadiene **61**, 2,3-dimethylbutadiene (DMB) and isoprene] are reported in Table 6.5. Some of the above catalysts have been also prepared on a solid support [50, 51], but the results in term of selectivity cannot be compared with those obtained under homogeneous Lewis-acid conditions.

Looking at the different aspects of the catalyzed DA reactions reported in Table 6.5, the efficiency of nearly every catalyst is the most important point for the synthetic organic chemist. Nevertheless, another important aspect concerns the relationship between the structure of the catalysts and the selectivity promoted in the reaction, because this may lead to an understanding of the modifications required for the catalyst, after coordination of a reactant, to be transformed into the complex involved in the catalytic cycle. These mechanistic investigations require either a crystal structure, or deep spectroscopic investigations, or high level theoretical calculations on the catalyst. These data are available for some of the above-described catalysts.

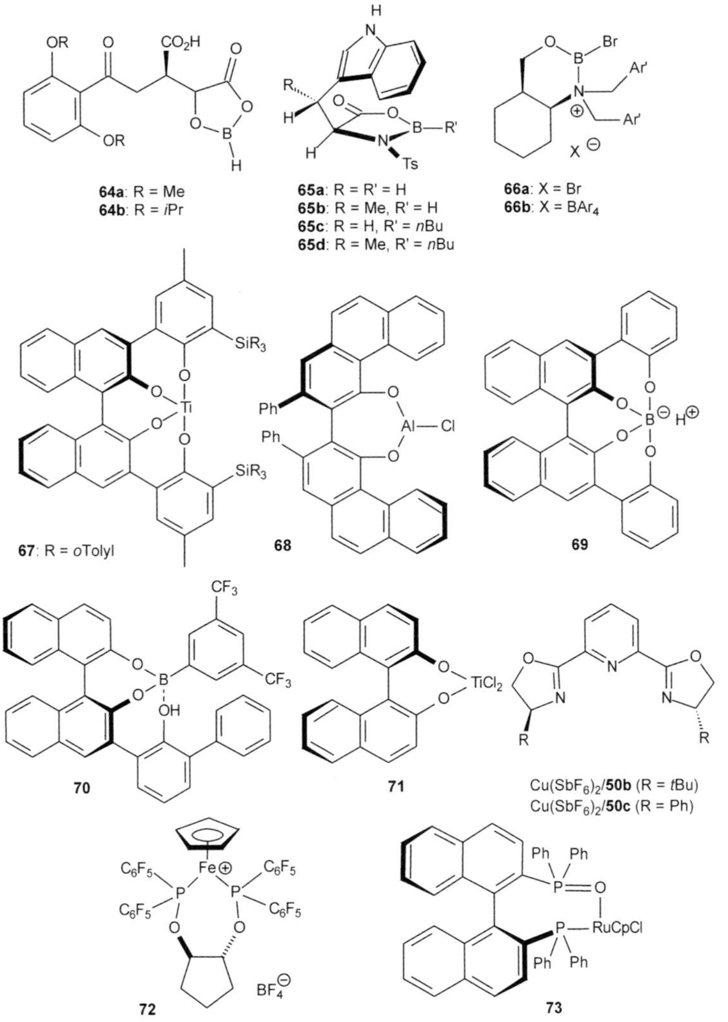

64a: R = Me
64b: R = *i*Pr

65a: R = R' = H
65b: R = Me, R' = H
65c: R = H, R' = *n*Bu
65d: R = Me, R' = *n*Bu

66a: X = Br
66b: X = BAr$_4$

67: R = *o*Tolyl

68

69

70

71

Cu(SbF$_6$)$_2$/**50b** (R = *t*Bu)
Cu(SbF$_6$)$_2$/**50c** (R = Ph)

72

73

Fig. 6.2 The optically active catalysts tested on the DA reaction of **34a,b** with different dienes (see Table 6.5).

The working model of the reacting complex **74** of the DA reaction between 2-bromoacrolein **34b** and cyclopentadiene catalyzed by the (*S*)-tryptophan-derived oxazaborolidine **65c**, and producing the *exo* adduct (*R*)-**63b**, is depicted in Fig. 6.3. This model is supported by physical and chemical studies, mainly based on deep NMR and UV spectroscopic investigations of the complexes between **65c,d** and 2-methylacrolein, as well as on the results of structural modifications of the ligand [52].

Starting from the X-ray structure of [**50b** Cu(H$_2$O)$_2$](SbF$_6$)$_2$, the reacting pybox complex **75** was proposed by replacing water with 2-bromoacrolein as ligand (Fig. 6.3). The complex has square-planar geometry with the coordinated dienophile in

Table 6.5 Enantioselective catalyzed DA reaction between
34a,b and three dienes.

Entry	Catalyst	Dienophile	Diene	endo:exo	exo-63 ee (%) (config.)	Ref.
1	64a	34a	61	11:89	96 (*R*)	32
2	64b	34a	61	–	87 (*R*)	33, 34
3	64b	34b	61	6:94	95 (*S*)	33, 34
4	64b	34b	DMB	–	95	33, 34
5	64b	34b	Isoprene	–	87 (*S*)	33, 34
6	65c	34b	61	4:96	>99 (*R*)	35
7	65c	34b	Isoprene	–	92 (*R*)	35
8	66a	34a	61	12:88	90 (*S*)	37
9	66a	34b	61	6:94	95 (*R*)	37
10	66a	34b	Isoprene	–	96 (*R*)	37
11	66b	34a	61	11:89	87 (*R*)	37
12	66b	34b	61	9:91	98 (*R*)	37
13	67	34a	61	1:99	94 (*S*)	38
14	68	34a	61	2:98	98 (*R*)	39
15	69	34a	61	1:99	99 (*R*)	40
16	69	34b	61	1:99	99 (*S*)	40
17	69	34b	Isoprene	–	98 (*S*)	40
18	70	34a	61	1:99	99 (*S*)	41, 42
19	70	34a	DMB	–	>99	41, 42
20	70	34a	Isoprene	–	>99	41, 42
21	70	34b	61	10:90	>99 (*R*)	41, 42
22	70	34b	DMB	–	91	41, 42
23	70	34b	Isoprene	–	>99 (*R*)	41, 42
24	71	34b	Isoprene	–	78	43, 44
25	50b	34a	61	3:97	92 (*S*)	45, 46
26	50b	34b	61	2:98	96 (*R*)	45, 46
27	50c	34a	61	4:96	49 (*R*)	46
28	72	34a	61	3:97	90 (*R*)	47
29	72	34b	61	5:95	95 (*S*)	47
30	72	34b	DMB	–	97	47
31	72	34b	Isoprene	–	97	47
32	73	34a	61	1:99	94 (*S*)	48
33	73	34b	61	2:98	76	48

the *s-cis* conformation; the *Si*-face of the reacting alkene is therefore blinded by the *tert*-butyl group of the ligand, and this rationalizes the formation of (*R*)-**63b** through the *exo*-favored transition state [46].

Although several structures for the reacting complex involved in processes with catalysts **50,64–73** (derived from different degrees of approximation in the investigation) have been proposed, not all results can be easily rationalized. Table 6.5 (entries 25 and 27) shows that Cu(II)-based catalysts derived from **50b** (*t*Bu-pybox) and **50c** (Ph-pybox), with the same configuration at the chiral center of the

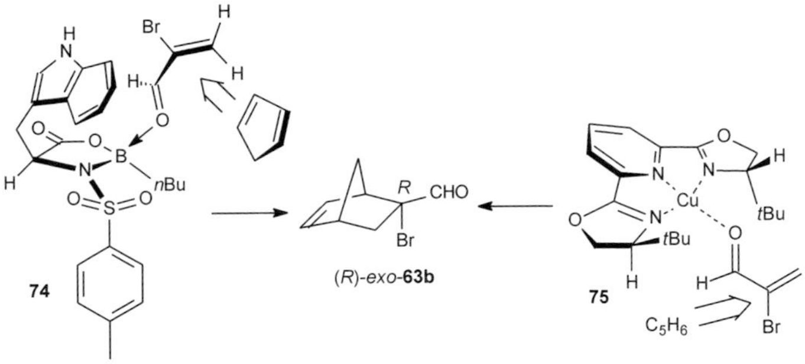

Fig. 6.3 Working models of the reacting complex of the DA reaction between cyclopentadiene and 2-bromoacrolein **34b** catalyzed by oxazaborolidine **65c** and by the Cu/pybox **50b** complex.

ligand, produce the opposite enantiomers of *exo*-**63a**. This means that different reacting complexes are involved, whose structures must be influenced, in this and in other different enantioselective catalytic processes [53], by the substituents of the ligand.

The concepts developed from the catalysis of the *exo*-enantioselective DA cycloadditions with 2-methyl or 2-bromoacrolein found a useful application in the total synthesis of some natural products. In 1992 Marshall and Xie [54] used oxazaborolidine **65a** (unsubstituted at the cyclic boron atom) as catalyst in the DA reaction of diene **76** and bromoacrolein **34b**. The adduct **77** was obtained in excellent yield and good *ee* (Scheme 6.16), but the dienophile, which under Corey–Loh catalysis gave mainly the *exo* adduct, with diene **76** gave the *endo* product exclusively. Therefore, to convert **77** into the spirotetronate subunit of kijanolide **78**, it was necessary to invert the quaternary stereocenter configuration.

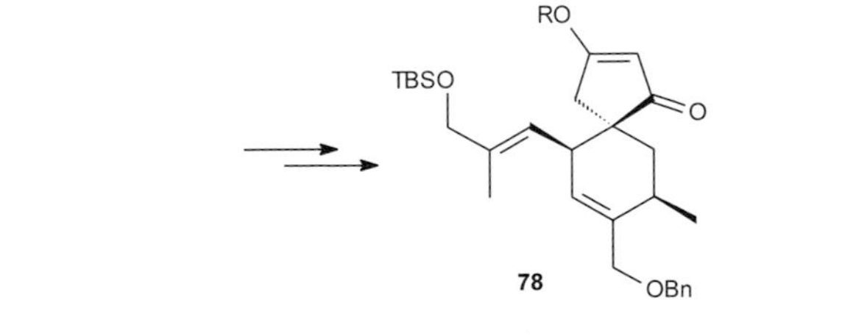

Scheme 6.16

Oxazaborolidine **65b**, unsubstituted at the cyclic boron atom, but with a methyl group in the tryptophan chain, reverted to being *exo*-selective in the reaction of bromoacrolein **34b** with 2-(2-bromoallyl)-1,3-cyclopentadiene **79** [55], and the adduct **80**, converted with excellent yields to **81**, allowed the first catalytic enantioselective synthesis of the gibberellic acid key intermediate **82** (Scheme 6.17).

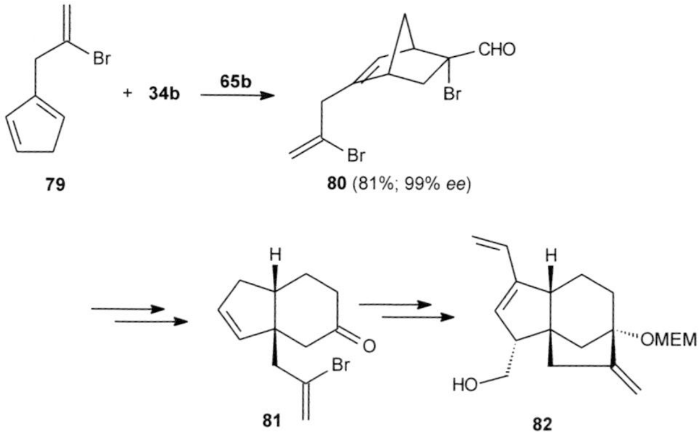

Scheme 6.17

After some initial disappointing results, a set of critical modifications both on the oxazaborolidine catalyst (**65b** was the best structure) and on the diene [(triiso-propylsilyl)oxy (TIPSO) was the crucial substituent in position 2 of **83**] allowed the DA reaction to run with 2-methylacrolein **34a** to give adduct **84** in 83% yield and 97% *ee* (Scheme 6.18). Again, as for the reaction reported in Scheme 6.16, this DA is strongly *endo*-selective and this seems to mark a sharp difference [55] between butadienes (*endo*-selectivity) and cyclopentadiene (*exo*-selectivity). The synthon **84** was converted in three steps into the antiulcer substance cassiol **85**.

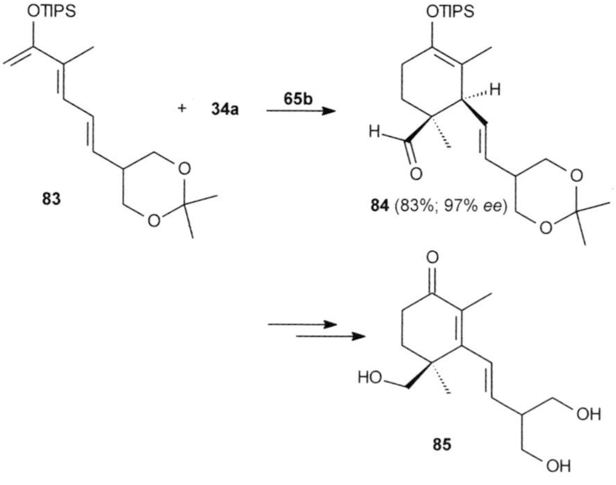

Scheme 6.18

The success obtained in the cassiol synthesis encouraged Corey to approach the synthesis of Eunicenone A **88**, a terpene molecule of marine origin, using the asymmetric DA reaction of bromoacrolein **34b** and polyene **86**, catalyzed by oxazaborolidine **65b** [56]. The challenge to face was the diene structure with six double bonds, but the catalyst was so efficient and selective that adduct **87** was obtained in 85% yield. Again the reaction was strongly *endo*-selective (>98:2), and the *ee* was 97%. The conversion of **87**, through a sequence of reactions, to euni-cenone A **88** with an overall yield of about 30%, allows the accomplishment of a synthesis that is both stereo- and regiocontrolled (Scheme 6.19).

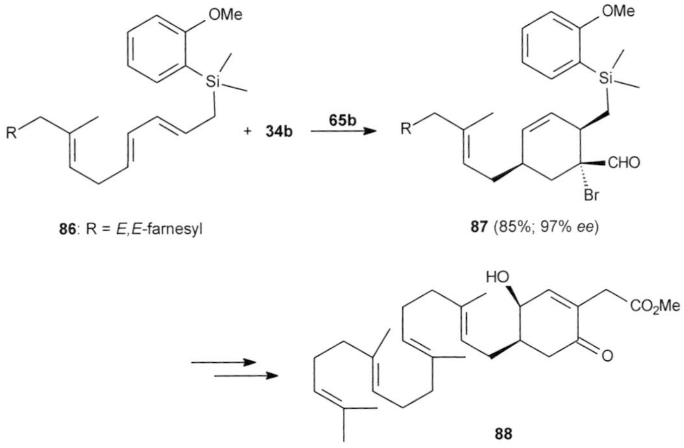

86: R = *E,E*-farnesyl   **87** (85%; 97% *ee*)

88

Scheme 6.19

A further development of oxazaborolidine-based catalysts was proposed by Corey employing a proline-derivative as precatalyst and triflic acid as the catalyst generator [57, 58]. The nature of the boron substituent at the aryl groups was optimized and catalysts **89a,b** were the results of this choice. The main difference with previous oxazaborolidines was the behavior of these catalysts as bi-coordinating agents: aldehydes coordinate their carbonyl oxygen atom to boron, and the proton forms a hydrogen bond with the oxygen atom of the catalyst. The results of the DA reactions between cyclopentadiene **61** and cyclohexadiene **90** with **34a,b** (Scheme 6.20), catalyzed by **89a,b**, are given in Table 6.6.

The stereochemical course of the reaction can be rationalized only by assuming the *s-trans* conformation of the coordinated enals, with the *Si*-face of the double bond screened by one of the aryl groups. The important result in terms of expectations for future applications is the strong diastereoselectivity of the reactions: *exo* with cyclopentadiene (in analogy with the results reported in Table 6.5), *endo* with cyclohexadiene (in analogy with the results previously described for substituted butadienes).

The DA method for obtaining quaternary carbon stereocenters is not confined to 2-substituted enals. Catalyst **93**, analogous to **89a**, but with $(CF_3SO_2)_2N^-$ as counterion, catalyzes the DA reaction between various dienes (cyclopentadiene, cyclohexadiene, butadiene, 2,3-dimethylbutadiene) [58] and 2,5-dimethylbenzoquinone, giving quantitative yields of *endo* adducts with, in general, almost complete enantioselectivity. The activation of quinone with **93** raised the question

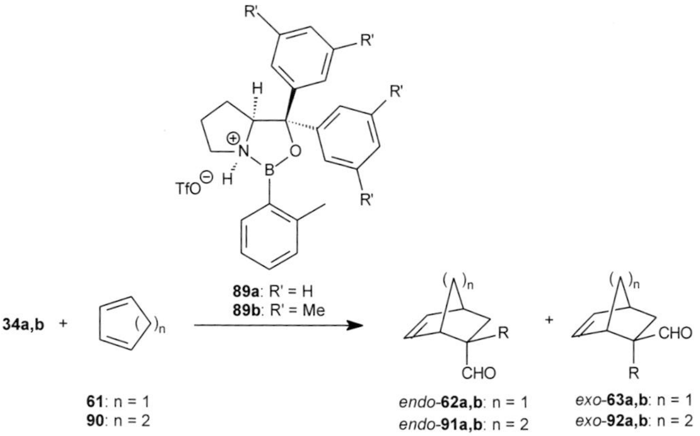

Scheme 6.20

of the type of coordination involved, since the expected ligation through the carbonyl lone pair *anti* to methyl would activate both C=C double bonds with the formation of regioisomers in the reaction with unsymmetrical dienes (as in fact occurs with isoprene, for example). Corey solved the problem by using 3-iodo-2,5-dimethyl-1,4-benzoquinone **94** as a dienophile synthetically equivalent to 2,5-dimethyl-1,4-benzoquinone. Iodine deactivates the C=C to which is attached, transferring the coordination to the sterically more available lone pair to give **95**, and Fig. 6.4 shows the DA reaction result with 2-triisopropylsilyloxy-1,3-butadiene **96**, which produces **97** as a single cycloadduct with complete control of regio- and stereoselectivity.

The relatively inert diene 1-vinyl-2,2,6-trimethylcyclohexene **98** is the starting material for the DA approach to drimane sesquiterpenes, but even if dienophile **99** can be activated by bi-coordination to a catalyst, the reaction is not easy. The problem was solved by using Cu(II)triflate/(1*R*,2*R*)-*N,N'*-bis(2,6-dichlorobenzylidene)-diaminocyclohexane **100** as catalyst (Scheme 6.21), and by running the reaction under 12–15 kbar pressure. Under these conditions, **101** was obtained as the predominant product (93:7 *dr*, >60% *ee*) [59].

Table 6.6 Enantioselective DA reactions between **34a,b** and two cyclic dienes, catalyzed by **89a,b** [57].

Entry	Catalyst	Dienophile	Diene	*endo:exo*	*endo ee* (%) (config.)	*exo ee* (%) (config.)
1	89a	34a	61	9:91	–	91 (*S*)
2	89a	34a	90	95:5	92 (*S*)	–
3	89a	34b	61	9:91	–	92 (*R*)
4	89a	34b	90	94:6	92 (*R*)	–
5	89b	34a	61	9:91	–	96 (*S*)
6	89b	34a	90	94:6	92 (*S*)	–
7	89b	34b	61	9:91	–	96 (*R*)
8	89b	34b	90	93:7	92 (*R*)	–

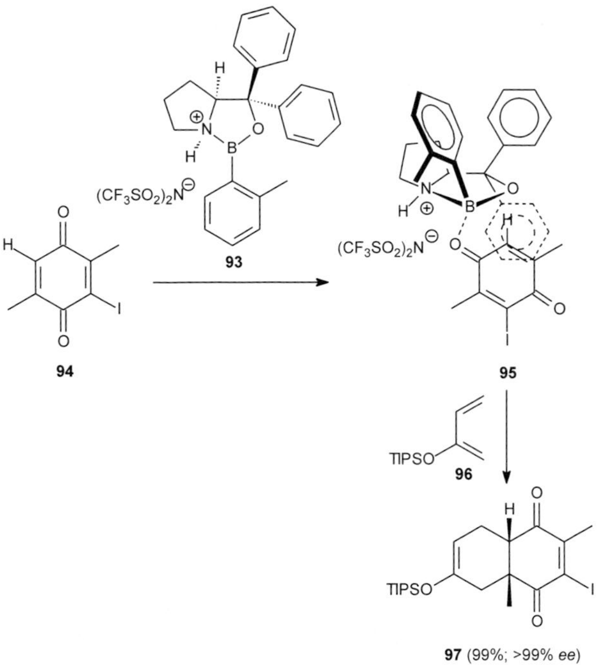

Fig. 6.4 The DA reaction of complex **95** derived from **93** and **94** with butadiene **96**.

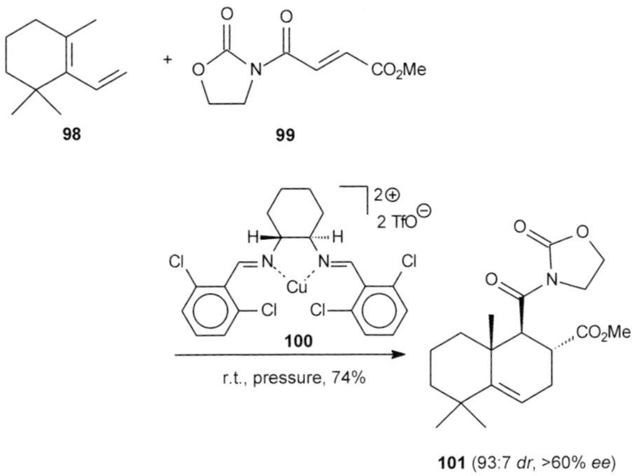

Scheme 6.21

Among the different chiral catalysts used in enantioselective DA reactions, those based on box ligands are certainly the most widely used for their high efficiency and flexibility [60] (Scheme 6.22).

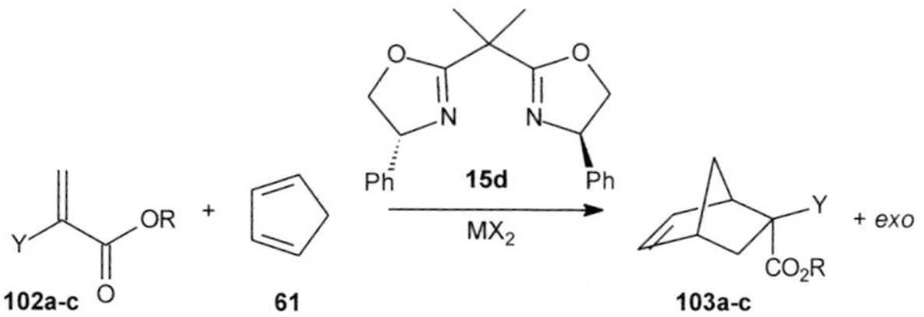

Scheme 6.22

Acrylates **102**, substituted in position 2 with a group able to cooperate with the carbonyl in a bidentate binding of dienophile, react with cyclopentadiene in a strongly *endo*-selective DA reaction, catalyzed by (*R*,*R*)-diphenyl-box **15d** and Mg(II) or Cu(II) (Table 6.7) [61–63]. The absolute configuration of adduct **103a** (entry 1 – from benzoyl acrylate **102a**) was determined by its conversion into the corresponding bromolactone, whose structure was established by X-ray crystallography [61].

Box-based complexes have been used as catalysts for the enantioselective DA reaction between cyclopentadiene and 2-acetyl-1,4-naphthoquinone [64], but the level of enantioselection with box and copper(II) triflate cannot be compared with the *ee* obtained with Gd(III) triflate/pybox complexes [65]. Several substituted pentadienes **104a–c** and 1,4-benzo- or naphthoquinone-2-methylcarboxylates **105a–c** smoothly react with (*S*,*S*)-diphenyl-pybox (**50c**: R = Ph, R' = H) and pybox derived from (2*S*,3*S*)-norephedrine (**50d**: R = Me, R' = Ph), Scheme 6.23. Different triflates have been tested: samarium and gadolinium were excellent cations, with Gd better than Sm. Table 6.8 reports the results with Gd(III) and in all cases only the *endo* adducts **106** were formed with almost complete enantioselectivity.

6.6
Hetero-Diels–Alder Reactions

The presence of one or more heteroatoms in either diene or dienophile is the specific feature of the hetero-Diels–Alder (HDA) reaction. Before discussing the catalytic asymmetric version of such a cycloaddition, as an ideal bridge between DA and HDA, we describe the sequence of reactions that convert the

Table 6.7 DA reactions between acrylates **102a–c** and cyclopentadiene, catalyzed by complexes of box **15d**.

Entry	Dienophile	R	Y	MX$_2$	*endo:exo*	103 *ee* (%) (config.)	Ref.
1[a]	102a	Et	COPh	MgI$_2$	>99:1	89 (*R*)	61, 62
2	102b	Et	SPh	Cu(SbF$_6$)$_2$	94:6	>95	63
3	102c	CH$_2$CF$_3$	SPh	Cu(SbF$_6$)$_2$	93:7	>95	63

a The diethyl-box was used.

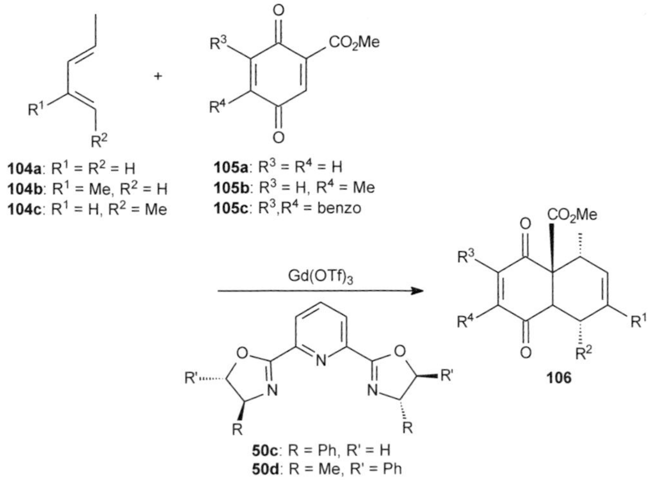

104a: $R^1 = R^2 = H$
104b: $R^1 = Me, R^2 = H$
104c: $R^1 = H, R^2 = Me$

105a: $R^3 = R^4 = H$
105b: $R^3 = H, R^4 = Me$
105c: $R^3, R^4 = benzo$

Gd(OTf)$_3$

50c: R = Ph, R' = H
50d: R = Me, R' = Ph

Scheme 6.23

highly functionalized polyenic β-keto ester **107**, first to macrocycle **108** by ring closure, then by tandem transannular DA/HDA reaction into **109** (a single stereoisomer in 40% yield, Scheme 6.24). The pentacyclic product has seven new contiguous stereocenters suitably placed to allow its easy conversion into FR182877, a biologically active product whose potency is comparable to that of the anticancer drug Taxol [66]. Even if only one of the new stereocenters of **109** is quaternary, it properly derives from the HDA part of this excellent example of how stereospecific the sequential process based on [4+2] + [4+2] cycloadditions can be.

A review by Jørgensen, an author active in the field [67], details catalytic asymmetric HDA reactions in cases in which either a carbonyl group behaves as heterodienophile, or the carbonyl group is part of an α,β-unsaturated system behaving as a heterodiene. Since the development of quaternary stereocenters by this route is in general specific for ketones, Jørgensen focused a second review [68] on the catalysis of the HDA reactions of ketones.

Table 6.8 DA reactions between quinones **105a-c** and dienes **104a-c**, catalyzed by pybox **50c,d**/Gd(OTf)$_3$ complexes [65].

Entry	Diene	Dienophile	Pybox	106 *ee* (%)
1	104a	105a	50c	99
2	104b	105a	50c	99
3	104c	105a	50c	99
4	104a	105b	50c	99
5	104b	105b	50c	98
6	104a	105c	50d	91
7	104c	105c	50d	>99

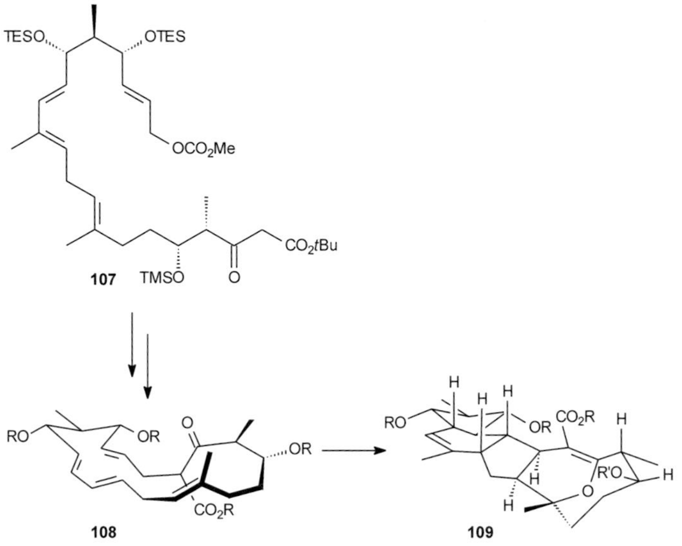

Scheme 6.24

6.6.1
The Carbonyl Group as Dienophile

The chiral HDA reaction with the C=O group acting as dienophile was first performed in 1987, when the reaction between 1-substituted butadienes and diethyl ketomalonate **110** was catalyzed either with menthyloxyaluminum dichloride (the Koga catalyst) [31], or with the chiral europium catalysts Eu(hfc)₃ [69]. The yields were sometimes excellent, but the *ee* was always poor.

More than ten years later [70], box-based complexes were found to be useful catalysts for the reaction of **110** with 1,3-cyclohexadiene **90** to give **111** (Scheme 6.25) and the overall best conditions for yield and selectivity are reported in Table 6.9. Four box ligands (**15b**, **15d**, **112** and **113**) were tested, Cu(II) and Zn(II) were the cations of choice with (*R,R*)-diphenyl-box **15d**, and the reaction was found to be unusually solvent dependent (entries 3–10). A further improvement of the enantioselectivity of the reaction was obtained with the chiral catalyst derived from (*S,S*)-bis(sulfoximine) **114** and copper triflate [71].

In order to rationalize the stereochemical outcome, theoretical calculations of the various modes of coordination of **110** to the Cu(II) and Zn(II) complexes of **15d** were run, suggesting that the HDA proceeds via the five-membered ring asymmetric intermediate **115** involving two adjacent C=O groups (Fig. 6.5). The alternative C₂-symmetric six-membered ring intermediate, involving the coordination of both C=O ester groups, was excluded for the absence of any facial discrimination [70].

Obviously the quaternary carbon atom generated in this reaction is not stereogenic, but its formation is the specific condition for the development of chirality in a product (**111**) whose ring opening gives a chiral diol, an important optically active building block.

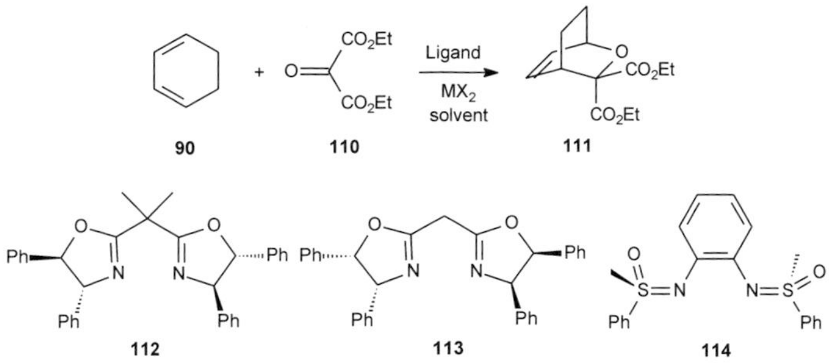

Scheme 6.25

Conversely, a simple example of how a stereogenic nonquaternary carbon easily becomes a quaternary stereocenter is the total synthesis of (R)-actinidiolide **119**, obtained by the HDA reaction of 2,6,6-trimethyl-1,3-cyclohexadiene **116** and ethyl glyoxylate **117**, catalyzed by the [Cu(II)/box **15a**] complex [72] (Scheme 6.26). The product **118**, obtained in 90% yield and excellent diastereomeric and enantiomeric excess, can be converted in three steps into the target pheromone.

An enantioselective catalyzed HDA reaction giving a quaternary stereocenter is the reaction between α-keto esters or α-diketones **120** and electron-rich butadienes [usually *trans*-1-methoxy-3-(trimethylsilyloxy)-1,3-butadiene (Danishefsky's diene, **121a**); Scheme 6.27].

A variety of combinations of different ligands, cations, dienes, and keto esters or diketones have been tested [73–76], and the most significant results are reported

Table 6.9 HDA reactions between diethyl ketomalonate **110** and 1,3-cyclohexadiene **90**, catalyzed by box (**15b**, **15d**, **112** and **113**)- and bis(sulfoximine) **114**-based complexes.

Entry	Ligand	MX$_2$	Solvent	Yield (%)	(%) ee (config.)	Ref.
1	15b	Cu(OTf)$_2$	CH$_2$Cl$_2$	20	40 (1R,4S)	70
2	15b	Cu(SbF$_6$)$_2$	CH$_2$Cl$_2$	11	64 (1R,4S)	70
3	15d	Zn(OTf)$_2$	CH$_2$Cl$_2$	39	86 (1R,4S)	70
4	15d	Zn(OTf)$_2$	Et$_2$O	94	91 (1R,4S)	70
5	15d	Zn(OTf)$_2$	PhMe	68	90 (1R,4S)	70
6	15d	Zn(OTf)$_2$	THF	15	86 (1R,4S)	70
7	15d	Zn(OTf)$_2$	MeCN	17	23 (1R,4S)	70
8	15d	Cu(OTf)$_2$	Et$_2$O	64	93 (1R,4S)	70
9	15d	Cu(OTf)$_2$	CH$_2$Cl$_2$	70	72 (1R,4S)	70
10	15d	Cu(OTf)$_2$	THF	70	72 (1R,4S)	70
11	112	Zn(OTf)$_2$	CH$_2$Cl$_2$	64	34 (1R,4S)	70
12	113	Zn(OTf)$_2$	CH$_2$Cl$_2$	66	75 (1R,4S)	70
13	114	Cu(OTf)$_2$	CH$_2$Cl$_2$	92	98 (1S,4R)	71

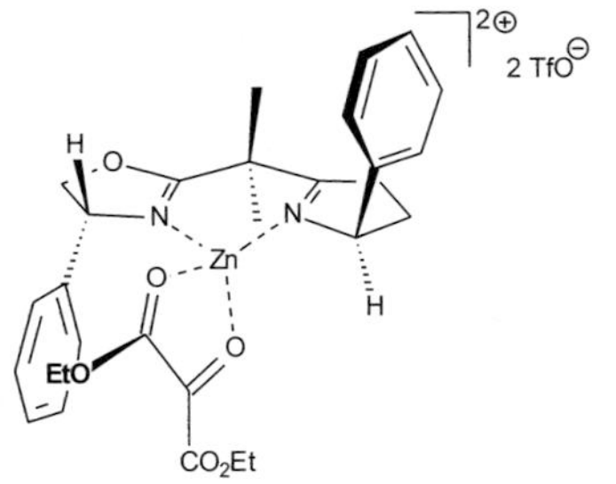

115

Fig. 6.5 Working model of the reacting complex of the HDA reaction between cyclohexadiene **90** and diethyl ketomalonate **110**, catalyzed by the complex of box **15d** with Zn(II).

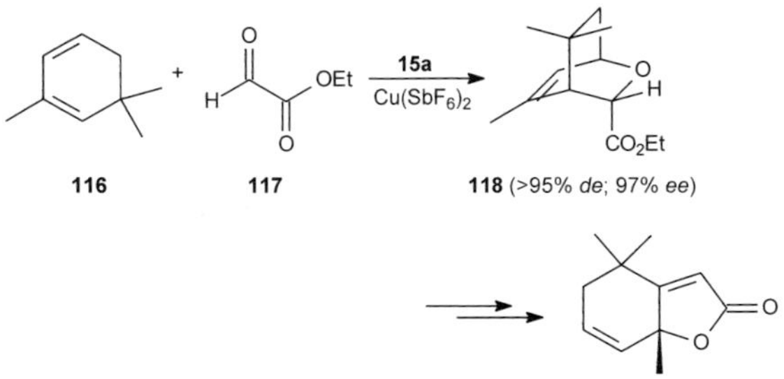

Scheme 6.26

in Table 6.10. Some general trends can be outlined:

- Among heterocyclic ligands, box gives better catalysts than those derived from pybox and the results of some experiments run with the latter ligands [73, 74] have not been reported in the table because the best *ee* was 63%.
- Copper(II) was by far the best cation and triflate was its anion of choice, better than hexafluoroantimonate (entries 1, 3 *vs* 2, 4).
- Although the configuration of box ligands **15a** and **15d** is opposite, the same absolute configuration of **122** (found to be (*S*) by the X-ray analysis of a suitable derivative prepared from product obtained under the conditions in entry 5) was induced (entries 1, 2 *vs* 3, 4).
- In α-keto esters (R′ = OMe or OEt) the increased steric hindrance of R reduces the enantioselectivity (entries 4–7).
- With box **112**, Zn(II) is better than Cu(II) (entries 8, 9).
- The enantioselectivity observed with α-diketones (entries 10–12) is not affected in the same way as noted for α-keto esters.
- Substituents in diene **121** influence the yield more than the enantioselectivity (entries 13, 14).

In addition to box ligands, very good enantioselectivities were obtained with the catalysts based on ligand **124**, obtained from the Schiff bases of (1*S*,2*S*)-diphenyldiaminoethane with cycloalkanones (in the ratio 1:1), and copper triflate (entries 19–21) [76].

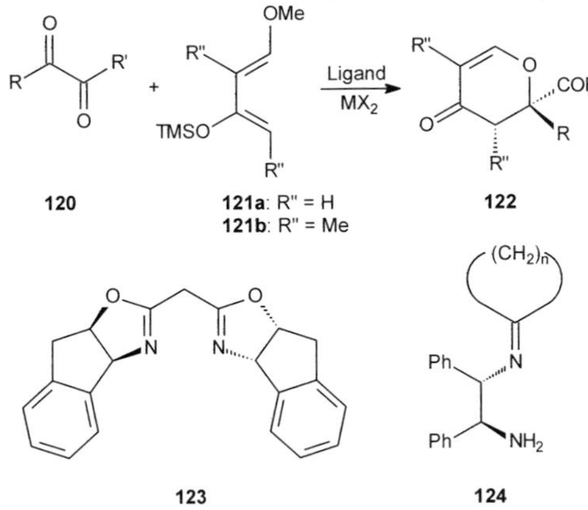

Scheme 6.27

Table 6.10 HDA reactions of α-keto esters and α-diketones **120** with dienes **121a,b** catalyzed by box **15a,15d,112,123** and diamine Schiff-base **124** complexes.

Entry	R	R$'$	Diene	Ligand	MX$_2$	Yield (%)	ee (%)	Ref.
1	Me	OEt	121a	15a	Cu(SbF$_6$)$_2$	37	89	73, 74
2	Me	OEt	121a	15a	Cu(OTf)$_2$	78	99	73, 74
3	Me	OEt	121a	15d	Cu(SbF$_6$)$_2$	24	23	74
4	Me	OEt	121a	15d	Cu(OTf)$_2$	85	26	74
5	Et	OMe	121a	15a	Cu(OTf)$_2$	80	94	74
6	i.Pr	OEt	121a	15a	Cu(OTf)$_2$	42	37	74
7	Ph	OEt	121a	15a	Cu(OTf)$_2$	77	77	74
8	Me	OEt	121a	112	Cu(OTf)$_2$	4	23	74
9	Me	OEt	121a	112	Zn(OTf)$_2$	86	63	74
10	Me	Me	121a	15a	Cu(OTf)$_2$	90	94	73, 74
11	Me	Et	121a	15a	Cu(OTf)$_2$	77	98	74
12	Me	Ph	121a	15a	Cu(OTf)$_2$	95	94	74
13	Me	OMe	121b	15a	Cu(OTf)$_2$	75	96	74
14	Me	Me	121b	15a	Cu(OTf)$_2$	60	91	74
15	Me	OMe	121a	15a	Cu(OTf)$_2$	66	99	75
16	Me	OEt	121a	123	Cu(OTf)$_2$	62	99	75
17	(CH$_2$)$_8$Me	OEt	121a	15a	Cu(OTf)$_2$	77	47	75
18	(CH$_2$)$_8$Me	OEt	121a	123	Cu(OTf)$_2$	73	56	75
19	Me	OEt	121a	124 ($n = 6$)	Cu(OTf)$_2$	72	70	76
20	Me	OEt	121a	124 ($n = 5$)	Cu(OTf)$_2$	85	91	76
21	Me	OEt	121a	124 ($n = 4$)	Cu(OTf)$_2$	85	94	76

These basic concepts allowed fruitful applications and **122** (R″ = H), obtained under the conditions of entry 2 in Table 6.10 with 99% *ee*, was the starting product of Danishefsky's synthesis [77] of **127**, the seven-stereocenter core of phomactin A (Scheme 6.28). A complex conversion of **122** to **125** gave the diene for a DA reaction with maleic anhydride **126** and it is remarkable that the stereoselectivity was complete and a single stereoisomer **127** was obtained with the correct stereochemical arrangements between functional groups necessary for conversion to phomactin A.

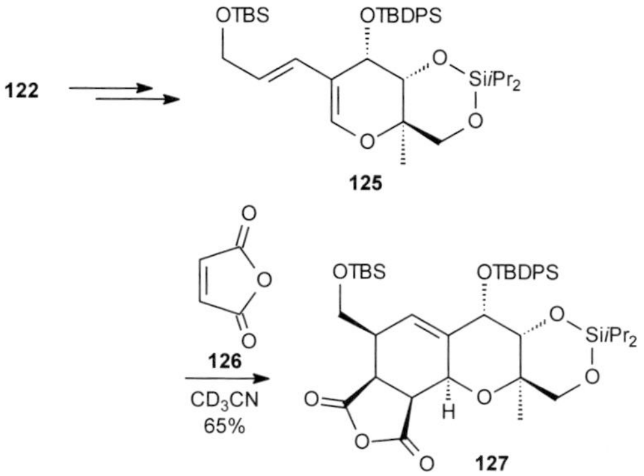

Scheme 6.28

6.6.2
α,β-Unsaturated Carbonyl Derivatives as Heterodienes

The α,β-unsaturated carbonyls may behave as electron-poor heterodienes when they react with electron-rich alkenes, ordinarily vinyl ethers, to give 3,4-dihydro-2*H*-pyrans [67, 68, 78]. This behavior is facilitated by an electron-withdrawing group bound to the carbonyl carbon atom, either an ester or a phosphonato group, which has two main effects: (i) to decrease the LUMO energy of the heterodiene lowering the FMO separation with the HOMO of the dienophile, increasing the rate of the reaction; (ii) to allow a bis-coordination to the chiral catalyst increasing the rigidity of the reacting complex and favoring the face discrimination of the coordinated reagent.

The HDA reaction of γ-substituted-β,γ-unsaturated α-keto esters **128** with the *exo*-cyclic vinyl ether **129** is enantioselectively catalyzed by the box **15a**/Cu(II) triflate complex to give **130**, which are useful synthons for the preparation of optically active spiro-carbohydrates and spiro-amino sugars (Scheme 6.29). When R is an amino residue [79], the reaction seems to be *exo* selective (Table 6.11 – entry 1), when R is an alkoxy group [80], *endo* selectivity is preferred (Table 6.11 – entries 2, 3); in all cases the *ee* of **130** are very good.

The reaction of α,β-unsaturated acyl phosphonates **131** and vinyl ethers, either when the heterodiene is β,β-disubstituted, or when the vinyl ether **132** is α-substituted, is an excellent example of the enantioselective synthesis of

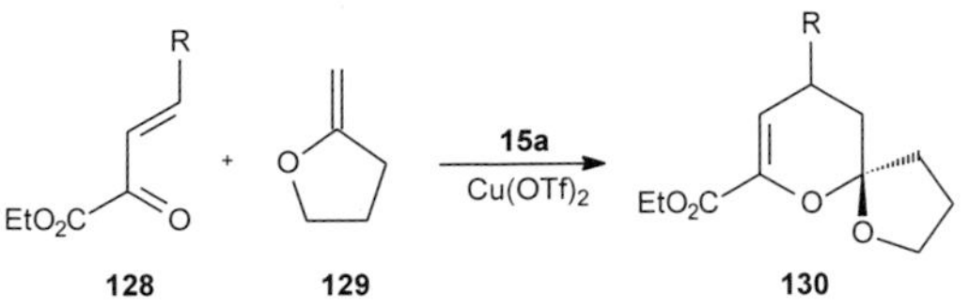

Scheme 6.29

dihydropyrans **133** and **134** carrying at least one quaternary stereocenter [81], whose control is developed by a copper triflate/box catalyst (Scheme 6.30, Table 6.12).

A novel asymmetric tandem transesterification(TE)-intramolecular version of HDA was proposed by Wada and coworkers [82, 83] starting from methyl (*E*)-4-methoxy-2-oxo-3-butenoate **135** and 5-methyl-4-hexen-1-ol **136**. Even if no quaternary stereocenter is formed in this cycloaddition, the example is an interesting case of an intramolecular HDA reaction that should be a subject of future development. The catalyzed TE gives **137**, which undergoes intramolecular-HDA cyclization to give the *trans*-fused pyranopyrans **138** (Scheme 6.31). First experiments catalyzed by Ti-taddolates gave good yields, but enantioselectivity was absent. By testing different Cu(II)/box-based catalysts (Table 6.13), *t*Bu-box **15a** was found to be better than Ph-box *ent*-**15d** (entries 5, 6 *vs* 7, 8), the SbF$_6$ anion was better than triflate (entries 5, 6 *vs* 2, 3), and 4-Å or, better, 5-Å molecular sieves (MS) were the necessary additive to develop enantioselectivity (entries 1, 4 *vs* 2, 5 and 3, 6). Similar results were obtained running the reaction on pre-formed **137**, or via its *in situ* formation (entry 6 *vs* 9).

Double asymmetric induction experiments have been done on the tandem TE/HDA reaction of **135** with racemic 6-methyl-5-hepten-2-ol, catalyzed by Cu(SbF$_6$)$_2$/**15a** [83], and excellent kinetic resolution (up to 95% selectivity), diastereoselectivity (up to 92% *de*) and enantioselectivity (up to 97% *ee*) were observed.

6.6.3
Imine Derivatives as Heterodienophiles

Compared to the great achievements with oxygen-containing dienes or dienophiles, asymmetric HDA reactions with their nitrogen analogues remained

Table 6.11 Results for the HDA reaction of α-keto esters **128** with the *exo*-cyclic vinyl ether **129**, catalyzed by the Cu(II)/**15a** complex.

Entry	R	Yield (%)	endo:exo	endo-130 ee (%)	exo-130 ee (%)	Ref.
1	NPht	94	25:75	–	80	79
2	OEt	84	75:25	74	84	80
3	OBn	52	77:23	76	95	80

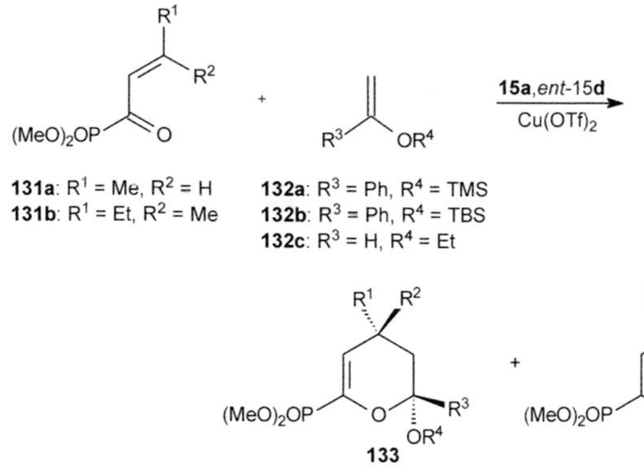

Scheme 6.30

unavailable for a long time and this field only began to be investigated in about 2002 [84] and the use of chiral catalysts to control the stereochemistry of reactions involving achiral reagents is an argument far to be defined. Since much of the work concerns the reaction of aldimines, very few examples are known of HDA reactions with a C=N group acting as dienophile for the generation of quaternary stereocenters.

The diastereoselective reaction with the enantiopure aziridine **139** bearing 8-phenylmenthol as chiral auxiliary was performed [85] with a set of dienes (Scheme 6.32): Danishefsky's diene **121a**, 1,3-cyclohexadiene **90**, 2-trimethylsilyloxy-1,3-cyclohexadiene **140** and cyclopentadiene **61**. Several Lewis acids were tested (Table 6.14), and the configuration of the major diastereoisomer, illustrated in the scheme, can be rationalized if the bidentate Lewis acid (see entries 1–4 *vs* 5,6) locks the aziridine ring in an *s-cis* conformation so that its *Re*-face is shielded by the phenyl group of the auxiliary. An attempt to develop an enantioselective variant with chiral Lewis acids gave unsatisfactory yields and enantioselectivities.

Table 6.12 Catalyzed HDA reactions of substituted acryloyl phosphonates **131a,b** with enol ethers **132a–c** [81].

Entry	Box	diene	dienophile	Yield (%)	133:134	133 ee (%) (config.)	134 ee (%) (config.)
1	15a	131a	132a	62[a]	92:8	99 (2R,4R)	74 (2R,4S)
2	15a	131a	132b	99[a]	93:7	99 (2R,4R)	82 (2R,4S)
3	ent-15d	131a	132a	20	90:10	99 (2R,4R)	9 (2S,4R)
4	ent-15d	131a	132b	100	60:40	96 (2R,4R)	99 (2S,4R)
5	ent-15d	131b[b]	132c	75[c]	83:17	93 (2S,4S)	–

a A Michael adduct is also formed.
b **131** is a 2.7:1 mixture of (*E*) and (*Z*) isomers.
c The product is a 4.8:1 mixture of diastereoisomers.

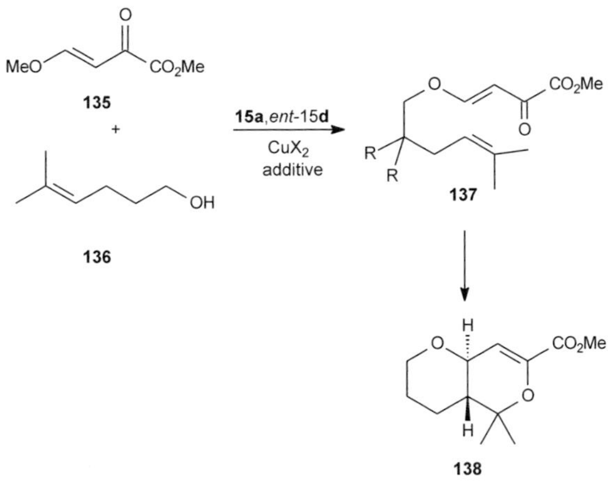

Scheme 6.31

6.7
Consecutive Cycloaddition Reactions

A sequence of cycloaddition reactions that follow one another, with or without including short intervening steps, obviously offers the organic chemist the possibility of adding the stereoselectivity gained from the first reaction to the second one, and so on, with a positive synergistic effect that strongly amplifies the stereoselectivity of the entire process. This topic, developed in the 1990s, briefly became very popular, mainly in the variant taking place without the agency of additional components or reagents, a variant known world-wide by the term "tandem cascade cycloadditions (TCCs)". Two review papers in 1996 defined the state of the art for DA cycloadditions [86], and [4+2]/[3+2] cycloadditions of nitroalkenes [87].

In this section we shall discuss not only TCCs but also a few consecutive reactions in which the strategy developed for a significant target incorporating

Table 6.13 Cu(II)/box-catalyzed transesterification–intramolecular-HDA reaction of **135** and **136** at room temperature [82].

Entry	Ligand	X⁻	Mol. sieve (Å)	Yield (%)	138 ee (%)
1[a]	15a	OTf	–	75	0
2[a]	15a	OTf	4	50	73
3[a]	15a	OTf	5	83	55
4[a]	15a	SbF_6	–	82	3
5[a]	15a	SbF_6	4	12	93
6[a]	15a	SbF_6	5	63	96
7[a]	ent-15d	SbF_6	4	51	13
8[a]	ent-15d	SbF_6	5	51	14
9[b]	15a	SbF_6	5	76	97

a Reaction run on pre-formed **137**. b Reaction run on the mixture of **135** and **136**.

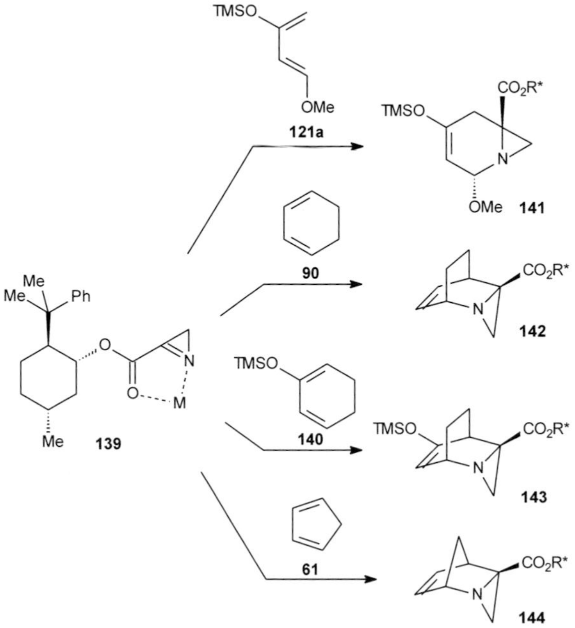

Scheme 6.32

one or more quaternary stereocenters involves a sequence of cycloadditions giving rise to results personally evaluated as elegant and admirable, such as the Nicolaou total synthesis of (–)-colombiasin A **145** [88, 89] (Scheme 6.33). The first step is the (*S*)-BINOL/TiCl$_2$-catalyzed DA between (*E*)-2-(*tert*-butyldimethylsilyloxy)penta-1,3-diene **146** and 2-methoxy-3-methyl-1,4-quinone **147** to give **148** with excellent *ee*. This is converted in a few steps into **149**, which still has no quaternary chiral carbon. Heating this in toluene at 180°C gives first a cheletropic extrusion of SO$_2$, then an intramolecular DA involving the just-generated tethered diene and the most easily approachable internal double bond. Hence, through **150**, the *endo* adduct **151** is obtained stereoselectively with two quaternary stereocenters that are of fundamental importance in the structure of the target **145**.

Table 6.14 Lewis-acid-catalyzed aza-HDA of chiral aziridines **139** with four dienes [85].

Entry	Diene	Lewis acid	Product	Yield (%)	% de
1	121a	MgBr$_2$·OEt$_2$	141	56	96
2	90	ZnCl$_2$·OEt$_2$	142	99	80
3	140	MgBr$_2$·OEt$_2$	143	99	97
4	61	MgBr$_2$·OEt$_2$	144	88	85
5	61	–	144	99	8
6	61	BF$_3$·OEt$_2$	144	0[a]	–

a All aziridine consumed.

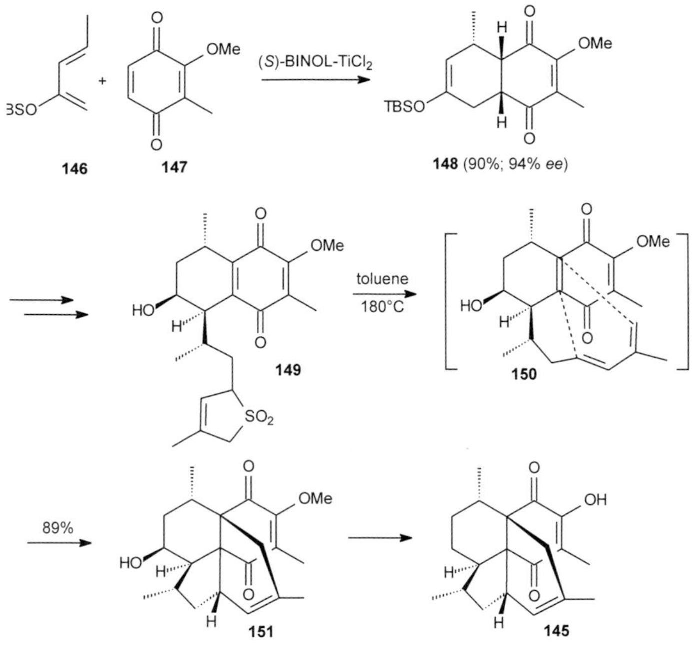

Scheme 6.33

Among TCCs, the most studied class is the [4+2]/[3+2] cycloaddition of nitroalkenes [87]. The components involved are an α,β-unsaturated nitroalkene, behaving as a heterodiene, which reacts with an electron-rich dienophile in an HDA reaction (the [4+2] step) to give a cyclic nitrone able to react with an electron-poor dipolarophile (the [3+2] step) to provide the bicyclic product. There are four different permutations arising from the combinations of inter- and intramolecular processes, and Scheme 6.34 illustrates the possible tethers between the reaction components. In the absence of the dipolarophile, the cyclic nitrone becomes the reaction product, and its isolation is useful to infer the mechanism of the different steps involved in the whole sequence.

Since the leading author in the field published a didactically excellent review [87], the same organization will be followed: first the "inter/inter" and then the "inter/intra" reactions, limiting the discussion in this section to the reactions giving rise to quaternary stereocenters.

The standard method of inducing enantioselectivity in a TCC involves the use of chiral vinyl ethers able to distinguish the enantiofaces of the achiral nitroalkene in the HDA step, so that the resulting nitrone can transfer the acquired stereochemical information to the 1,3-dipolar cycloaddition characterizing the second step of the process.

An example of inter/inter TCC is the SnCl₄-catalyzed reaction of (*Z*)-1-nitro-2-(3,4-dimethoxyphenyl)-1-butene **152** and (*Z*)-1-acetyloxy-2-[(*R*)-2,2-diphenylcyclopentoxy]ethene **153** [90] to give **154** with excellent diastereoselectivity (Scheme 6.35).

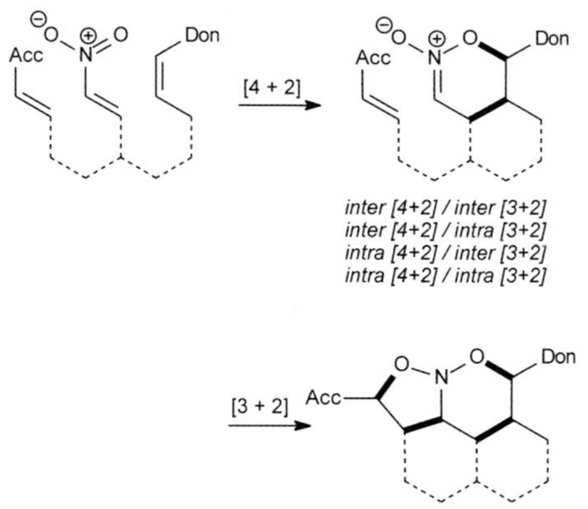

Scheme 6.34

The nitrone may further react with methyl acrylate to give the nitroso acetal **155** in a 7:1 *dr* (the major diastereoisomer is illustrated), whose structure was determined by X-ray analysis. Both **154** and **155** are useful synthons: hydrogenation gives ring contraction of oxazine, the pyrrolidine **156** and the hydroxy-pyrrolizinone derivative **157** are respectively obtained, the chiral auxiliary **158a** is recovered, and the overall balance of the reaction is the formation of four new stereocenters.

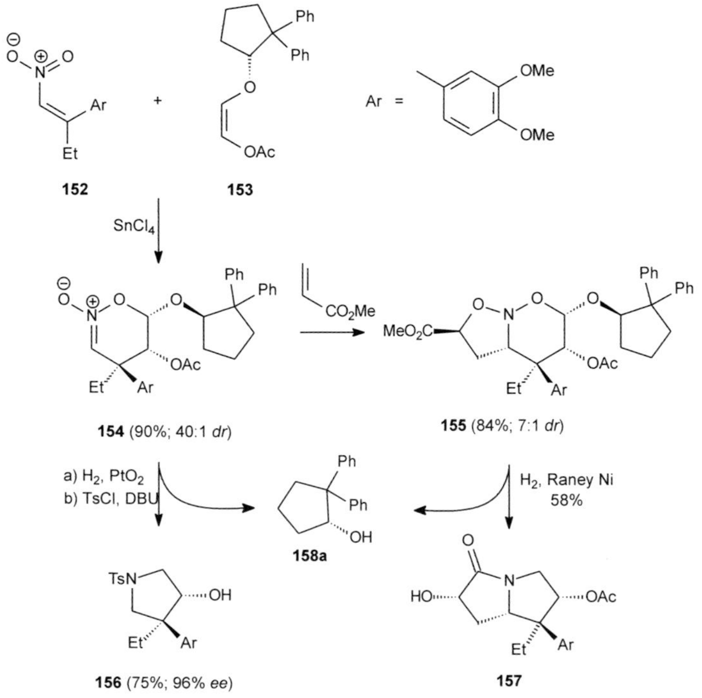

Scheme 6.35

If the above sequence is strictly considered in the logic of TCCs, the reaction could be run "one pot", with nitroalkene, vinyl ether and acrylate used "together", because FMOs dictates the sequence with the [4+2] step controlled by LUMO(nitroalkene), which therefore may only interact with HOMO(vinyl ether); the [3+2] step, being under HOMO(nitrone)/LUMO(dipolarophile) control, may complete the TCC with acrylate, no different sequence being possible.

In fact, for sake of simplicity, the reaction is generally run in two different steps, the inter[4+2]/intra[3+2] variant of TCC simply requires the addition of an optically active vinyl ether to a reagent that has the dipolarophile function tethered to nitroalkene. The reactions of methyl (E,E)-7-nitro-2,6-octadienoate **161** with the vinyl ethers **158–160b** derived from (R)-1,1-diphenyl-2-cyclopentanol **158a**, (1R,2S)-2-phenylcyclohexanol **159a** and (2S,3R)-2-(2,2-dimethylpropoxy)-3-bornanol **160a** [91–93] were catalyzed with different Lewis acids (TiCl$_2$(OiPr)$_2$, methyl aluminum bis(2,6-di-*tert*-butyl-4-methylphenoxide) (MAD) and methyl aluminum bis-(2,6-diphenylphenoxide) (MAPh); (Scheme 6.36). The product was **162**, again reduced to the tricyclic hydroxylactam **163** with recover of the chiral auxiliary **158–160a**. The influence of different variants on the stereoselectivity of **162** and **163** is given in Table 6.15 and the stereochemical course of the reaction largely depends on the nature of the Lewis acid: TiCl$_2$(OiPr)$_2$ (independent of the nature of the vinyl ether) gives (1S)-**163** (entries 1–3), MAPh gives the opposite enantiomer (entries 4–6), and MAD is not the Lewis acid of choice (entries 7, 8). The interesting inversion in enantioselectivity can be explained as a consequence of the reversal in the *endo/exo* selectivity of the [4+2] cycloaddition. The coordination of **161** to TiCl$_2$(OiPr)$_2$ gives a complex where the attack of the vinyl ether occurs *endo* selectively on the *Re*-face of the heterodienophile, while the coordination of MAPh favors the *exo* approach to the *Si*-face.

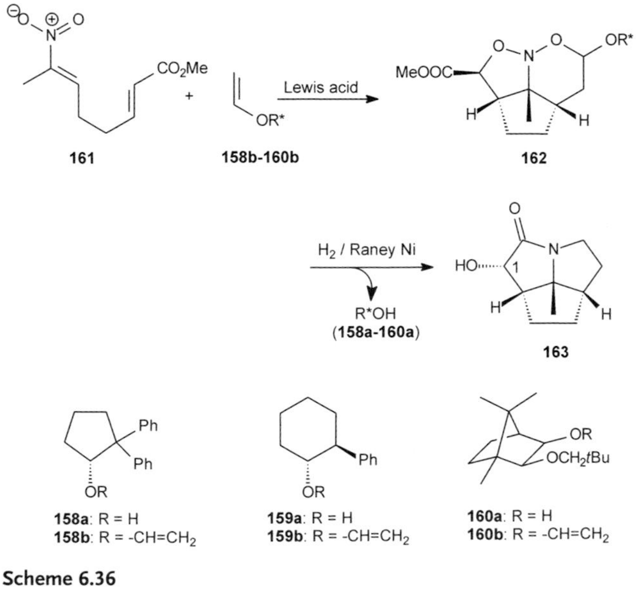

Scheme 6.36

Table 6.15 Influence of the chiral vinyl ether and the Lewis acid on the inter[4+2]/intra[3+2] TCC of **161**.

Entry	Chiral vinyl ether	Lewis acid	162 Yield (%)	162 dr	R*OH Yield (%)	163 Yield (%)	163 ee (%) (config.)	Ref.
1	158b	TiCl$_2$(OiPr)$_2$	88	98:2	82	70	98 (1S)	87
2	159b	TiCl$_2$(OiPr)$_2$	73	99:1	92	76	98 (1S)	93
3	160b	TiCl$_2$(OiPr)$_2$	82	81:19	92	86	98 (1S)	91–93
4	158b	MAPh	86	–	94	74	93 (1R)	87
5	159b	MAPh	85	2:98	97	76	79 (1R)	93
6	160b	MAPh	83	–	86	88	99 (1R)	92, 93
7	159b	MAD	88	58:42	78	73	72 (1S)	93
8	160b	MAD	86	–	81	67	2 (1S)	93

Three variants of inter[4+2]/intra[3+2] TCC demonstrate the enormous flexibility of these reactions.

The first mode (the bridged mode) [94] has the dienophile and the dipolarophile on the same molecule, a 1,4-pentadiene with the chiral alkoxyde at the 2-position. This first reacts as a vinyl ether under mild catalyzed conditions with nitroalkene, then undergoes intramolecular 1,3-dipolar cycloadditions under more severe conditions. If (E)-2-tert-butyldimethylsilyloxy-1-nitroethene **164** reacts with 2-[(1S,2R)-2-(1-methyl-1-phenylethyl)cyclohexyloxy]penta-1,4-diene **165** in toluene at –78°C, the [4+2] cycloaddition proceeds with exo selectivity to give **166**, then, under reflux, the intramolecular [3+2] step gives ring closure to **167**. This tricyclic adduct can be hydrogenated and its ring opening (with recovery of the chiral auxiliary) gives first **168** [95] and finally the protected aminocarba-sugar **169** (Scheme 6.37).

The second mode [96] has again both dienophile and dipolarophile on the same molecule, but differs from the first strategy because the chiral alkoxy group is in the 1-position of the 1,4-pentadiene **171**. This reacts with nitroalkene **170** and if the Lewis acid is SnCl$_4$, a mixture of exo diastereoisomers **172** and **173** in the ratio 11:1 is obtained in excellent yield. If the Lewis acid is MAPh, again the reaction is exo selective, but the diastereofacial selectivity is opposed and the ratio [**172**]:[**173**] now becomes 1:8 (Scheme 6.38). Each diastereoisomer, through intramolecular [3+2] cycloaddition, provides a tricyclic adduct (**174** and **175**, respectively) that can be reduced and protected, each giving a single enantiomer with excellent enantiomeric purity: (1S,2R,3S,5R)-**176** and (1R,2S,3R,5S)-**176**, respectively.

The third cycloaddition mode (the so-called spiro mode) [97] starts from methyl (E)-7-nitro-2,7-octadienoate **177**, a structure carrying both heterodiene and dipolarophile, which reacts with the chiral vinyl **159b**. The stereoselectivity was optimized with MAPh as Lewis acid since spiro tricyclic adduct **178** was obtained in 98% yield (19.5:1 dr). Scheme 6.39 describes the stereochemical course of this TCC, the reductive cleavage of the adduct with recovery of the chiral auxiliary **159a**, and the formation of the tricyclic lactam **179**. The observed asymmetric

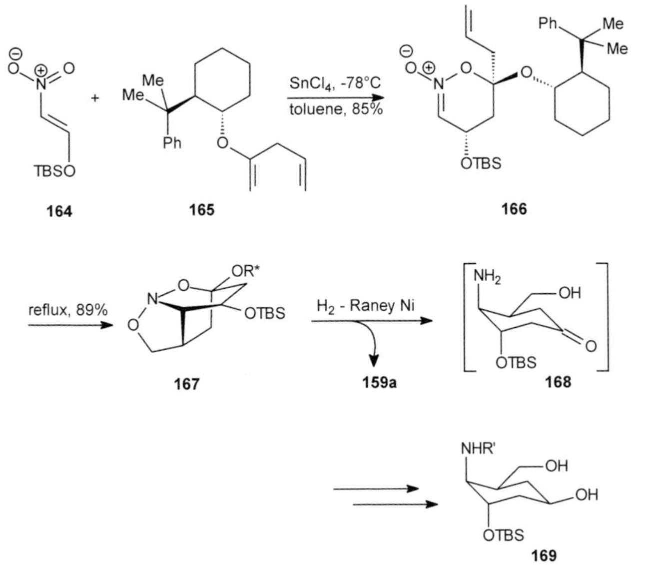

Scheme 6.37

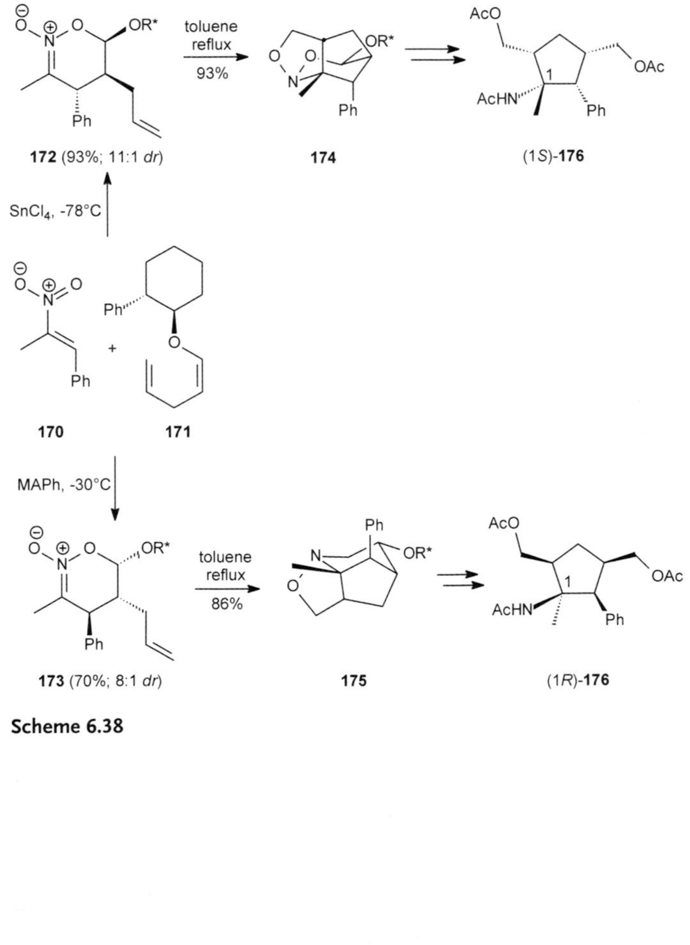

Scheme 6.38

induction can be rationalized as being due to a combination of face selectivity induced by the vinyl ether and preference for a distal fold of the tether (in the intramolecular 1,3-dipolar cycloaddition step) far from the substituent on the anomeric carbon.

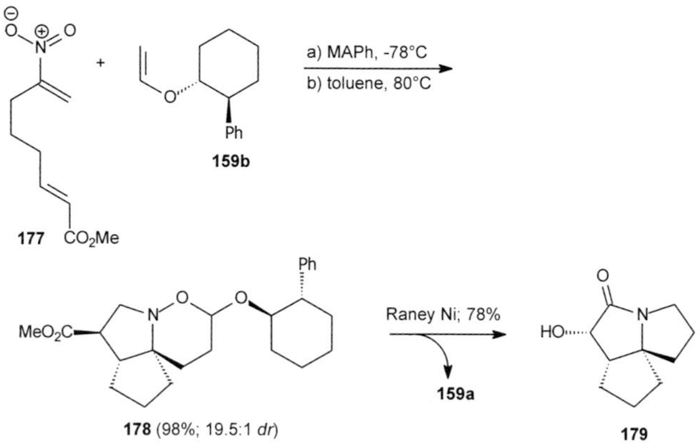

Scheme 6.39

The intra[4+2]/intra[3+2] TCC process has also been developed [98] but, for the moment, only on achiral substrates.

With this ample section devoted to TCC reactions, discussion of the generation of quaternary stereocenters through cycloadditions could be closed, but, again as a personal choice, we shall discuss an example showing the variety of synthetic possibilities made available to the organic chemist by cycloaddition methodology. Shair has reported an elegant total synthesis of (–)-longithorone A **180**, a cytotoxic marine natural product, through an intermolecular Lewis-acid-catalyzed *endo* DA reaction [99]. The two protected [12]-paracyclophanes **181** and **182**, each divergently prepared from the same precursor, behave as diene and dienophile to give the "dimeric" structure **183**, which has the first quaternary stereocenter required for the target. The deprotection of the paracyclophanes gives the bis-quinonic structure **184**, ready for the *in situ* intermolecular and transannular DA that directly affords longithorone in 90% yield (Scheme 6.40).

The synthesis of the complex architecture of a natural product, characterized by the simultaneous presence of two forms of chirality (six stereogenic centers and the atropisomerism deriving from the hindered rotation of a quinone ring) is the first example of inter[4+2]/intra[4+2] DA reaction with transfer of the chirality from atropisomerism to stereogenic centers in the product, and can be the icon of this chapter.

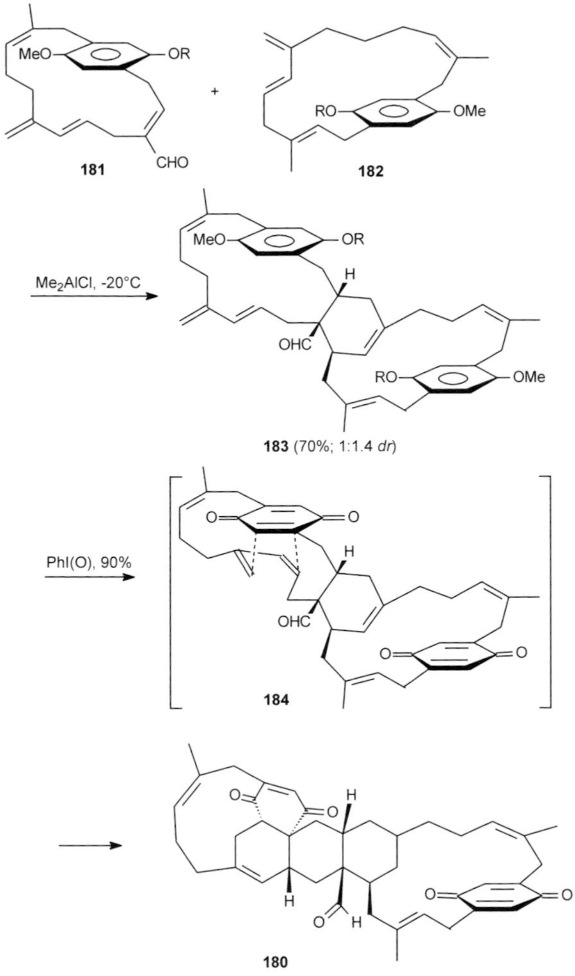

Scheme 6.40

References

1 For recent reviews on asymmetric syntheses of quaternary carbon centers see: (a) K. Fuji, *Chem. Rev.* **1993**, *93*, 2037–2066; (b) E. J. Corey, A. Guzman-Perez, *Angew. Chem. Int. Ed.* **1998**, *37*, 388–401; (c) J. Christoffers, A. Mann, *Angew. Chem. Int. Ed.* **2001**, *40*, 4591–4597.

2 The analogy between a catalyst and a molecular machine assembling the achiral reagents into a chiral supramolecular device was first stressed by E. J. Corey in his Nobel lecture dedicated to the logic of chemical synthesis: *Angew. Chem. Int. Ed. Engl.* **1991**, *30*, 455–465.

3 H. Nozaki, S. Moriuti, H. Takaya, R. Noyori, *Tetrahedron Lett.* **1966**, 5239–5242.

4 T. Aratani, *Pure Appl. Chem.* **1985**, *57*, 1839–1844.

5 M. P. Doyle, M. N. Protopopova, *Tetrahedron* **1998**, *54*, 7919–7946.

6 H. M. L. Davies, P. R. Bruzinski, D. H. Lake, N. Kong, M. J. Fall, *J. Am. Chem. Soc.* **1996**, *118*, 6897–6907.

7 P. Müller, G. Bernardinelli, Y. F. Allenbach, M. Ferri, H. D. Flack, *Org. Lett.* **2004**, *6*, 1725–1728.

8 P. Müller, S. Grass, S. P. Shahi, G. Bernardinelli, *Tetrahedron* **2004**, *60*, 4755–4763.

9 M. P. Doyle, R. E. Austin, A. S. Bailey, M. P. Dwyer, A. B. Dyatkin, A. V. Kalinin, M. M. Y. Kwan, S. Liras, C. J. Oalmann, R. J. Pieters, M. N. Protopopova, C. E. Raab, G. H. P. Roos, Q.-L. Zhou, S. F. Martin, *J. Am. Chem. Soc.* **1995**, *117*, 5763–5775.

10 M. P. Doyle, Q.-L. Zhou, *Tetrahedron: Asymmetry* **1995**, *6*, 2157–2160.

11 M. P. Doyle, C. S. Peterson, Q. Zhou, H. Nishiyama, *Chem. Commun.* **1997**, 211–212.

12 P. Müller, C. Bolèa, *Synlett* **2000**, 826–828.

13 R. Tokunoh, H. Tomiyama, M. Sodeoka, M. Shibasaki, *Tetrahedron Lett.* **1996**, *37*, 2449–2452.

14 D. H. Rogers, E. C. Yi, C. D. Poulter, *J. Org. Chem.* **1995**, *60*, 941–945.

15 T. G. Gant, M. C. Noe, E. J. Corey, *Tetrahedron Lett.* **1995**, *36*, 8745–8748.

16 P. Müller, C. Fruit, *Chem. Rev.* **2003**, *103*, 2905–2919.

17 K.-S. Yang, K. Chen, *J. Org. Chem.* **2001**, *66*, 1676–1679.

18 H. Ito, M. Hasegawa, Y. Takenaka, T. Kobayashi, K. Iguchi, *J. Am. Chem. Soc.* **2004**, *126*, 4520–4521.

19 K. V. Gothelf, K. A. Jørgensen, *Chem. Rev.* **1998**, *98*, 863–909.

20 S. Kanemasa, *Synlett* **2002**, 1371–1387.

21 F. Viton, G. Bernardinelli, E. P. Kündig, *J. Am. Chem. Soc.* **2002**, *124*, 4968–4969.

22 M. Shirahase, S. Kanemasa, Y. Oderaotoshi, *Org. Lett.* **2004**, *6*, 675–678.

23 M. Shirahase, S. Kanemasa, M. Hasegawa, *Tetrahedron Lett.* **2004**, *45*, 4061–4063.

24 D. Carmona, M. P. Lamata, F. Viguri, R. Rodriguez, L. A. Oro, A. I. Balana, F. J. Lahoz, T. Tejero, P. Merino, S. Franco, I. Montesa, *J. Am. Chem. Soc.* **2004**, *126*, 2716–2717.

25 M. R. Mish, F. M. Guerra, E. M. Carreira, *J. Am. Chem. Soc.* **1997**, *119*, 8379–8380.

26 R. Grigg, *Tetrahedron: Asymmetry* **1995**, *6*, 2475–2486.

27 H. Suga, K. Inoue, S. Inoue, A. Kakehi, *J. Am. Chem. Soc.* **2002**, *124*, 14836–14837.

28 R. B. Woodward, F. Sondheimer, D. Taub, K. Heusler, W. M. McLamore, *J. Am. Chem. Soc.* **1952**, *74*, 4223–4251.

29 K. C. Nicolaou, S. A. Snyder, T. Montagnon, G. Vassilikogiannakis, *Angew. Chem. Int. Ed.* **2002**, *41*, 1668–1698.

30 W. R. Roush, R. J. Sciotti, *J. Am. Chem. Soc.* **1998**, *120*, 7411–7419.

31 S. Hashimoto, N. Komeshima, K. Koga, *J. Chem. Soc., Chem. Comm.* **1979**, 437–438.

32 K. Furuta, S. Shimizu, Y. Miwa, H. Yamamoto, *J. Org. Chem.* **1989**, *54*, 1481–1483.

33 K. Ishihara, Q. Gao, H. Yamamoto, *J. Am. Chem. Soc.* **1993**, *115*, 10412–10413.

34 K. Ishihara, Q. Gao, H. Yamamoto, *J. Org. Chem.* **1993**, *58*, 6917–6919.

35 E. J. Corey, T.-P. Loh, *J. Am. Chem. Soc.* **1991**, *113*, 8966–8967.

36 E. J. Corey, T.-P. Loh, *Tetrahedron Lett.* **1993**, *34*, 3979–3982.

37 Y. Hayashi, J. J. Rohde, E. J. Corey, *J. Am. Chem. Soc.* **1996**, *118*, 5502–5503.

38 K. Maruoka, N. Murase, H. Yamamoto, *J. Org. Chem.* **1993**, *58*, 2938–2939.

39 J. Bao, W. D. Wulff, A. L. Rheingold, *J. Am. Chem. Soc.* **1993**, *115*, 3814–3815.

40 K. Ishihara, H. Yamamoto, *J. Am. Chem. Soc.* **1994**, *116*, 1561–1562.

41 K. Ishihara, H. Kurihara, H. Yamamoto, *J. Am. Chem. Soc.* **1996**, *118*, 3049–3050.

42 K. Ishihara, H. Kurihara, M. Matsumoto, H. Yamamoto, *J. Am. Chem. Soc.* **1998**, *120*, 6920–6930.

43 Y. Motoyama, M. Terada, K. Mikami, *Synlett* **1995**, 967–968.

44 K. Mikami, Y. Motoyama, M. Terada, *J. Am. Chem. Soc.* **1994**, *116*, 2812–2820.

45 D. A. Evans, J. A. Murry, P. von Matt, R. D. Norcross, S. J. Miller, *Angew. Chem. Int. Ed. Engl.* **1995**, *34*, 798–800.

46 D. A. Evans, D. A. Barnes, J. S. Johnson, T. Lectka, P. von Matt, S. J.

Miller, J. A. Murry, R. D. Norcross, E. A. Shaughnessy, K. R. Campos, *J. Am. Chem. Soc.* **1999**, *121*, 7582–7594.

47 E. P. Kündig, B. Bourdin, G. Bernardinelli, *Angew. Chem. Int. Ed. Engl.* **1994**, *33*, 1856–1858.

48 J. W. Faller, B. J. Grimmond, D. G. D'Alliessi, *J. Am. Chem. Soc.* **2001**, *123*, 2525–2529.

49 E. J. Corey, *Angew. Chem. Int. Ed.* **2002**, *41*, 1650–1667.

50 J. M. Fraile, J. I. Garcia, J. A. Mayoral, A. J. Royo, *Tetrahedron: Asymmetry* **1996**, *7*, 2263–2276.

51 J. M. Fraile, J. A. Mayoral, A. J. Royo, R. V. Salvador, B. Altava, S. V. Luis, M. I. Burguete, *Tetrahedron* **1996**, *52*, 9853–9862.

52 E. J. Corey, T.-P. Loh, T. D. Roper, M. D. Azimioara, M. C. Noe, *J. Am. Chem. Soc.* **1992**, *114*, 8290–8292.

53 G. Desimoni, G. Faita, P. Quadrelli, *Chem. Rev.* **2003**, *103*, 3119–3154.

54 J. A. Marshall, S. Xie, *J. Org. Chem.* **1992**, *57*, 2987–2989.

55 E. J. Corey, A. Guzman-Perez, T.-P. Loh, *J. Am. Chem. Soc.* **1994**, *116*, 3611–3612.

56 T. W. Lee, E. J. Corey, *J. Am. Chem. Soc.* **2001**, *123*, 1872–1877.

57 E. J. Corey, T. Shibata, T. W. Lee, *J. Am. Chem. Soc.* **2002**, *124*, 3808–3809.

58 D. H. Ryu, E. J. Corey, *J. Am. Chem. Soc.* **2003**, *125*, 6388–6390.

59 J. Knol, A. Meetsma, B. L. Feringa, *Tetrahedron: Asymmetry* **1995**, *6*, 1069–1072.

60 A. K. Ghosh, P. Mathivanan, J. Cappiello, *Tetrahedron: Asymmetry* **1998**, *9*, 1–45 and references therein.

61 Y. Honda, T. Date, H. Hiramatsu, M. Yamauchi, *Chem. Commun.* **1997**, 1411–1412.

62 M. Yamauchi, T. Aoki, M.-Z. Li, Y. Honda, *Tetrahedron: Asymmetry* **2001**, *12*, 3113–3118.

63 V. K. Aggarwal, E. S. Anderson, D. E. Jones, K. B. Obierey, R. Giles, *Chem. Commun.* **1998**, 1985–1986.

64 M. A. Brimble, J. F. McEwan, *Tetrahedron: Asymmetry* **1997**, *8*, 4069–4078.

65 D. A. Evans, J. Wu, *J. Am. Chem. Soc.* **2003**, *125*, 10162–10163.

66 D. A. Vosburg, C. D. Vanderwal, E. J. Sorensen, *J. Am. Chem. Soc.* **2002**, *124*, 4552–4553.

67 K. A. Jørgensen, *Angew. Chem. Int. Ed.* **2000**, *39*, 3558–3588.

68 K. A. Jørgensen, *Eur. J. Org. Chem.* **2004**, 2093–2102.

69 M. Quimpère, K. Jankowski, *J. Chem. Soc., Chem. Commun.* **1987**, 676–677.

70 S. Yao, M. Roberson, F. Reichel, R. G. Hazel, K. A. Jørgensen, *J. Org. Chem.* **1999**, *64*, 6677–6687.

71 C. Bolm, O. Simic, *J. Am. Chem. Soc.* **2001**, *123*, 3830–3831.

72 S. Yao, M. Johannsen, R. G. Hazel, K. A. Jørgensen, *J. Org. Chem.* **1998**, *63*, 118–121.

73 M. Johannsen, S. Yao, K. A. Jørgensen, *Chem. Commun.* **1997**, 2169–2170.

74 S. Yao, M. Johannsen, H. Audrain, R. G. Hazell, K. A. Jørgensen, *J. Am. Chem. Soc.* **1998**, *120*, 8599–8605.

75 A. K. Ghosh, M. Shirai, *Tetrahedron Lett.* **2001**, *42*, 6231–6233.

76 P. I. Dalko, L. Moisan, J. Cossy, *Angew. Chem. Int. Ed.* **2002**, *41*, 625–628.

77 S. R. Chemler, U. Iserloh, S. J. Danishefsky, *Org. Lett.* **2001**, *3*, 2949–2951.

78 G. Desimoni, G. Tacconi, *Chem. Rev.* **1975**, *75*, 651–692.

79 W. Zhuang, J. Thorhauge, K. A. Jørgensen, *Chem. Commun.* **2000**, 459–460.

80 H. Audrain, J. Thorhauge, R. G. Hazel, K. A. Jørgensen, *J. Org. Chem.* **2000**, *65*, 4487–4497.

81 D. A. Evans, J. S. Johnson, E. J. Olhava, *J. Am. Chem. Soc.* **2000**, *122*, 1635–1649.

82 E. Wada, H. Koga, G. Kumaran, *Tetrahedron Lett.* **2002**, *43*, 9397–9400.

83 H. Koga, E. Wada, *Tetrahedron Lett.* **2003**, *44*, 715–719.

84 P. Müller, C. Fruit, *Chem. Rev.* **2003**, *103*, 2905–2919.

85 A. S. Timén, P. Somfai, *J. Org. Chem.* **2003**, *68*, 9958–9963.

86 J. D. Winkler, *Chem. Rev.* **1996**, *96*, 167–176.

87 S. E. Denmark, A. Thorarensen, *Chem. Rev.* **1996**, *96*, 137–165.

88 K. C. Nicolaou, G. Vassilikogiannakis, W. Mägerlein, R. Kranich, *Angew. Chem. Int. Ed.* **2001**, *40*, 2482–2486.

89 K. C. Nicolaou, G. Vassilikogiannakis, W. Mägerlein, R. Kranich, *Chem. Eur.J.* **2001**, *7*, 5359–5371.

90 S. E. Denmark, M. E. Schnute, *J. Org. Chem.* **1994**, *59*, 4576–4595.

91 S. E. Denmark, C. B. W. Senanayake, G.-D. Ho, *Tetrahedron* **1990**, *46*, 4857–4876.

92 S. E. Denmark, C. B. W. Senanayake, *J. Org. Chem.* **1993**, *58*, 1853–1858.

93 S. E. Denmark, M. E. Schnute, C. B. W. Senanayake, *J. Org. Chem.* **1993**, *58*, 1859–1874.

94 S. E. Denmark, V. Guagnano, J. A. Dixon, A. Stolle, *J. Org. Chem.* **1997**, *62*, 4610–4628.

95 S. E. Denmark, M. Juhl, *Helv. Chim. Acta* **2002**, *85*, 3712–3736.

96 S. E. Denmark, J. A. Dixon, *J. Org. Chem.* **1998**, *63*, 6178–6195.

97 S. E. Denmark, D. S. Middleton, *J. Org. Chem.* **1998**, *63*, 1604–1618.

98 S. E. Denmark, L. Gomez, *J. Org. Chem.* **2003**, *68*, 8015–8024.

99 M. E. Layton, C. A. Morales, M. D. Shair, *J. Am. Chem. Soc.* **2002**, *124*, 773–775.

7
Asymmetric Cross-coupling and Mizoroki–Heck Reactions

Louis Barriault and Effiette L. O. Sauer

7.1
The Asymmetric Heck Reaction

7.1.1
Introduction

In the early 1970s, Mizoroki [1] and Heck [2] each reported the discovery of a new palladium(0)-mediated coupling of an alkene with an aryl halide (Eq. 1) [3]. In the following years, however, the potential for this reaction's use in natural product synthesis lay dormant, perhaps because of the problems associated with the regiocontrol of unsymmetrical alkenes and a general lack of mechanistic understanding.

$$R \overset{X}{\underset{}{\bigcirc}} \quad + \quad \diagup \diagdown^{R'} \quad \xrightarrow[nBu_3N]{Pd(0) \text{ cat.}} \quad \bigcirc \diagup\diagdown^{R'} \quad + \quad nBu_3NHX \tag{1}$$

Interest in this powerful reaction grew rapidly following seminal work on the asymmetric Heck reaction, independently reported by Shibasaki [4] and Overman [5] in 1989. They described the asymmetric formation of tertiary and quaternary carbon centers using $Pd(OAc)_2$ as a precatalyst and a chiral phosphine ligand such as (R)-BINAP or (R,R)-DIOP (Scheme 7.1). Despite the fact that low enantiomeric excesses were observed in both cases, their work represented a significant break-through in the development of catalytic asymmetric methods for the synthesis of quaternary carbon centers [6].

7.1.2
Mizoroki–Heck Reaction Mechanism

The basic reaction mechanism for the Heck coupling is illustrated in Scheme 7.2. This simplistic representation, however, fails to account for the differences in enantioselectivity that have been observed between the Heck reactions of alkenyl

Quaternary Stereocenters: Challenges and Solutions for Organic Synthesis. Edited by Jens Christoffers, Angelika Baro
Copyright © 2005 WILEY-VCH Verlag GmbH & Co. KGaA, Weinheim
ISBN: 3-527-31107-6

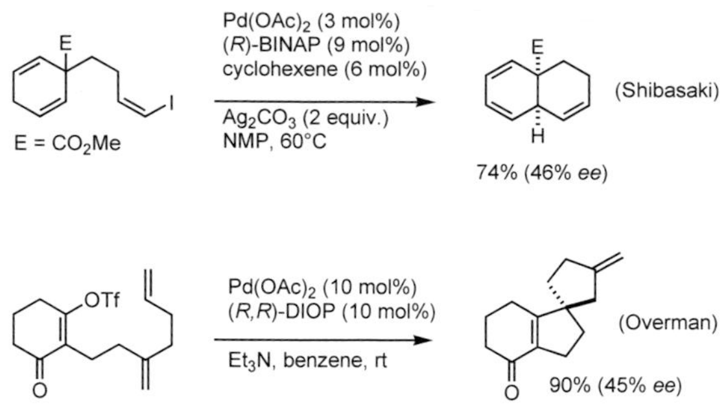

Scheme 7.1 First reported examples of the asymmetric Mizoroki–Heck reaction.

halides compared with alkenyl triflates. To this end, two distinct mechanisms have been proposed, cationic and neutral [7].

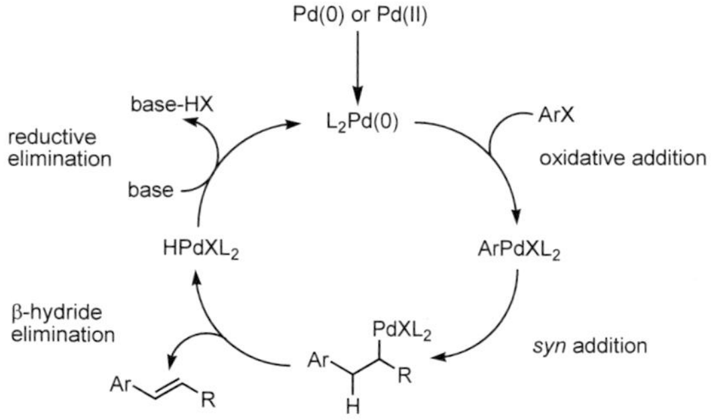

Scheme 7.2 General mechanism of the Mizoroki–Heck reaction.

Independent work by Cabri [8] and Hayashi [9] described a cationic pathway for the Heck reaction of aryl triflates in the presence of bidentate phosphine ligands (Scheme 7.3). Today, this mechanism is typically used when discussing the asymmetric Heck reactions of unsaturated triflates, or those of halides in the presence of halide scavengers such as Ag(I) and Tl(I) (Scheme 7.3). Oxidative insertion of Pd(0) gives intermediate **1** which can either undergo triflate dissociation or halide abstraction in the presence of Ag(I) or Tl(I) thus liberating a coordination site. This allows the alkene to complex onto the Pd(II) to form intermediate **2** which undergoes a *syn* addition to give complex **3**. In all steps, the diphosphine ligand remains attached to the metal, thus ensuring a high transfer of chirality.

In the absence of halide scavengers, a neutral pathway has been invoked to describe the Heck reaction of aryl or vinyl halides (Scheme 7.4). In these cases, modest enantioselectivities are typically observed suggesting partial dissociation of the diphosphine ligand during the palladium–alkene complex formation (**6**→**9**). Some notable exceptions to this trend, however, have been observed. In 1992, Overman et al. reported examples of a highly enantioselective asymmetric

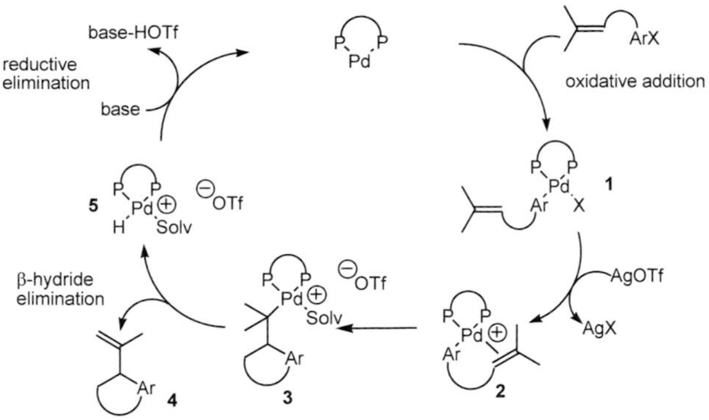

Scheme 7.3 Proposed mechanism for the cationic pathway.

Pd/BINAP-catalyzed Heck reaction using aryl halides *without the addition of halide scavengers* [10]. These results suggest that both phosphorus atoms remain bound to the metal during the chirality-transfer step. Additionally, the fact that different results are often obtained under neutral and cationic conditions rules out the direct conversion of **6** to **8**. Alternatively, the involvement of a pentacoordinate Pd(II) intermediate, **7**, has been suggested, owing to the associative nature by which substitution chemistry at square planar Pd(II) complexes is known to occur [11]. Finally, both calculations [12] and experimental evidence [13] indicate that the direct conversion of **7** to **10** ought to be an energetically unfavorable process. Taken all together, these suppositions point to a neutral pathway which proceeds from **6** to **7** to **8** to **10**.

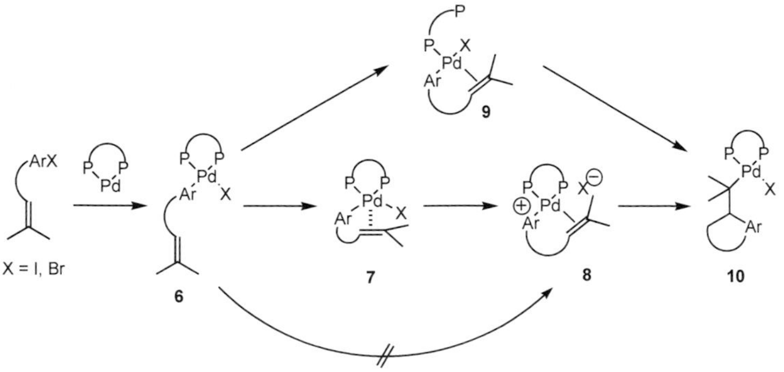

Scheme 7.4 Overman's proposed mechanism for the neutral pathway.

7.1.3
Asymmetric Formation of Quaternary Carbon Centers

Since the pioneering work of Overman and Shibasaki on the asymmetric construction of quaternary carbon centers using an intramolecular Mizoroki–Heck reaction, several groups have devoted considerable effort to this area [14]. Through their collective research, they have clearly demonstrated the power and versatility of the intramolecular asymmetric Mizoroki–Heck reaction as a valuable synthetic tool. As testament to its worth, the ongoing research in this area has culminated in the elegant total synthesis of several biologically active natural products [14a].

In 1996, Keay and co-workers reported the first total synthesis of (+)-xestoquinone 14, a potent cardiotonic agent (Scheme 7.5) [15]. The quaternary carbon center as well as the C and D rings of 14 were created via two asymmetric palladium-catalyzed Heck reactions in tandem. The authors reported poor enantiomeric excesses (5–13%) when naphthyl bromide 11 (X = Br) was used under either neutral or cationic conditions. However, replacement of the bromide by a triflate 12 dramatically improved the cyclization process yielding the pentacycle 13 in 82% yield and 68% *ee* [Method A: $Pd_2(dba)_3$ (2.5 mol%), (*S*)-BINAP (10 mol%), PMP (8 equivalents), toluene, reflux].

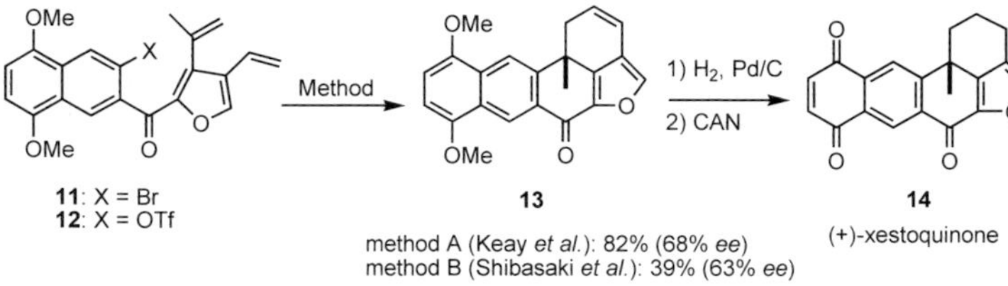

11: X = Br
12: X = OTf

13

method A (Keay *et al.*): 82% (68% *ee*)
method B (Shibasaki *et al.*): 39% (63% *ee*)

14

(+)-xestoquinone

Scheme 7.5 Synthesis of (+)-xestoquinone via a tandem asymmetric Heck sequence.

Although Keay et al. had demonstrated the potential for naphthyl bromides as precursors to the xestoquinone framework, the low enantiomeric excesses obtained (*vide supra*) prompted Shibasaki and co-workers to search for improved reaction conditions [16]. Soon after, they reported that the enantiomeric excess of this asymmetric Heck reaction cascade could be improved by the use of Ag-exchanged zeolite in place of Ag_3PO_4 as a halide scavenger. In fact, treatment of 11 with $Pd_2(dba)_3$-$CHCl_3$, (*S*)-BINAP, $CaCO_3$ and Ag-exchanged zeolite (1 equivalent) in NMP at 80°C gave 13 in 39% yield and 63% *ee* (Method B). Furthermore, it was discovered that the quantity of zeolite used was crucial to obtaining good enantiomeric excesses. Indeed, when more than one equivalent was used (2–6 equivalents), the reaction yield and enantiomeric excess were significantly diminished.

The versatility of the intramolecular Mizoroki–Heck reaction has been widely exploited in the development of asymmetric methods for the synthesis of indole rings bearing a quaternary center at C-3. Overman et al. investigated the asymmetric cyclizations of various *Z*-anilides and *E*-anilides [17]. Exposure of 15 and 16 to $Pd_2(dba)_3$-$CHCl_3$ (5 mol%) and (*S*)-BINAP (12 mol%) in DMA in the presence of additives such as PMP or Ag_3PO_4 gave the corresponding 3,3-bisalkyloxindole ring 17 in yields ranging from 53 to 85% with 38–92% *ee* (Table 7.1) [17a,d]. In the case of *Z*-2-triflateanilides 18–28, $Pd(OAc)_2$ as a precatalyst and (*R*)-BINAP in THF at 80°C were found to be the optimal conditions to convert 18–28 to 3-alkyl-3-aryloxindoles 29–39 (Table 7.2) [17g].

Table 7.1 Asymmetric cyclizations of Z- and E-anilides.

Entry	Substrate	Additive	Yield (%)	ee (%) (config.)
1	(Z)-**15**	Ag$_3$PO$_4$	53	78 (R)
2	(Z)-**15**	PMP	80	92 (R)
3	(E)-**16**	Ag$_3$PO$_4$	80	45 (S)
4	(E)-**16**	PMP	85	38 (S)

Following these reports, the potential of the asymmetric Heck reaction for the preparation of 3,3-disubstituted oxindoles was further established through the

Table 7.2 Asymmetric cyclizations of Z-2-triflateanilides.

Entry	Ar	Pd(OAc)2, (mol%)	(R)-BINAP, (mol%)	Product (yield %)	ee (%)
1	Ph (**18**)	10	20	**29** (86)	84
2	Ph (**18**)	5	10	**29** (86)	82
3	4-MeO-C$_6$H$_4$ (**19**)	10	20	**30** (86)	77
4	4-MeO-C$_6$H$_4$ (**19**)	5	10	**30** (81)	79
5	3-pyridyl (**20**)	10	20	**31** (76)	90
6	3-pyridyl (**20**)	5	10	**31** (77)	88
7	4-AcNHC$_6$H$_4$ (**21**)	10	20	**32** (92)	72
8	4-AcNHC$_6$H$_4$ (**21**)	5	10	**32** (91)	71
9	1-naphthyl (**22**)	10	20	**33** (95)	94
10	1-naphthyl (**22**)	5	10	**33** (92)	92
11	2-NO$_2$C$_6$H$_4$ (**23**)	30	60	**34** (60)	89
12	2-NO$_2$-3-MeC$_6$H$_3$ (**24**)	40	80	**35** (48)	87
13	(**25**)	10	20	**36** (86)	95
14	(**26**)	10	20	**37** (86)	86
15	2-BocNHC$_6$H$_4$ (**27**)	10	20	**38** (91)	98
16	2-BnNHC$_6$H$_4$ (**28**)	10	20	**39** (24)	0

completion of several elegant total syntheses of complex alkaloids including (–)-esermethole (**40**) [17d], (–)-physostigmine (**41**) [17a], (–)-physovenine (**42**) [17e] and (+)-asperazine (**43**) [18] (Scheme 7.6).

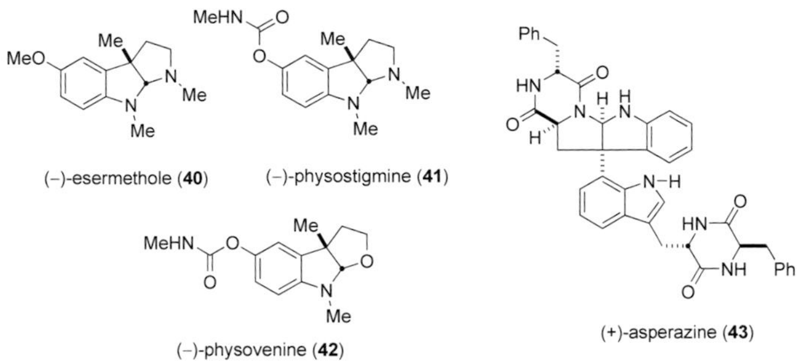

Scheme 7.6 Alkaloids containing 3,3-disubstituted oxindoles synthesized via the asymmetric Heck reaction.

A handful of synthetic methods are available for creating vicinal quaternary carbon centers with complete stereocontrol [19]. The challenge lies primarily in the steric congestion of such architectural motifs, which plagues an efficient transfer of chirality. Recently, Overman and collaborators reported a highly stereoselective double intramolecular Heck reaction that overcomes this problem [20]. Treatment of **44** with $(Ph_3P)_2PdCl_2$, Et_3N, DMA, 100°C gave the pseudo *meso*-bispyrroloindoline **45** in 71% yield (Scheme 7.7). The latter was then transformed into *meso*-chimonanthine **46** in a few steps [21].

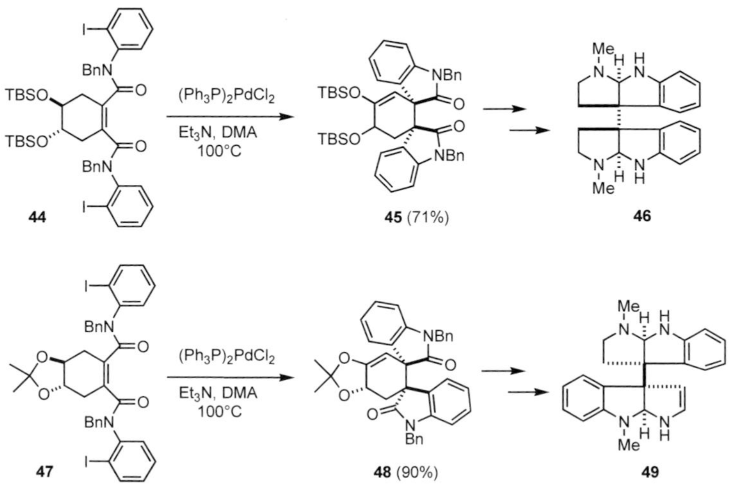

Scheme 7.7 Preparation of vicinal quaternary carbon centers by a double intramolecular Heck reaction.

Replacement of the TBS group by an acetonide had a drastic effect on the asymmetric Heck reaction. In fact, the palladium-catalyzed cyclization of **47** now afforded exclusively the (C-2)-symmetric bispyrroloindoline **48** in 90% yield. With desymmetrized **48** in hand, (–)-chimonanthine **49** was thus obtained following the reaction sequence used to prepare *meso*-chimonanthine **46**. In a later report, Overman et al. also described the desymmetrization of anilide **50** using a stoichiometric quantity of Pd(OAc)$_2$ and (R)-tolyl-BINAP (2 equivalents) in acetonitrile (Scheme 7.8) [22]. The two 3-alkyl-3-aryloxindoles in **51** were simultaneously formed in 62% yield (90% ee) and 21% of *meso* isomers were collected. Removal of both benzyl and tosyl groups gave quadrigemine C, **52**.

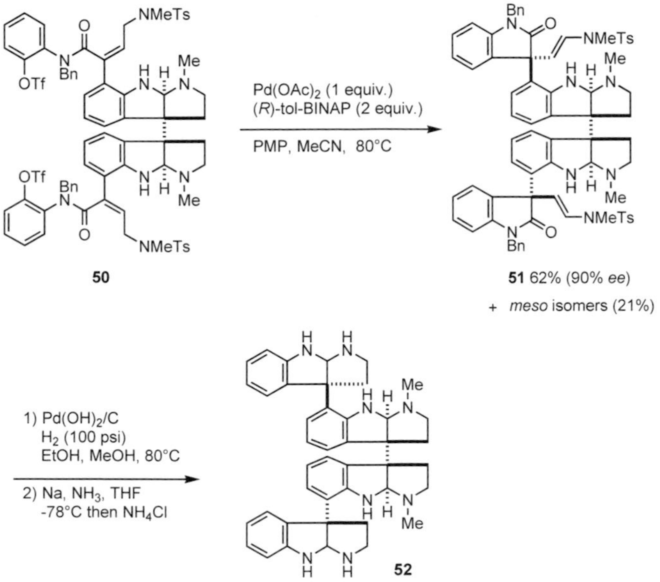

Scheme 7.8 Synthesis of quadrigemine C by a desymmetrizing Heck reaction.

Numerous strychnos indole alkaloids such as (–)-aspidospermine **53** and (–)-strychnine **54** contain a 3-alkyl-indoline subunit **55** in their backbone (Scheme 7.9). The indoline framework of these alkaloids can be readily obtained from an intramolecular Mizoroki–Heck cyclization of cyclohexylamine. In 2003, Mori and co-workers demonstrated the effectiveness of this approach by accomplishing the total synthesis of (–)-strychnine **54**, (–)-dehydrotubifoline **56** and (–)-tubifoline **57** [23]. Cyclohexylamine **58** was stirred in the presence of Pd(OAc)$_2$ (5 mol%), PPh$_3$ (10 mol%) and DIPEA (2 equivalents) in DMF at 90°C to give the corresponding indolines **59**, **60** and starting material **58** in 26%, 13% and 31% yield, respectively (Table 7.3).

After the examination of various ligands, bases and solvents, they found that Me$_2$PhP (10 mol%) and AgCO$_3$ (2 equivalents) in DMSO at 90°C constituted the best reaction conditions to convert **58** into the corresponding indoline **59** (47%) (Table 7.3, entry 7). To complete the synthesis of strychnine, **58** was transformed into the corresponding nitrile derivative **61** which underwent a palladium-catalyzed

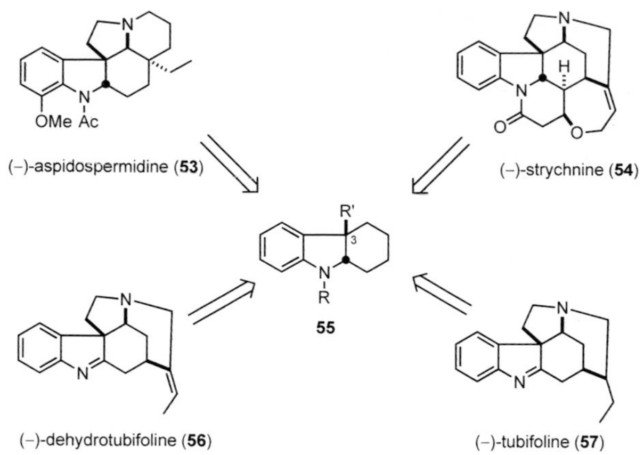

(−)-aspidospermidine (**53**)

(−)-strychnine (**54**)

55

(−)-dehydrotubifoline (**56**)

(−)-tubifoline (**57**)

Scheme 7.9 Strychnos indole alkaloids containing a
3-alkyl-indoline subunit.

Mizoroki–Heck reaction to give the key intermediate indoline **62** in 87% yield
(Scheme 7.10).

Trost and co-workers reported a modular approach to the total synthesis of
furaquinocin A **65**, B **66** and E **67** [24]. Their strategy took advantage of a reductive
Heck cyclization of **63** to construct the quaternary carbon center at C-9. They
discovered that replacing triethylamine by a more hindered base, such as PMP,
greatly improved the yield of the 5-*exo*-product **64** (Scheme 7.11).

Hirama and collaborators disclosed a stereocontrolled synthesis of the ABC
ring framework of zoanthenol [25]. The congested benzylic quaternary carbon

Table 7.3 Asymmetric Heck reaction of a cyclohexylamine to
yield 3-alkyl-indolines.

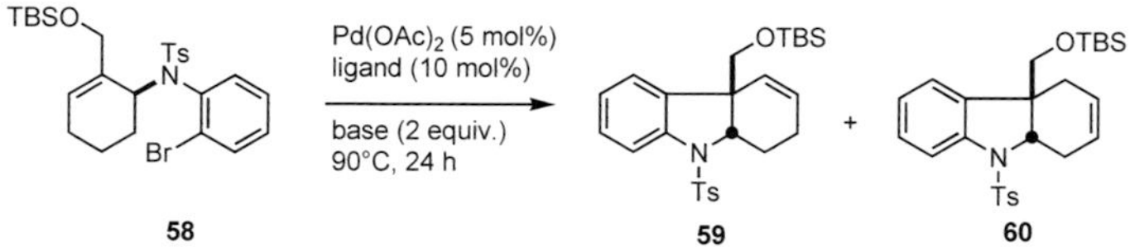

Entry	Ligand	Solvent	Base	59 (%)	60 (%)	58 (%)
1	PPh$_3$	DMF	iPr$_2$NEt	26	13	31
2	dppb	DMF	iPr$_2$NEt	13	1	76
3	PPh$_3$	DMF	Ag$_2$CO$_3$	42	8	33
4	PMePh$_2$	DMF	Ag$_2$CO$_3$	52	13	29
5	PMe$_2$Ph	DMF	Ag$_2$CO$_3$	56	19	20
6	PMe$_2$Ph	C$_3$H$_7$CN	Ag$_2$CO$_3$	17	1	71
7	PMe$_2$Ph[a]	DMSO	Ag$_2$CO$_3$	47	0	38

a Reaction temperature 105°C

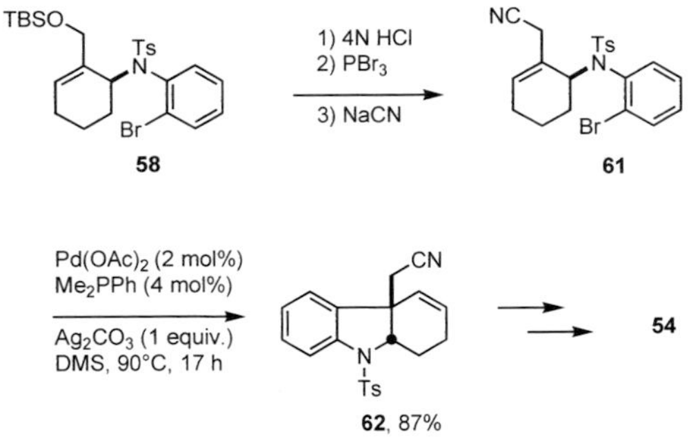

Scheme 7.10 Preparation of a 3-alkyl-indoline by an asymmetric Heck reaction: application to the synthesis of (–)-strychnine.

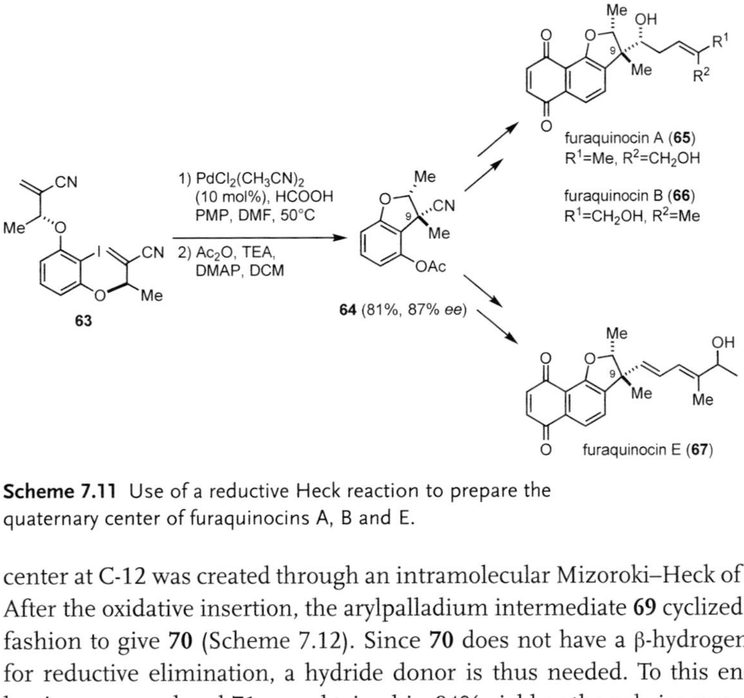

Scheme 7.11 Use of a reductive Heck reaction to prepare the quaternary center of furaquinocins A, B and E.

center at C-12 was created through an intramolecular Mizoroki–Heck of enone **68**. After the oxidative insertion, the arylpalladium intermediate **69** cyclized in a 6-*exo* fashion to give **70** (Scheme 7.12). Since **70** does not have a β-hydrogen available for reductive elimination, a hydride donor is thus needed. To this end, triethylamine was used and **71** was obtained in 84% yield as the sole isomer.

As revealed by the previous examples, the choice of ligand can be crucial to obtaining a highly stereoselective process. Several groups have devoted efforts toward the development of new ligands [26]. In one such endeavor, Busacca and co-workers probed the electronic effects of various phosphinoimidazole ligands (BIPI) in the asymmetric Mizoroki–Heck reaction [27]. They demonstrated that the enantioselectivity is linearly dependent on the phosphine's electronic density and that the variation in ligand basicity affects the facial selectivity. When comparing the enantiomeric excesses for the asymmetric formation of 3-spiroindole **73**, the optimized BIPI ligand **76** proves to be a superior chiral ligand to that of the commonly used BINAP (**75**) (Scheme 7.13).

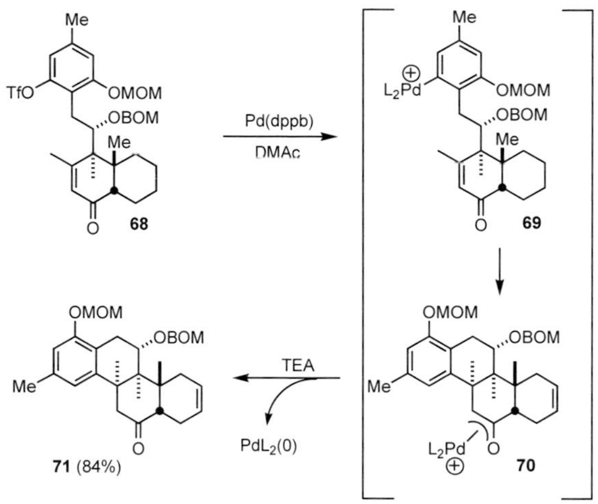

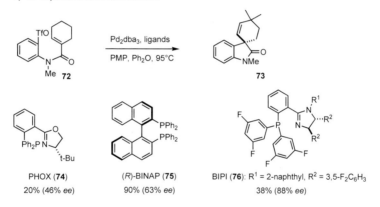

Scheme 7.12 Synthesis of the ABC ring system of zoanthenol by an asymmetric Heck reaction.

PHOX (**74**)
20% (46% *ee*)

(*R*)-BINAP (**75**)
90% (63% *ee*)

BIPI (**76**): R^1 = 2-naphthyl, R^2 = 3,5-F$_2$C$_6$H$_3$
38% (88% *ee*)

Scheme 7.13 Effect of different phosphinoimidazole ligands on the asymmetric Heck reaction.

While the majority of asymmetric Heck reactions rely on ligands as their means of chiral induction, Hallberg and co-workers have reported a novel asymmetric Heck arylation using a chiral auxiliary to control the stereoselective formation of cyclopentanones bearing α-quaternary carbon centers [28]. The observed stereocontrol is presumed to arise from the coordination of palladium to the pyrrolidine nitrogen, thus preferentially directing the carbopalladation to one face of vinyl ether **77** via an *N*-chelated π-complex such as **78** (Scheme 7.14). Using different aryl iodides and bromides, a variety of 2,2-disubstituted cyclopentanones **80** were prepared in 45 to 78% yield with enantiomeric excesses ranging from 90 to 98% (Table 7.4).

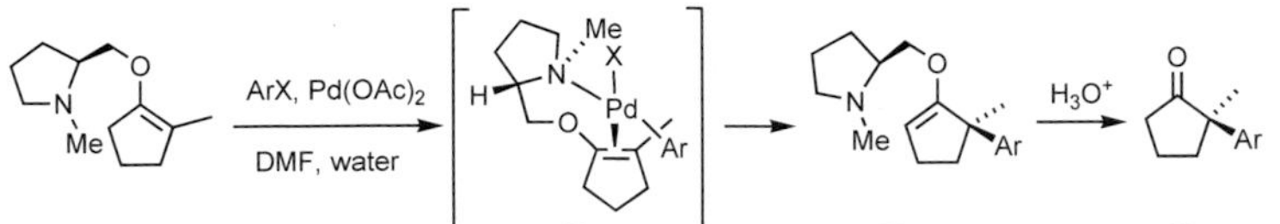

Scheme 7.14 Auxiliary-controlled asymmetric Heck arylation.

7.2
Metal-catalyzed Cross-coupling Reactions

7.2.1
Palladium-catalyzed α-Arylation

In addition to the widely used asymmetric Heck reaction for the formation of new carbon–carbon bonds, several other cross-coupling reactions have emerged as viable methods for the enantioselective synthesis of quaternary carbon centers. One such alternative that has seen widespread use is the palladium-catalyzed

Table 7.4 Auxiliary-controlled asymmetric Heck arylations.

Entry	Aryl halide	*T* (°C), time (h)	Yield (%)	*ee* (%)
1	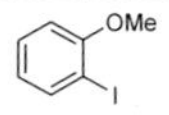	70, 24	67	98
2	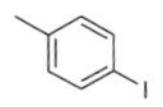	70, 18	54	93
3	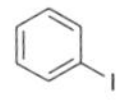	80, 30	50	94
4	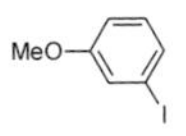	70, 18	68	93
5	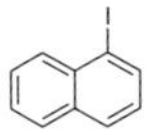	80, 48	45	90
6	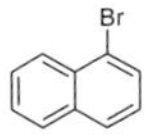	100, 48	49	91
7	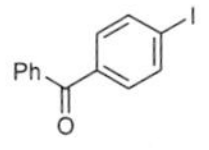	80, 24	47	97
8	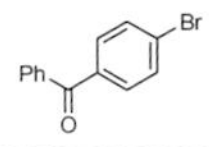	100, 24	78	94

All components (1.3 equiv. **77**, 1 equiv. ArX, 3% Pd catalyst) were added to aqueous DMF, the vessel sealed under air, and heated for 18–68 h.

α-arylation of ketone [29] and lactone [30] enolates reported by Buchwald and coworkers. In Buchwald's initial report, enolates of α-methyl substituted tetralones **81** were treated with several different aryl bromides in the presence of NaO*t*Bu and a palladium(0)/BINAP complex to afford α-disubstituted tetralones **82** in 56 to 74% yield with enantioselectivities ranging from 73 to 88% (Table 7.5).[31].

Treatment of the α′-blocked α-methyl cyclopentanone **83** under similar conditions showed a significant improvement to both yield and enantioselectivity (Table 7.6). Unfortunately, the benzylidine α-blocking group proved impossible to remove from the product. In addition, high catalyst loadings and the need for elevated temperatures prompted the search for improved reaction conditions.

In order to overcome the aforementioned shortcomings, a new catalytic system was introduced [32]. The benzylidene moiety was replaced by a 2-methyl-5-*N*-methyl-anilinomethylene blocking group to give ketone **85**. Treatment of this new starting material with various aryl bromides in the presence of NaO*t*Bu and a Pd(0)/BINAP complex afforded products **86** with enantiomeric excesses varying from 43 to 89%. To further improve the reaction conditions, a variety of ligands was screened. Ligand **87** was found to offer the best results, raising the enantiomeric excesses as high as 94% (Table 7.7).

Table 7.5 Asymmetric α-arylation of tetralone derivatives.

Entry	R	ArBr	Yield (%)	*ee* (%)
1	H		66	73
2	H		73	88
3	MeO		56	77
4	H		74	84
5	H		40[a]	61[a]

a The reaction was carried out at 70°C using 5 equiv. of halide and 5 equiv. of NaO*t*Bu.

Table 7.6 Asymmetric α-arylation of cyclopentanone **83** using a benzilidine α-blocking group.

Entry	ArBr	Conditions	Yield (%)	*ee* (%)
1		NaHMDS, 10 mol% Pd$_2$(dba)$_3$/BINAP	86	95
2		NaOtBu, 20 mol% Pd(OAc)$_2$/BINAP	80	94
3		NaOtBu, 20 mol% Pd(OAc)$_2$/BINAP	75	98

In addition to the improved yields and enantiomeric excesses, the new reaction could be performed at room temperature with catalyst and ligand loadings of 1.0 mol% and 2.5 mol%, respectively. Lastly, the blocking group could be easily removed by acid hydrolysis and subsequent retro-Mannich reaction under basic conditions to afford ketone **89** (Scheme 7.15).

7.2.2
Palladium-catalyzed α-Vinylation

Efforts to extend Buchwald's methodology to include the α-vinylation of ketones initially looked at the treatment of **85** with the previously reported [29a] palladium(0)/BINAP system [33]. These reactions, however, were low yielding and afforded poor enantiomeric excesses. Following an extensive screening of various biaryl ligands, optimized reaction conditions were found using the electron-rich aminophosphine ligand **90** [34]. Treatment of **85** with various vinyl bromides in the presence of 2.5–5.0 mol% Pd$_2$(dba)$_3$, 6.5 mol% **90**, and NaOtBu in toluene at ambient temperature resulted in the preparation of α-vinyl ketones **91–95** in 84–95% yield and 71–92% *ee* (Table 7.8).

7.2.3
Intramolecular Palladium-catalyzed α-Arylation of Amides

Analogous to the intermolecular α-arylation of carbonyls cited above, Hartwig and co-workers reported an intramolecular α-arylation of amides to form α,α-disubstituted oxindoles [35]. The initial search for a set of conditions to convert **96**

Table 7.7 Asymmetric α-arylation of cyclopentanone **85** using an N-methyl-anilinomethylene blocking group.

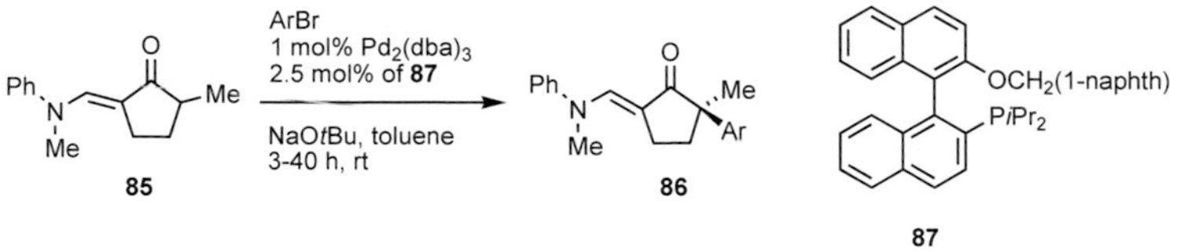

Entry	ArBr	Yield (%)	ee (%)
1	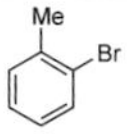 (2-Me-phenyl)Br	43	22
2	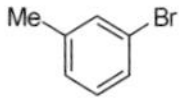 (3-Me-phenyl)Br	85	94
3	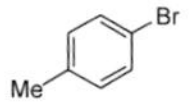 (4-Me-phenyl)Br	84	93
4	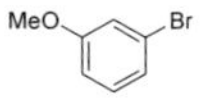 (3-MeO-phenyl)Br	80	89
5	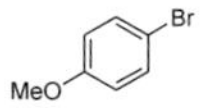 (4-MeO-phenyl)Br	80	94
6	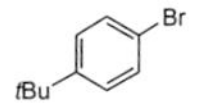 (4-tBu-phenyl)Br	84	93

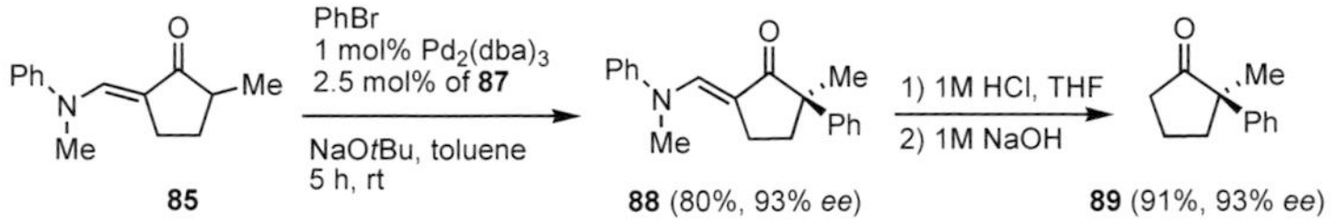

Scheme 7.15 Removal of the *N*-methyl-anilinomethylene blocking group.

to **97** began with the screening of numerous commercially available mono- and bisphosphines. These reactions, however, provided only modest enantiomeric excess and were often slow despite catalyst loadings of up to 10 mol%. Consequently, a new ligand system was pursued. Several heterocyclic carbene ligands were investigated, from which **98** and **99** emerged as promising leads. In particular, the complementary orientations of the amine and α-methyl groups offered the potential for the pair of ligands to provide opposite absolute stereochemistry. Indeed, treatment of **96** with Pd(dba)$_2$ and ligands **98** and **99** afforded **97** in good yield and moderate *ee*, with an opposite preference for absolute stereochemistry being displayed by the two ligands (Scheme 7.16).

Table 7.8 Palladium-catalyzed asymmetric vinylation of cyclopentanone derivative **85**.

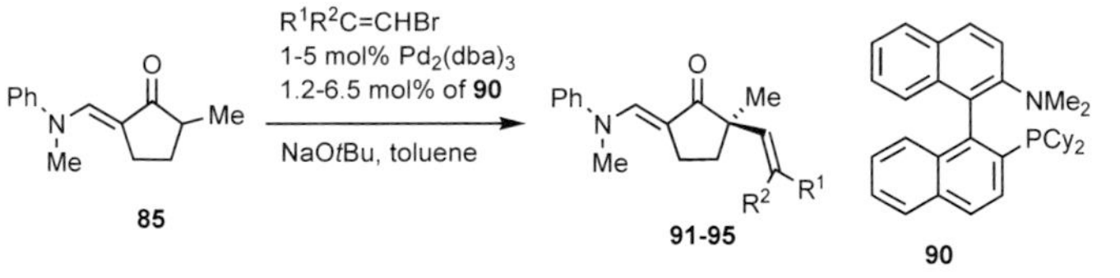

Entry	R^1	R^2	Product	Yield (%)	ee (%)
1	Me	H	91	95	90
2	H	H	92	94	92
3	Ph	H	93	92	89
4	Me	Me	94	95	71
5	H	Me	95	84	76

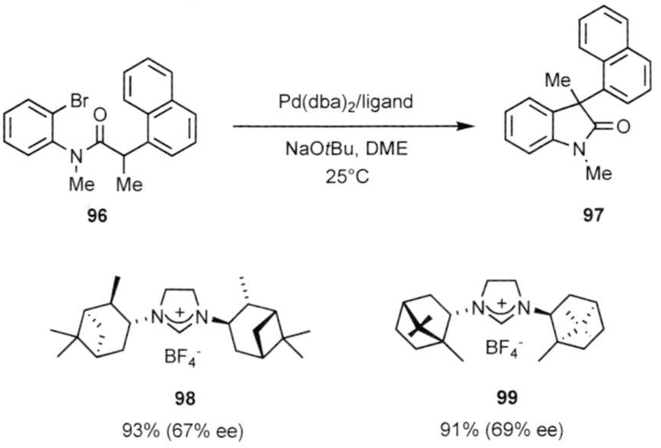

Scheme 7.16 New heterocyclic carbene ligands for an asymmetric intramolecular α-arylation reaction.

7.2.4
Palladium-catalyzed Rearrangements

An alternative route to α-disubstituted carbonyls was reported by Yoshida and Ihara in 2002 wherein allenylcyclobutanols **100**, **103–105** were readily transformed via a palladium-catalyzed arylation/ring expansion reaction sequence [36]. The results, summarized in Tables 7.9 and 7.10, show the reaction to be tolerant of both electron-withdrawing (entries 5, 6) and electron-donating groups (entries 2–4) on the aryl moiety. The use of aryl bromides, however, was found to be detrimental, leading to slow, low-yielding reactions (entries 7, 8). While the reaction with α-bromostyrene (entry 9) suffered from a similarly low yield (26%), it is nonetheless interesting to note the formation of the corresponding diene **101**. The authors also discovered that elevated temperatures enhanced the

Table 7.9 Palladium-catalyzed rearrangement of allenylcyclobutanol **100**.

Entry	RX	Time (h) (%)	101:102 ratio	Yield
1		3	**101** only	80
2	MeO—	1.5	**101** only	72
3	Me—	1	94:6	89
4		3	84:16	79
5	O₂N—	1.5	**101** only	66
6		4	**101** only	77
7	MeO——Br	30	**101** only	37(52)[a]
8		24	84:16	26
9		30	**101** only	26

a based on recovered starting material

diastereo-selectivity of the process. In fact, poor selectivities were observed at low temperatures. Assuming that the ratio of products is a reflection of the two transition states **110** and **113**, Yoshida and Ihara propose that at high temperatures the equilibrium shifts towards the more stable transition state **110** thereby favoring the formation of **111** over **114** (Scheme 7.17).

7.2.5
Desymmetrizing Suzuki Couplings of *meso*-Substrates

The use of the Suzuki cross-coupling reaction for the stereoselective formation of new quaternary carbons is complicated by the fact that no new stereogenic

Table 7.10 Palladium-catalyzed rearrangement of allenylcyclobutanols **103–105**.

Entry	Allenylcyclobutanol	Product	Yield (%)
1	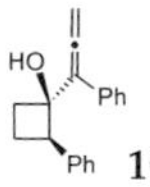 **103**	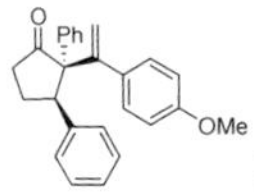 **106**	20
2	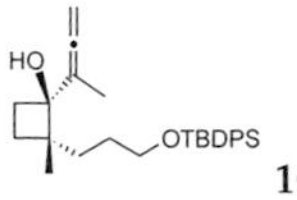 **104**	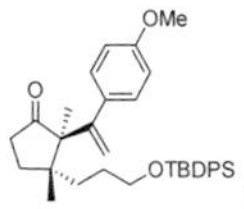 **107**	81
3	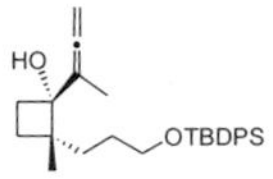 **105**	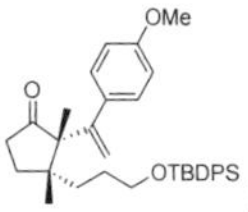 **108**	91

In all cases 5 mol% each of $Pd(PPh_3)_4$ and Ag_2CO_3 in toluene at 80°C.

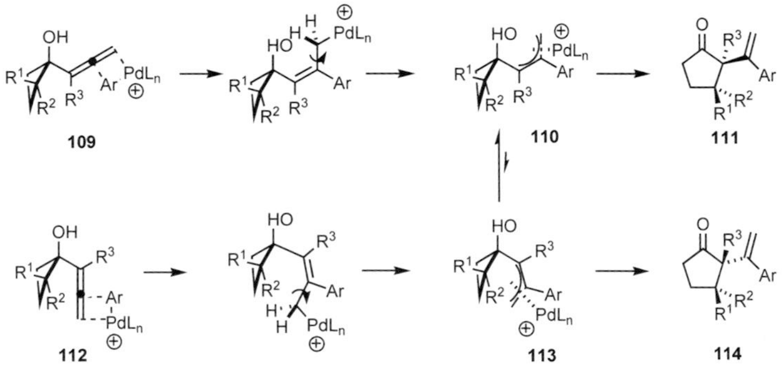

Scheme 7.17 Proposed mechanism for the palladium-catalyzed rearrangement of allenylcyclobutanols.

sp³ centers are typically produced in the reaction. An interesting solution to this problem has been reported involving a desymmetrization strategy wherein products contained newly stereodefined sp³ carbon centers [37].

meso-Ditriflates such as **115** are readily available from the corresponding cyclic 1,3-diketone. Treatment of such achiral substrates with a chiral palladium catalyst should yield a newly desymmetrized vinyl–palladium species, **116**, via a selective oxidative insertion reaction, which could be subsequently coupled with an aryl boronic acid (Scheme 7.18).

Following some initial optimization of the reaction conditions, a catalytic system derived from $Pd(OAc)_2$ and ligand **117** was found to induce the desired coupling of **115** with a variety of aryl boronic acids to give products **118–126** in moderate to good yields and good enantiomeric excesses (Table 7.11). Positional isomers of the acetylbenzene and formylbenzene boronic acids showed similar

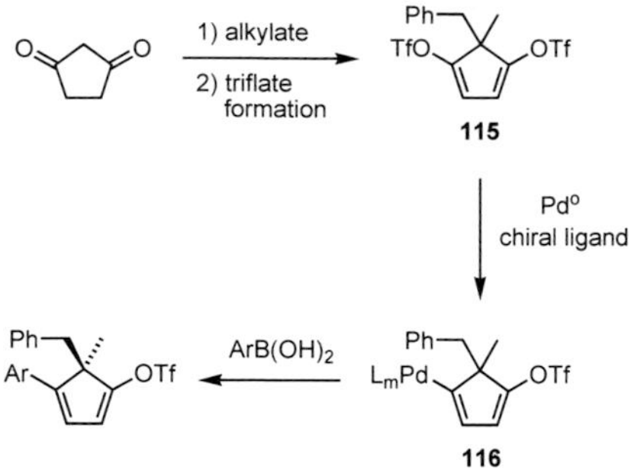

Scheme 7.18 Desymmetrization-based enantioselective Suzuki coupling.

reactivity and enantiomeric excess (entries 1–6), with the exception of the *ortho*-substituted acetyl substrate, which presumably suffered from steric factors (entry 3). The reaction also proceeded well in the presence of an unprotected hydroxyl group (entry 7) and was similarly tolerant of heterocyclic boronic acids (entries 8, 9).

In addition to illustrating a unique strategy for the preparation of compounds containing newly defined quaternary carbon centers, the products of this reaction possess an additional vinyl triflate unit allowing for further substrate manipulation. As an example, product **119** was subjected to a second Suzuki coupling to yield bisaryl **127** in 93% yield (Scheme 7.19).

Table 7.11 Enantioselective Suzuki coupling of a *meso*-ditriflate.

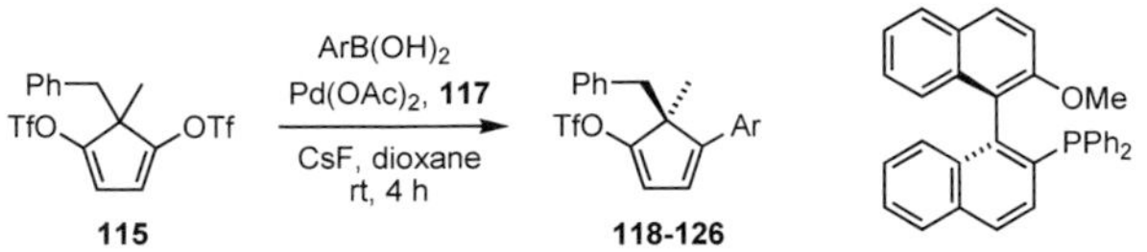

115 118-126 117

Entry	Ar-B(OH)$_2$	Product	Yield (%)	*ee* (%)
1	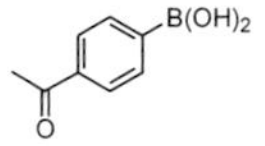	**118**	46	77
2	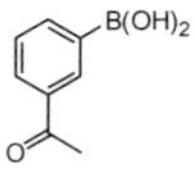	**119**	51	86
3	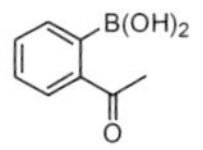	**120**	<5	–
4	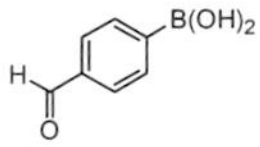	**121**	47	73
5	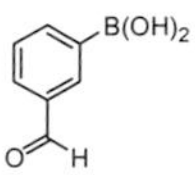	**122**	57	80
6	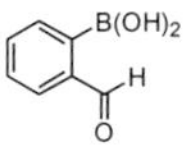	123	41	82
7	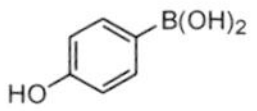	**124**	43	74
8	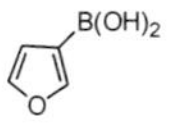	**125**	53	72
9[a]	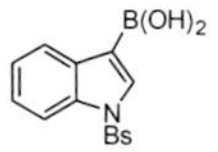	**126**	66	85

a Bs = benzenesulfonyl.

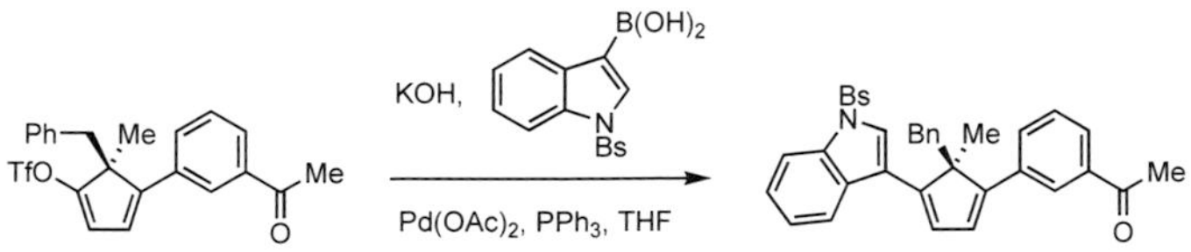

119 93% 127

Scheme 7.19 Functionalization of the monotriflate via a second Suzuki reaction.

7.3
Summary

As illustrated by the preceding examples, asymmetric cross-coupling reactions have emerged as powerful synthetic tools for the stereoselective formation of quaternary carbon centers. In particular, the Mizoroki–Heck reaction has received detailed attention from the synthetic community. Following extensive research on the effects of additives and ligand choice, the asymmetric variant of this reaction has developed into a reliable means of constructing even the most challenging of quaternary centers. Consequently, its use in natural-product synthesis has become widespread, making it a valuable addition to the existing body of organic reactions.

Complementary to the Mizoroki–Heck reaction are several additional cross-coupling reactions, including the α-arylation of carbonyl compounds, the rearrangement of allenylcyclobutanols and a unique desymmetrizing Suzuki reaction, which together form a valuable array of methods for the stereoselective formation of quaternary carbon centers.

References

1 T. Mizoroki, K. Mori, A. Ozaki, *Bull. Chem. Soc. Jpn.* **1971**, *44*, 581.

2 R. F. Heck, J. P. Nolley, Jr., *J. Org. Chem.* **1972**, *37*, 2320.

3 For review, see: (a) A. de Meijere, F. E. Meyer, *Angew. Chem. Int. Ed. Engl.* **1994**, *33*, 2379. (b) S. Bräse, A. de Meijere, *Metal-Catalyzed Cross Coupling Reactions*, Wiley-VCH, Weinheim, **1998**, Ch. 3. (c) J. T. Link, L. E. Overman, *Metal-Catalyzed Cross Coupling Reactions*, Wiley-VCH, Weinheim, **1998**, Ch. 6. (d) J. T. Link, *Organic Reactions*, Vol. 60, Wiley, Hoboken, NJ, **2002**, Ch. 2.

4 Y. Sato, M. Sodeoka, M. Shibasaki, *J. Org. Chem.* **1989**, *54*, 4753.

5 N. E. Carpenter, D. J. Kucera, L. E. Overman, *J. Org. Chem.* **1989**, *54*, 5846.

6 For reviews on the synthesis of quaternary carbon centers, see: (a) C. J. Douglas, L. E. Overman, *Proc. Natl. Acad. Sci. U.S.A.* **2004**, *101*, 5363. (b) L. Barriault, I. Denissova, *Tetrahedron* **2003**, *59*, 10105. (c) J. Christoffers, A. Baro, *Angew. Chem. Int. Ed.* **2003**, *42*, 1688. (d) J. Christoffers, A. Mann, *Angew. Chem. Int. Ed.* **2001**, *40*, 4591. (e) E. J. Corey, A. Guzman-Perez, *Angew. Chem. Int. Ed.* **1998**, *110*, 402. (f) K. Fuji, *Chem. Rev.* **1993**, *93*, 2037. (g) S. F. Martin, *Tetrahedron* **1980**, *36*, 419.

7 (a) I. P. Beletskaya, A. V. Cheprakov, *Chem. Rev.* **2000**, *100*, 3009. (b) C. Amatore, A. Jutand, *J. Organomet. Chem.* **1999**, *576*, 254.

8 W. Cabri, I. Cadiani, S. DeBernardis, F. Francalanci, S. Penco, *J. Org. Chem.* **1991**, *56*, 5796.

9 F. Ozawa, A. Kubo, T. Hayashi, *J. Am. Chem. Soc.* **1991**, *113*, 1417.

10 A. Ashimori, L. E. Overman, *J. Org. Chem.* **1992**, *57*, 4571.

11 R. J. Cross, *Adv. Inorg. Chem.* **1989**, *34*, 219.

12 D. L. Thorn, R. Hoffmann, *J. Am. Chem. Soc.* **1978**, *100*, 2079.

13 E. G. Samsel, J. R. Norton, *J. Am. Chem. Soc.* **1984**, *106*, 5505.

14 For excellent reviews, see: (a) A. B. Dounay, L. E. Overman, *Chem. Rev.* **2003**, *103*, 2945. (b) Y. Donde, L. E. Overman, *Catalytic Asymmetric Synthesis*, Wiley-VCH, New York, **2000**, Ch. 8G. (c) M. Shibasaki, E. M. Vogl, *J. Organomet. Chem.* **1999**, *576*, 1. (d) M. Shibasaki, C. Boden, A. Kojima, *Tetrahedron* **1997**, *53*, 7371.

15 S. P. Maddaford, N. G. Andersen, W. A. Cristofoli, B. A. Keay, *J. Am. Chem. Soc.* **1996**, *118*, 10766.

16 F. Miyazaki, K. Uotsu, M. Shibasaki, *Tetrahedron* **1998**, *54*, 13073.

17 (a) A. Ashimori, T. Matsuura, L. E. Overman, D. J. Poon, *J. Org. Chem.* **1993**, *58*, 6949. (b) L. E. Overman, D. J. Poon, *Angew. Chem. Int. Ed. Engl.* **1997**, *36*, 518. (c) A. Ashimori, B. Bachand, L. E. Overman, D. J. Poon, *J. Am. Chem. Soc.* **1998**, *120*, 6477. (d) A. Ashimori, B. Bachand, M. A. Calter, S. P. Govek, L. E. Overman, D. J. Poon, *J. Am. Chem. Soc.* **1998**, *120*, 6488. (e) T. Matsuura, L. E. Overman, D. J. Poon, *J. Am. Chem. Soc.* **1998**, *120*, 6500. (f) L. E. Overman, M. D. Rosen, *Angew. Chem. Int. Ed.* **2000**, *39*, 4596. (g) A. D. Dounay, K. Hatanaka, J. J. Kodanko, M. Oestreich, L. E. Overman, L. A. Pfeifer, M. M. Weiss, *J. Am. Chem. Soc.* **2003**, *125*, 6261.

18 S. P. Govek, L. E. Overman, *J. Am. Chem. Soc.* **2001**, *123*, 9468.

19 Claisen rearrangement is often used, see Chapter 5, Section 5.2.1.

20 L. E. Overman, D. V. Paone, B. A. Stearns, *J. Am. Chem. Soc.* **1999**, *121*, 7702.

21 J. T. Link, L. E. Overman, *J. Am. Chem. Soc.* **1996**, *118*, 8166.

22 A. D. Lebsack, J. T. Link, L. E. Overman, B. A. Stearns, *J. Am. Chem. Soc.* **2002**, *124*, 9008.

23 M. Mori, M. Nakanishi, D. Kajishima, Y. Sato, *J. Am. Chem. Soc.* **2003**, *125*, 9801.

24 B. M. Trost, O. R. Thiel, H.-C. Tsui, *J. Am. Chem. Soc.* **2003**, *125*, 13255.

25 G. Hirai, Y. Koizumi, S. M. Moharram, H. Oguri, M. Hirama, *Org. Lett.* **2002**, *4*, 1627.

26 (a) E. Gorobets, G.-R. Sun, B. M. M. Wheatley, M. Parvez, B. A. Keay, *Tetrahedron Lett.* **2004**, *45*, 3597. (b) N. G. Andersen, M. Parvez, R. McDonald, B. A. Keay, *Can. J. Chem.* **2004**, *82*, 145. (c) R. Kuwano, M. Sawamura, J. Shirai, M. Takahashi, Y. Ito, *Bull. Chem. Soc. Jpn.* **2000**, *73*, 485. (d) H. Doucet, E. Fernandez, T. P. Layzell, J. M. Brown, *Chem. Eur. J.* **1999**, *5*, 1320. (e) B. C. Hamann, J. F. Hartwig, *J. Am. Chem. Soc.* **1998**, *120*, 3694. (f) Y. Crameri, J. Foricher, M. Scalone, R. Schmid, *Tetrahedron: Asymmetry*, **1997**, *8*, 3617. (g) A. Kojima, C. Boden, M. Shibasaki, *Tetrahedron Lett.* **1997**, *38*, 3459. (h) M. Cereghetti, W. Arnold, E. A. Broger, A. Rageot, *Tetrahedron Lett.* **1996**, *37*, 5347. (h) J.-P. Tranchier, V. Ratovelomanana-Vidal, J.-P. Genêt, S. Tong, T. Cohen, *Tetrahedron Lett.* **1997**,

38, 2951. (i) R. Schmid, E. A. Broger, M. Cereghetti, Y. Crameri, J. Foricher, M. Lalonde, R. K. Müller, M. Scalone, G. Schoettel, U. Zutter, *Pure Appl. Chem.* **1996**, *68*, 131. (j) T. Benincori, E. Brenna, F. Sannicolò, L. Trimarco, P. Antognazza, E. Cesarotti, F. Demartin, T. Pilati, *J. Org. Chem.* **1996**, *61*, 6244.

27 C. A. Busacca, D. Grossbach, R. C. So, E. M. O'Brien, E. M. Spinelli, *Org. Lett.* **2003**, *5*, 595.

28 P. Nilsson, M. Larhed, A. Hallberg, *J. Am. Chem. Soc.* **2003**, *125*, 3430.

29 J. Ahman, J. P. Wolfe, M. V. Troutman, M. Palucki, S. L. Buchwald, *J. Am. Chem. Soc.* **1998**, *120*, 1918. For reviews on catalyzed asymmetric arylation reactions see: (a) C. Bolm, J. P. Hildebrand, K. Muniz, N. Hermanns, *Angew. Chem. Int. Ed.* **2001**, *40*, 3284. (b) D. Culkin, J. F. Hartwig, *Acc. Chem. Res.* **2003**, *36*, 234.

30 D. J. Spielvogel, S. L. Buchwald, *J. Am. Chem. Soc.* **2002**, *124*, 3500.

31 For the racemic version, see: M. Palucki, S. L. Buchwald, *J. Am. Chem. Soc.* **1997**, *119*, 11108.

32 (a) T. Hamada, A. Chieffi, J. Ahman, S. L. Buchwald, *J. Am. Chem. Soc.* **2002**, *124*, 1261. (b) T. Hamada, S. L. Buchwald, *Org. Lett.* **2002**, *4*, 999.

33 A. Chieffi, K. Kamikawa, J. Åhman, J. M. Fox, S. L. Buchwald, *Org. Lett.* **2001**, *3*, 1897.

34 The authors have suggested that this ligand acts as a monodentate ligand coordinating through the phosphorus atom alone. Their claim is supported by several findings, including the observation that similar ligands bearing a non-coordinating alkyl group instead of the dimethylamino moiety display only slightly lower enantioselectivities. By contrast, bidentate ligands show significantly diminished enantioselectivities.

35 S. Lee, J. F. Hartwig, *J. Org. Chem.* **2001**, *66*, 3402.

36 M. Yoshida, K. Sugimoto, M. Ihara, *Tetrahedron* **2002**, *58*, 7839.

37 M. C. Willis, L. H. W. Powell, C. K. Claverie, S. J. Watson, *Angew. Chem. Int. Ed.* **2004**, *43*, 1249.

8
Alkylation of Ketones and Imines

Diego J. Ramón, and Miguel Yus

8.1
Introduction

After the disastrous incident of thalidomide, whose two enantiomers have totally different biological effects in humans [1], there is increasing public demand for the synthesis of chiral compounds. The responsibility of synthetic chemists is to prevent a repetition of this tragedy by providing highly efficient and reliable methods of asymmetric synthesis [2].

All chiral compounds have at least one stereogenic element, which is usually a stereogenic center. Nowadays, the generation of tertiary carbon-atom stereocenters (carbon atoms bearing one hydrogen atom) can be easily achieved in most cases by using the appropriate chiral auxiliary, reagent or catalyst from the vast number of synthetic approaches. However, the related approach to complex compounds bearing quaternary stereocenters is still a challenge for synthetic organic chemistry, and every asymmetric procedure for the construction of a fully substituted carbon center is of great value [3]. In fact, the simple extension of methods from the preparation of tertiary stereocenters to the preparation of quaternary ones is not straightforward; in some cases, the interval between the two could be more than ten years, and the development may involve the use of new and different approaches.

One of the most frequently occurring classes of compound in nature with a heterosubstituted quaternary center [4] is the corresponding tertiary alcohols and their related amine derivatives. Among the different approaches to the synthesis of this type of compound, such as kinetic resolution, asymmetric desymmetrization, oxidation processes, electrophilic alkylation and nucleophilic addition, those that involve stereoselective C–C bond formation are of particular interest, since one C–C bond and one stereoelement are created in a single synthetic step. In the case that engages our attention, the simplest conceptual approach for the preparation of chiral tertiary alcohols, and the relate amines, is the asymmetric 1,2-addition of organometallic reagents [5] to ketones, or to the corresponding imines. However, this approach has a central problem: the structural similarity between the two substituents around the electrophilic carbonyl group and therefore the

Quaternary Stereocenters: Challenges and Solutions for Organic Synthesis. Edited by Jens Christoffers, Angelika Baro
Copyright © 2005 WILEY-VCH Verlag GmbH & Co. KGaA, Weinheim
ISBN: 3-527-31107-6

stereodifferentiation step. The presence of two alkyl/aryl substituents on ketones makes the electrophilic character of the carbonyl group very low, compared to that of aldehydes, which narrows the range of possible nucleophiles able to react with them. Moreover, the difference between the hydrogen substituent and the carbon substituent in the case of aldehydes is always higher than that between the substituents in ketones. All these factors make the stereoselective addition of organometallics to ketones more difficult than the related reaction with aldehydes.

In this chapter, we describe briefly the different possibilities as well as the scope of these approaches. Other related topics, such as electrophilic α-alkylation of ketone derivatives, aldol, Claisen and radical processes covered in this book, are beyond the scope of this chapter, as are the asymmetric synthesis of cyanohydrins [6] and Friedel–Crafts processes [7]. The chapter is presented in three main sections according to the type of stereoselective addition – that is between diastereoselective or enantioselective approaches – and the final section is in turn divided between modulation and promotion processes. Stereoselective differentiation is due to the presence of chiral solvents, ligands or other compounds that only take part in the pertinent transition state, while enantioselective addition is defined as the transformation of two or more achiral reagents into one of the two possible enantiomers of a product [8]. In the diastereoselective approach at least one reagent is chiral.

8.2
Diastereoselective Additions

The diastereoselective approach for the synthesis of compounds with quaternary stereocenters implies the use of either a chiral nucleophile or a chiral ketone derivative. Both approaches have been tested with notable differences in the results.

8.2.1
Chiral Nucleophiles

The approach using chiral organometallics has been less studied than the alternative, as it generally yields inferior diastereoselectivity compared with that achieved using chiral ketones. That is the case of chiral organolithium intermediates **1** [9], **2** [10] and **3** [11] which, in their reaction with ketones, gave the expected tertiary alcohols as a nearly 1:1 mixture of the two possible diastereoisomers.

Structures 1-3

Despite the aforementioned tendency, there is one interesting result that must be pointed out. That is the case of the addition of the chiral stannane **4** (95% *ee*) to different ketones **5** (Scheme 8.1). The product **6** obtained from a methyl ketone derivative (R^2 = Me) was a unique diastereoisomer keeping the *ee* of the starting organometallic [12]. The configuration and *ee* values are the same as those achieved by direct use of a combination of butyllithium and sparteine [13], followed by transmetallation with triisopropoxytitanium chloride, indicating that the exchange of trialkylstannyl by trichlorotitanyl in the intermediate proceeds with *anti*-stereochemistry yielding the corresponding α-functionalized organometallic compound. This reaction has been successfully employed in the synthesis of 8-oxabicyclo[3.2.1]octane-6-carboxaldehyde derivatives [14], changing only the starting ketone to a 1,5-diketone derivative. Thus, after the diastereoselective homoaldol reaction, the resulting hydroxyketone of type **6** undergoes a final cycloaddition process catalyzed by a boron trifluoride–diethyl ether complex, yielding the corresponding tricyclic compound.

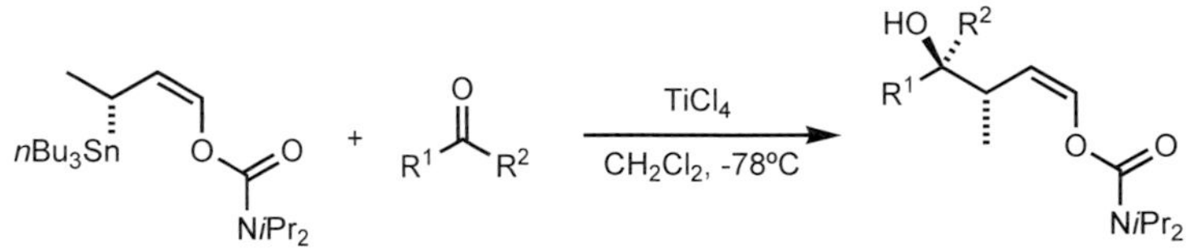

Scheme 8.1 Diastereoselective addition of chiral stannanes to ketones.

8.2.2
Chiral Electrophiles

The pioneer work of Tiffeneau [15] raised the possibility of using chiral ketones from natural products as starting materials in the asymmetric preparation of chiral tertiary alcohols. The work was performed using a camphonelic amide derivative. Its reaction with ethylmagnesium bromide, to obtain the corresponding ethyl ketone, followed by addition of phenylmagnesium bromide, gave a mixture of two alcohols in different proportions. When the reaction was performed changing the order of Grignard addition, the same products were isolated but in different ratios. Although the nature of these results was not clearly established at the time, this was the starting point for using chiral ketones from natural products as chiral partners in the asymmetric preparation of systems with tertiary alcohol moieties [16]. More recently, different simple chiral ketones, such as camphor, fenchone and menthone, have been used as starting materials for the preparation of chiral ligands bearing, among other functionalities, a stereogenic oxido-functionalized quaternary carbon atom [17]. A related process using more elaborate ketones such as camphorsulfonamide **7** is outlined in Scheme 8.2 [18]. The reaction gave only the diastereoisomer **9** which came from an *endo*-attack of nucleophile.

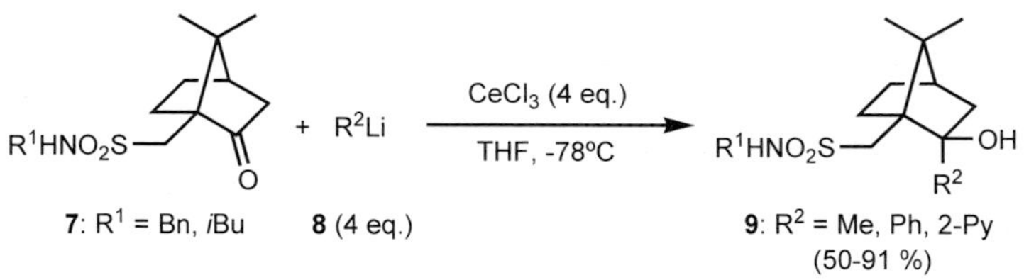

Scheme 8.2 Diastereoselective addition of organolithiums to chiral camphorsulfonamides.

Similar results, as far as the diastereoselectivity is concerned, were obtained using chiral estrone and cholesterone [19] and different chiral and achiral functionalized organolithium compounds [20]. This procedure permitted the preparation of a new family of steroids, which, in some cases, combined a sugar and a steroid moiety.

Also, different chiral ketones derived from carbohydrates have been used as starting materials in the preparation of chiral alcohols. Thus, for example, the glucose derivative **10** reacted with different fluoroalkyl nucleophiles to yield the expected tertiary alcohols [21] with good both chemical yields and diastereoselectivities, depending on the nature of the nucleophile: silyl derivatives yielded only one diastereoisomer (addition to the β-face of the carbonyl moiety) while magnesium derivatives gave diastereomeric excesses greater than 45%. When the same reaction was performed with functionalized organolithium intermediates of type **11**, which were prepared from the corresponding heterocycles phthalan or isochroman by a reductive ring-opening process [22] using lithium powder and a substoichiometric amount of an arene [23], the only isolated diol **12** came from a β-face attack [24] (Scheme 8.3). Similar results were obtained using a chiral ketone derived from fructose.

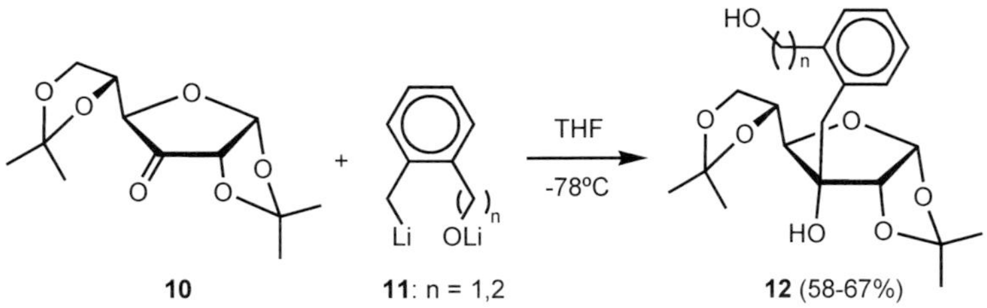

Scheme 8.3 Diastereoselective addition of functionalized organolithiums to a chiral carbohydrate derivative.

An interesting study has been performed with erythrulose **13** using different protecting groups. This study serves as an example of the difficulty of rationalizing the obtained results and illustrates the problems that have to be overcome. The obtained diastereoselectivity, and the absolute configuration, depend strongly on the protecting groups (R[1], R[2] and R[3] in Scheme 8.4). Thus, for highly chelating systems such as system **13a**, the diastereoselectivity is very high (nearly 99.5%)

yielding the compound **15** [25]. However, when the reaction is performed using a 3,4-acetonide derivative **13b**, the diastereoselectivity dropped considerably, the main diastereoisomer being the epimeric one. These results were rationalized assuming a Cram's cyclic model involving α-chelation for the first case and a Felkin–Anh model for the second case. This hypothesis has been confirmed by a theoretical study using semiempirical PM3 procedures and *ab initio* methods on the HF/3.21G basis set [26]. Theoretical calculations for magnesium derivatives showed that the first step in the addition process was the formation of a chelate complex, and that the inclusion of a second equivalent of the Grignard reagent decreases the energetic barrier for the nucleophilic addition step. All these facts were consistent with the experimental data. However, changing the metal (Li, Mg, Zn, Al and Ti) in the nucleophile had a random effect. When the reaction was performed using a 1,3-dioxane structure **13c**, the results were highly dependent on the nature of the metallic atoms, the main diastereoisomer being attributed to axial attack on the dioxanone derivative [27].

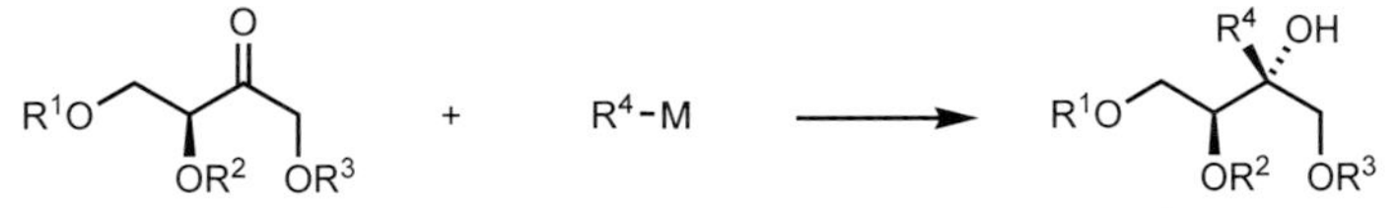

13a: $R^1 = R^2 = Bn$,
  $R^3 = H, SiPh_2tBu, Tr$
13b: $R^1, R^2 = Me_2C$,
  $R^3 = H, SiPh_2tBu, Bn, Tr$
13c: $R^1, R^2 = MeHC$,
  $R^3 = SiPh_2tBu, Bn, Tr$

14: M = Li, Mg, Cu,
  Ti, Zr, In, Ce, Yb

15: R^4 = alkyl, allyl,
  vinyl, 1-alkenyl

Scheme 8.4 Diastereoselective addition of organometallics to (*S*)-erythrulose derivatives.

The course of this addition is controlled not only by the nature of the protecting groups on the electrophile **13** and of the metal atom of nucleophile **14** but also by the actual nature of the nucleophile. This is the case for allylic nucleophiles, in which the reaction in aqueous media using organometallics derived from indium gave the expected compound **15** ($R^4 = CH_2=CHCH_2$) [28] while the reaction using allyl magnesium derivatives [29] gave the opposite epimer, which is in its turn the opposite one to that obtained using the related propylmagnesium halide, which goes to show the extreme difficulty of rationalizing all the results. Nevertheless, the allylation reaction has been used as the asymmetric key step in the synthesis of different natural lactones such as malyngolide and tanikolide [30]. This strategy, but using a xylulose derivative instead of an erythrulose derivative **13**, has been successfully used in the synthesis of secosyrins and syributins [31].

The diastereoselective addition of different organometallic reagents to other chiral ketose derivatives to give the corresponding chiral tertiary alcohol has been successfully used in the synthesis of highly substituted tetrahydrofurans [32] and 1-(1*H*-imidazol-4-yl)-1-(6-methoxy-2-naphthyl)-2-methylpropan-1-ol, which is a novel inhibitor of $C_{17,20}$-lyase, a key enzyme involved in androgen biosynthesis [33].

The utility of the tin moiety as a chelating group has been demonstrated in another example of diastereoselective addition of organolithium compounds to chiral ketones derived from natural products bearing a vinylstannane moiety, such as the carvone system **16** [34] (Scheme 8.5). *Ab initio* analysis of compound **16** predicts equatorial positions for both the methyl and the isopropenyl groups and an axial position for the stannylvinyl group, corroborated by coupling constants and nuclear Overhauser effect (NOE) data, which should inhibit axial and favor equatorial attack. However, only one diastereoisomer is obtained and this diastereoisomer comes from an axial attack. The role of the tin atom seems to be as a chelating director group for the organolithium intermediate favoring this topological pathway, excluding the possible formation of the corresponding hypervalent tin anion owing to the exchange of scrambling reactions when the alkyl groups of the organolithium and tin derivatives are different. Products related to system **17** (R = allyl) have been used as starting materials in an allyl-transfer reaction to aldehydes, yielding the corresponding homoallylic alcohols with total enantioselectivities [35].

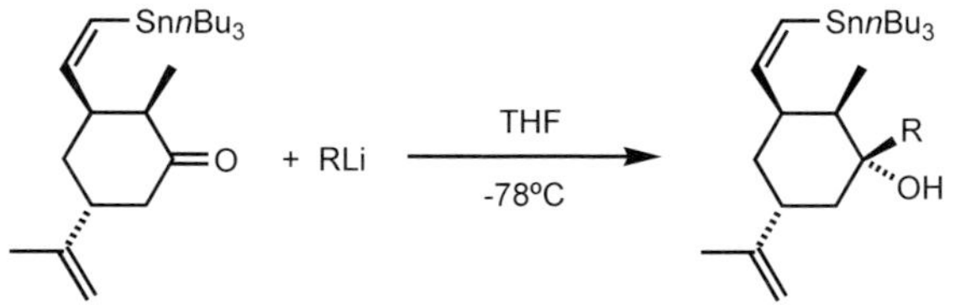

Scheme 8.5 Diastereoselective addition of organolithium reagents to a chiral vinylstannane ketone.

Rationalization of the topological pathway of the reactions in the above examples is in some cases very difficult owing to the presence of several stereogenic centers as well as extra coordinating atoms in the chiral ketone. However, for simple acyclic ketones with only one stereogenic center, the possible topological control and the main diastereoisomer was first described and predicted in the early 1950s, thanks to the work of Cram's group [36]. Although their reaction model has been further refined and improved by the work of groups led by Felkin, Anh and others [37], some aspects are still insufficiently clear. The origin of the diastereoselectivity on the nucleophilic addition of organometallics to different alkyl 1-phenylethyl ketones was not elucidated until 2004. *Ab initio* MO calculations on the conformational Gibbs energy showed two main rotamers whereby the alkyl moiety is synclinal to the phenyl group and the carbonyl group is nearly eclipsed. The more stable rotamer showed a stabilization effect due to a CH/phenyl hydrogen bond in a five-membered ring and the diastereoisomeric ratio was estimated on the basis of the ground-state rotamer distribution [38].

The presence of an extra chelating atom or functionality on the stereogenic center has a strong influence on the prediction of the main diastereomer of the addition. However, a modification of the possible main conformer overcomes this inconvenience. The modification is known as the "Cram chelated model" and it takes into account that in the main conformer a five-membered ring is formed by chelation from the oxygen of the carbonyl group and from the aforementioned extra functionality with the metal [39]. This strategy has been successfully used for the synthesis of the (C-1)–(C-11) subunit of 8-*epi*-fostriecin [40]. The addition of different alkenylmagnesium derivatives to the α-functionalized ketone 18 gave the expected alcohols 20 with diastereoselectivities higher than 95% (Scheme 8.6).

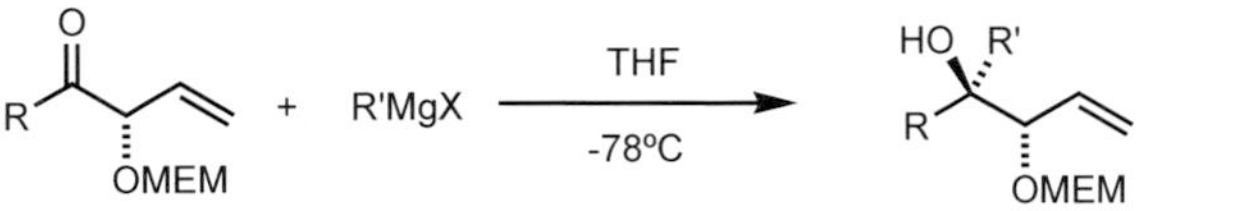

18: R = Me, Ph   **19**   **20**: R' = alkyl, vinyl, 1-alkynyl
                                                              (75-92%)

Scheme 8.6 Diastereoselective addition of Grignard reagents to a chiral α-functionalized ketone.

The extra chelating group can be not only an ether moiety but also an amino [41] group, the diastereomeric excess being strongly dependent on the nature of substituents at the nitrogen atom as well as on the nature of the organometallic reagents. This model has been extended to chiral β-hydroxyketones, the main conformer having a six-membered ring complex. However, the obtained diastereoselectivity was far less satisfactory than for the previous α-functionalized ketones, being always lower than 70% [42]. However, the diastereomeric addition of organoaluminum and organotin derivatives to chiral ketone complexes 21 yielded the expected alcohols 22 with diastereoselectivities higher than 95% (Scheme 8.7). This high level of diastereoselectivity was attributed to the bulkiness of the iron complex, which is able to bias the nucleophilic attack to the opposite side, even though the stereogenic center is far from the reactive center [43].

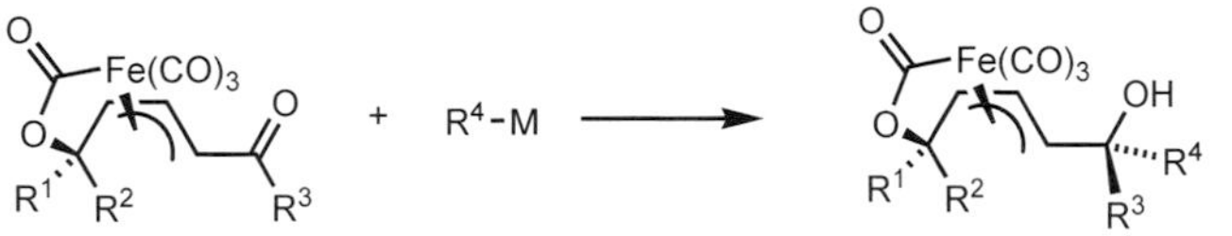

21: R^1, R^2 = H, alkyl   **14**: M = Al, Sn   **22**: R^4 = alkyl, 1-alkenyl, 1-alkynyl
    R^3 = Me, Ph   (46-98%)

Scheme 8.7 Diastereoselective addition of organometallics to chiral ketone complex derivatives.

High levels of diastereoselectivity have been reached using ketones with bicyclic structures not derived from natural products [44], which permitted the formal synthesis of ngaione, a toxic constituent of the stock-poisoning shrub ngaio tree (*Hyporum deserti*).

The results obtained using achiral amides, which crystallize in chiral crystals, merit separate comment (Scheme 8.8). Some achiral molecules can adopt chiral arrangement in the crystal structure without any outside source of chirality, and if the racemization in the solubilization process, usually due to the overcoming of a rotational barrier, is slower than the addition one, then the system is transitorily chiral and it is possible to obtain some stereodiscrimination in its reactions. This is the case for the imide **23** which is achiral in solution but in the solid state can be spontaneously resolved owing to the rotational barrier of the amide group with the bulky substituent. When the chiral crystals of this system, which present chiral imides, are dissolved at low temperature, the rotational barrier is high enough to keep transitorily the same rotamer as in the crystal structure and therefore to perform the addition of butyllithium **8a** to yield the expected tertiary alcohol **24** with an *ee* of up to 83% [45].

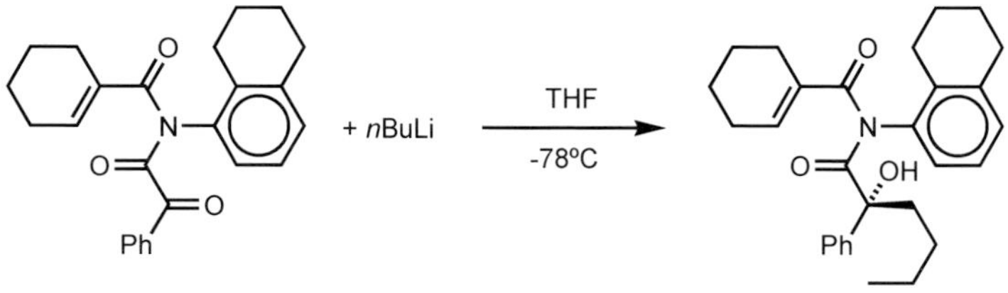

Scheme 8.8 Diastereoselective addition of butyllithium to a transitory chiral ketone.

The aforementioned approaches to the synthesis of chiral molecules containing tertiary alcohol moieties have the inconvenience of using natural products, which limits their general applicability enormously. A more flexible approach resulted from the work of Prelog [46]. The idea is to join a chiral auxiliary to the corresponding achiral ketone, yielding a new chiral ketone that is now able to react with the organometallic to yield diastereoselectively a compound with a quaternary stereocenter. Final removal of the initial chiral auxiliary liberates the corresponding chiral alcohol. This approach has been extensively used and different auxiliaries have been introduced for the purpose [47] showing its great flexibility and possibilities.

The first auxiliaries used were chiral alcohols, which were attached to the achiral ketone by an ester bond. The initial experiments were conducted with different derivatives of menthyl pyruvate [46], and more recently this has been revisited for the preparation of 2-naphthylpropionic acid by the corresponding diastereoselective addition of 2-naphthylmagnesium bromide [48], yielding the expected alcohols with a very low diastereoselectivity. Another chiral alcohol, 2-nitroxycyclohexan-1-ol [49], has also been proposed as a chiral auxiliary for the preparation of the corresponding phenylglyoxalate. Its reaction with different organozinc halides gave the expected alcohols with high diastereoselectivities.

An interesting result is outlined in Scheme 8.9, in which the diastereoselective nucleophilic addition and cleavage of the chiral binaphthol auxiliary take place in a single synthetic step. Nucleophilic addition of different Grignard reagents to ketone **25** in the presence of a Lewis acid, such as magnesium bromide, gave lactone **26** directly, with excellent chemical yield and enantioselectivity [50]. When the reaction is performed with an excess of nucleophile the addition continues, yielding the corresponding bis tertiary diol-type product instead of the lactone-type compound. The high stereoselectivity was attributed to the formation of a rigid *pseudo* macrocyclic magnesium complex composed of the podand oligoether group (R^1) and the keto ester with the magnesium bromide.

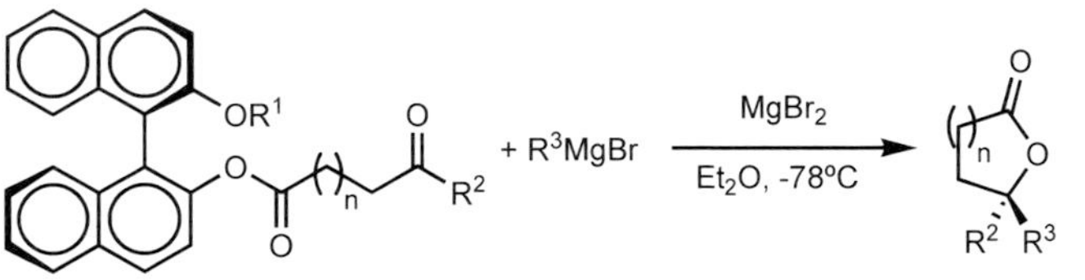

25: $R^1 = (CH_2)_kO(CH_2)_mOMe$, **19** **26**: R^3 = alkyl, aryl
k, m = 0, 2, 3 (<90%, <98% *ee*)
R^2 = Me, Ph
n = 1-6

Scheme 8.9 Diastereoselective addition of Grignard reagents to chiral ketoesters.

The chiral auxiliary can also be bonded to the ketone moiety as an ether functionality, as depicted in Scheme 8.10. The addition of different methyl metal derivatives to the α-ether-functionalized ketone **27** gave the expected tertiary alcohols **29** with moderate to excellent diastereoselectivities [51]. The diastereoselectivity and the relative configuration are a function of the identity of the metal (the best results being obtained with lithium) and the ketone. Different tetrahydropyranyl auxiliaries obtained from glutamic acid or arabinose have permitted the use of other organometallic derivatives in addition to methyl, giving similar results for the diastereoselectivity [52].

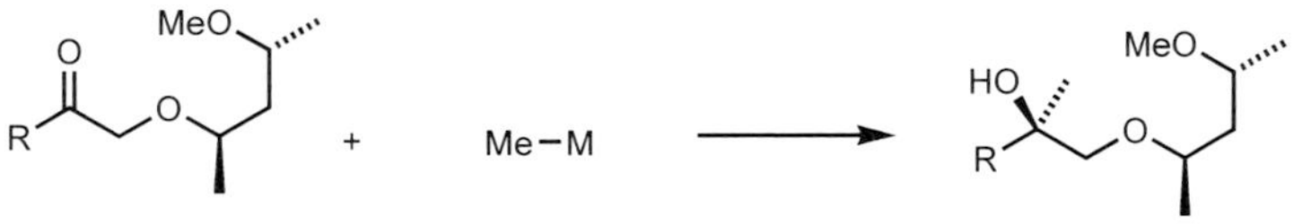

27: R = aryl **28**: M = Li, MgBr, TiCl$_3$, **29**
TiCl$_2$Me, Ti(O*i*Pr)$_3$ (<92%, <92% *de*)

Scheme 8.10 Diastereoselective addition of Grignard reagents to chiral α-alkoxy-functionalized ketones.

Other related systems such as α-amino-functionalized ketones derived from prolinol [53], α-sulfanyl- [54] and α-sulfinyl-functionalized ketones [55] have been proposed as alternatives to ether auxiliary linkages. However, although the diastereoselectivity was in some cases excellent, the removal of the auxiliary presented problems.

Another possibility for chiral auxiliary bonding to the ketone is through a ketal structure, such as that shown in Scheme 8.11, in which phenylglyoxal is transformed into its binaphthol ketal derivative **30**. Its reaction with different organometallics gave the expected tertiary alcohols with excellent diastereoselectivities [56], the best result being obtained when Grignard reagents were used as

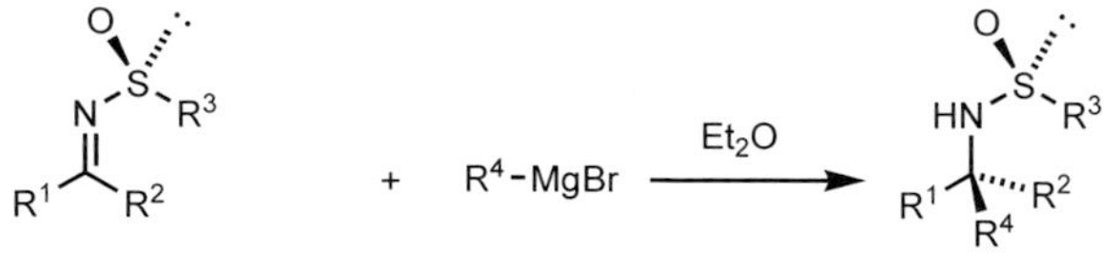

36a: R^1 = Me, R^2 = C(OEt)$_3$,   **19**   **37**: R^4 = Me, allyl
     R^3 = Tol   (<86%, <99% *de*)
36b: R^1-R^2 = (*R*)-(CH$_2$)$_3$CHMe,
     R^3 = *t*Bu

Scheme 8.14 Diastereoselective addition of organometallics to chiral sulfinamides.

Another chiral auxiliary that has been used in this strategy is 1-phenylethylamine [67]. Condensation of this chiral amine with different phenones yielded the expected imines. Their subsequent reaction with different organolithium or organomagnesium compounds gave the corresponding amines with good diastereoselectivities, but not as homogeneously as in the case depicted in Scheme 8.14, the best results being obtained when 2-imidoylphenol derivatives were used. There is another problem with this strategy for obtaining primary amines, namely the final selective cleavage of the auxiliary, which is usually performed by hydrogenolysis.

The final example is outlined in Scheme 8.15, and shows the great difficulty in the preparation of this type of molecule containing quaternary stereocenters. The SAMP-/RAMP-hydrazone technique is one of the most successfully used strategies in the preparation of different kinds of molecules [68], and although different nucleophiles have been 1,2-added to the corresponding aldehyde hydrazone derivative with excellent levels of both diastereoselectivity and chemical yield [69], there is only one example of the related 1,2-addition to a ketone hydrazone derivative [70]. The reaction of allenyllithium derivative **38** with the ketone hydrazone **39** gave, after hydrolysis, pyrroline **40** with excellent diastereoselectivity but in very low chemical yield. The formation of the final product **40** was explained as a result of an intramolecular cyclization of the corresponding lithium hydrazide intermediate after the expected 1,2-addition.

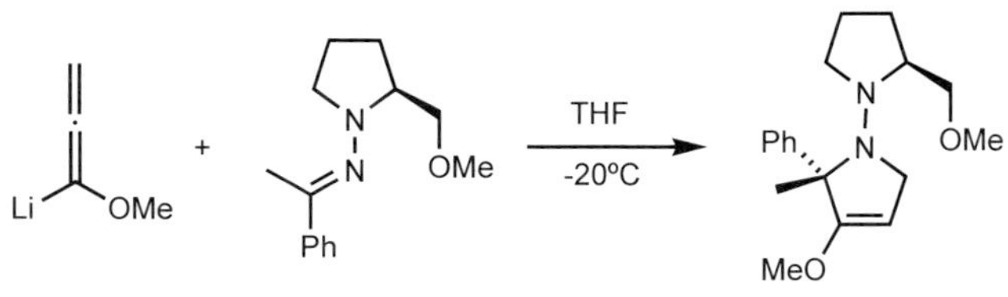

38   **39**   **40** (8%, 99% *de*)

Scheme 8.15 Diastereoselective addition of allenyllithium to the chiral SAMP-hydrazone of acetophenone.

8.3
Enantioselective Additions by Modulated Processes

Nowadays, all chemists recognize that an enantioselective reaction has several advantages compared to the related diastereomeric version. In the first case it is not necessary to bind covalently a chiral auxiliary to the achiral starting material (in the case of this section a ketone or an imine) or to the organometallic derivative, to perform the stereoselective addition, and therefore later to detach the previously attached auxiliary. However, the examples of enantioselective 1,2-addition of organometallics [5] to ketones are very scarce and, in some cases, limited to only one reagent. These examples can be collected in two main classes.

The first class includes those nucleophiles that can react with ketones (or imines) in the absence of any other component, the stereoselective addition being due to modulation of the high reactivity of the nucleophile by the ligand, permitting discrimination between the two heterotopic faces of the carbonyl or imine group. In these cases the use of one equivalent (or more) of the chiral modulator is essential to avoid nonstereoselective direct addition.

The second class includes those organometallics that are not nucleophilic enough to add to the ketone or imine derivative themselves and need the presence of another component to enhance either their own reactivity or the reactivity of the ketone (or both) to perform the addition. In these cases, as direct addition does not occur, the amount of chiral promoter can be reduced to substoichiometric amounts, giving an extra value to these reactions. The specific classification of a given reaction as a modulation or a promotion process is in some cases very controversial.

The old dream of asymmetric modulated addition of an achiral organometallic reagent to a prostereogenic ketone (or imine) started to be realized in 1953. This first example was the addition of Grignard reagents to ketones using a chiral solvent to yield the expected tertiary alcohol with very modest enantioselectivity [71]. However, this work opened up the possibility of using the new concept of asymmetric modulation. Since then, more examples have appeared that improve on the earlier results.

8.3.1
Alkylation Processes

After the first asymmetric modulations of the addition of Grignard reagents by the use of chiral solvents [71], attention turned to the use of some chiral additives as possible modulators of the reaction. The first compounds tested were different alcohols, such as the systems **41** [72] and **42** [73] which, although they produced only moderate enantioselectivities, represented a breakthrough for the new concept.

However, it is only with the use of the TADDol derivative **43** that the results have been extremely good (Scheme 8.16). The best results were obtained using primary organomagnesium derivatives and unfunctionalized aryl methyl ketones

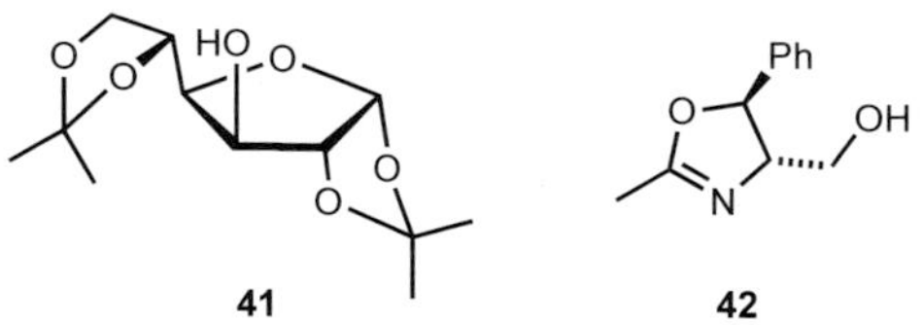
HO
O
Ph
OH
N

41 **42**

Structures 41-42

[74], with other substrate types giving lower enantiomeric ratios. To obtain these good results, the reaction must be performed using an equimolar amount of the chiral TADDol and the ketone at –105°C, which implies that it is a heterogeneous reaction. Mechanistic investigations demonstrated that the enantioselectivity is highly solvent dependent, THF being the medium of choice. In other solvents, even ethereal ones, the enantioselectivity falls and a racemic mixture is produced. The presence of magnesium salts does not influence the reaction. However, the counterion halide in the Grignard reagent is of great importance: bromide gives the best results and chloride the worst. Finally it is worth noting that the correlation between the *ee* of the chiral TADDol ligand **43** and the *ee* of the tertiary alcohol product is a straight line, which implies that only one TADDolate molecule is involved in the rate-determining step. On the contrary, the correlation between the chemical yield and the *ee* of the final alcohol shows that the *in situ* formed tertiary alcoholate deactivates its self-formation.

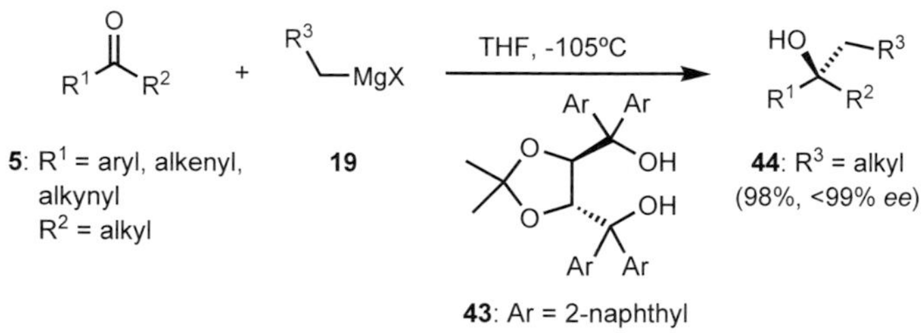
THF, -105°C
MgX
Ar Ar
OH
OH
Ar Ar
HO
R3
R1 R2
5: R1 = aryl, alkenyl, alkynyl
R2 = alkyl
19
43: Ar = 2-naphthyl
44: R3 = alkyl
(98%, <99% ee)

Scheme 8.16 Enantioselective addition of Grignard reagents to ketones modulated by TADDol.

As well as alcohols, diamines such as systems **45** and **46** can also be used as modulators for the organomagnesium intermediates. However, the enantioselectivities found were slightly lower [75].

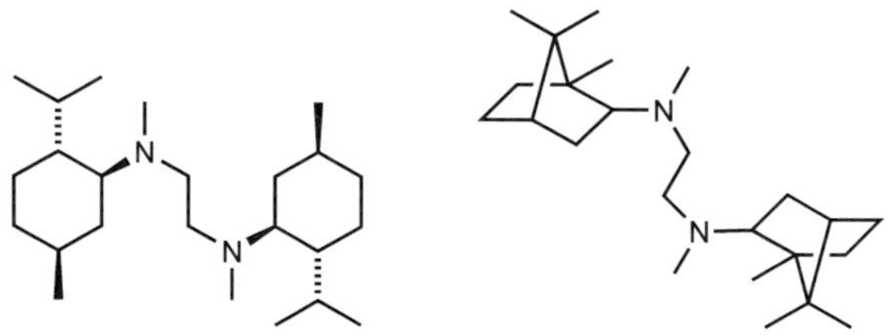
N
N

45 **46**

Structures 45–46

The Kulinkovich reaction [76] is a particular alkylation of carbonyl compounds, whereby esters are transformed into cyclopropanols using alkylmagnesium reagents (2 equivalent) and titanium alkoxides [77]. The mechanism of this process is still unclear but it seems to go through a dialkoxy dialkyltitanium species, which suffers a deprotonation to yield a dialkoxy titanacyclopropane derivative. This species reacts with an ester to yield first a titanium homoenolate and finally the corresponding cyclopropanolate derivative by intramolecular addition to the *in situ* formed ketone. The enantioselective version of this reaction has been performed using the spirotitanate derivative **47** [78]. Thus, the reaction of 2-phenylethylmagnesium bromide **19a** (2 equivalent) with ethyl acetate yielded, after hydrolysis, the expected chiral cyclopropanol derivative **48**, the reaction being completely diastereoselective (Scheme 8.17).

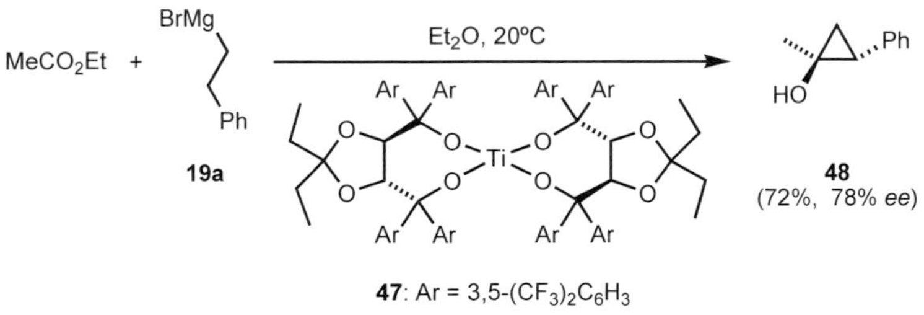

Scheme 8.17 The enantioselective Kulinkovich reaction.

Other organometallic compounds have been used in this type of alkylation. Thus, the reaction of lithium tetrabutylaluminate with acetophenone in the presence of ephedrine derivative **49** gave the expected tertiary alcohol with 56% chemical yield and 31% *ee* [79].

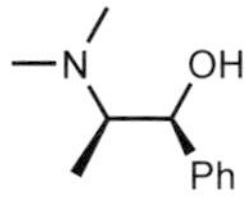

Structure 49

Organolithium intermediates can also be used as starting nucleophiles, the first reported example being the addition of methyllithium **8b** to different phenones (Scheme 8.18). In order to perform the asymmetric reaction, a stoichiometric amount of the C_3-symmetric chiral zirconium derivative **50** must be added [80].

As in previous sections, the number of examples of enantioselective addition of organometallics to ketoimines is rather smaller than the corresponding reaction with ketones. The first example is the reaction of the *in situ* formed metalimine **52** (prepared by addition of ethylmagnesium bromide to the corresponding nitrile) with methylmagnesium bromide **19b** in the presence of a stoichiometric amount of the titanium derivative **53** (Scheme 8.19). In this way, the corresponding primary amines **54** were obtained with good *ee* [81].

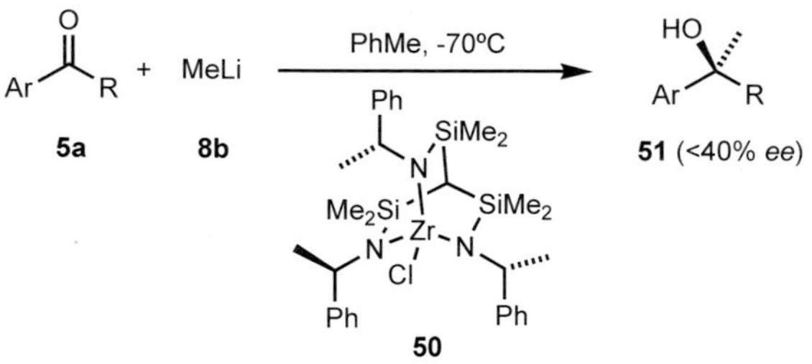

Scheme 8.18 Enantioselective addition of methyllithium to phenones.

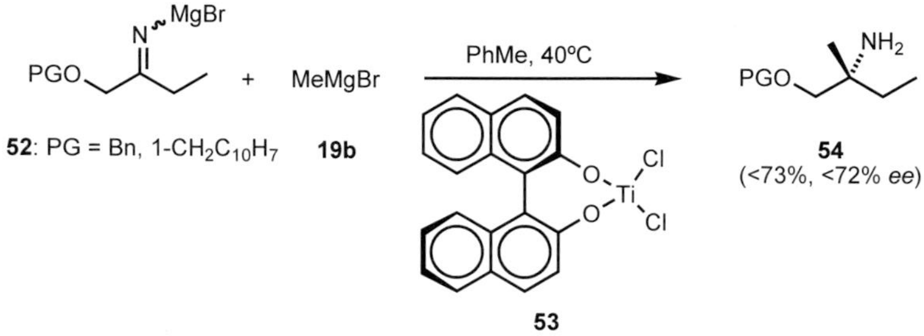

Scheme 8.19 Enantioselective addition of methylmagnesium bromide to imine derivatives.

Another less successful example is the enantioselective addition of different alkyllithium derivatives to 3-(2-naphthyl)-2*H*-azirine in the presence of stoichiometric amounts of (–)-sparteine which yielded, after hydrolysis, the expected α,α-disubstituted aziridines with very modest enantioselectivity [82].

8.3.2
Allylation Processes

Apart from the intrinsic value of carbon–carbon bond formation, the nucleophilic allylation of ketones and derivatives has an extra interest due to the transfer of a carbon–carbon double bond, which can be further converted into different functionalities. The example depicted in Scheme 8.20 shows the addition of allylmagnesium bromide **19c** to methyl ketone derivatives **5b** to yield the corresponding homoallylic alcohol **56** with moderate results [83].

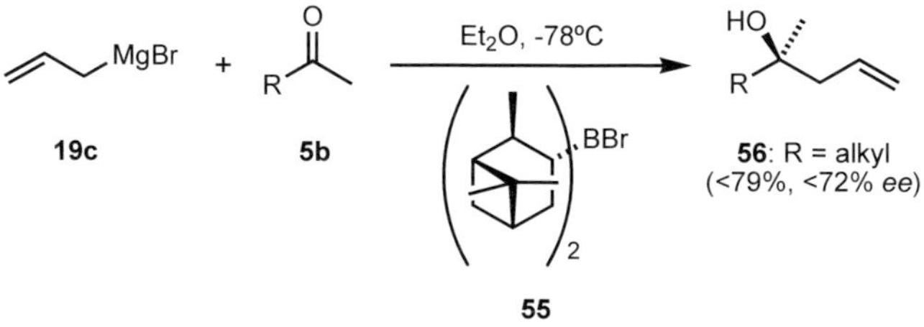

Scheme 8.20 Enantioselective addition of allylmagnesium bromide to methyl ketones.

A similar approach, which implied a transmetallation process between the corresponding allylmagnesium bromide and titanium species **57**, gave similar results [84].

Other organometallics can also be modulated in their addition to ketones as in the case of allylindium derivatives (*in situ* prepared from the corresponding allyl bromide derivative and indium metal), which were enantioselectively added to trifluoromethyl aryl ketones in the presence of a large excess of cinchonidine **58** [85]. The addition of different tertiary amines did not cause any change in the enantioselectivity, even using an excess of the additive.

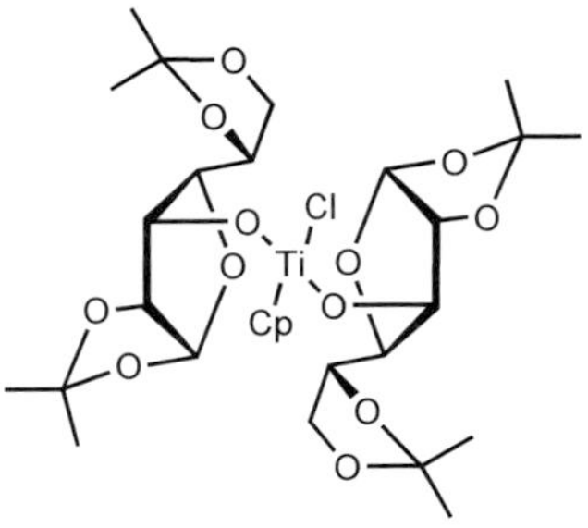
Cl
Ti
Cp

57

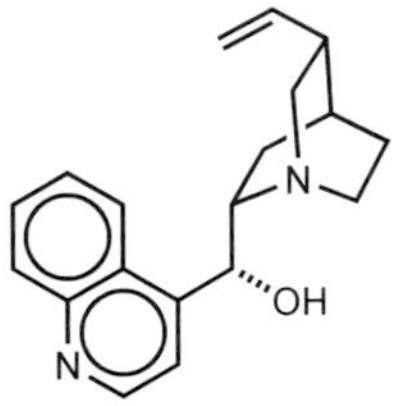
N
OH

58

Structures 57–58

The enantioselective allylation of different alkynyl ketones **59a** was performed using stoichiometric amounts of the chiral complex **61**, with the corresponding allylzinc derivative **60** (Scheme 8.21). The enantioselectivity strongly depended on the bulkiness of the ketone, giving the best result for adamantyl ethynyl ketone [86]. On the basis of the determined absolute stereochemistry, as well as by *ab initio* calculations, a transition-state model was proposed. The reaction has been extended to α-ketoester oxime derivatives **59b**, the reaction providing in this case different amino acids with an excellent *ee* [87].

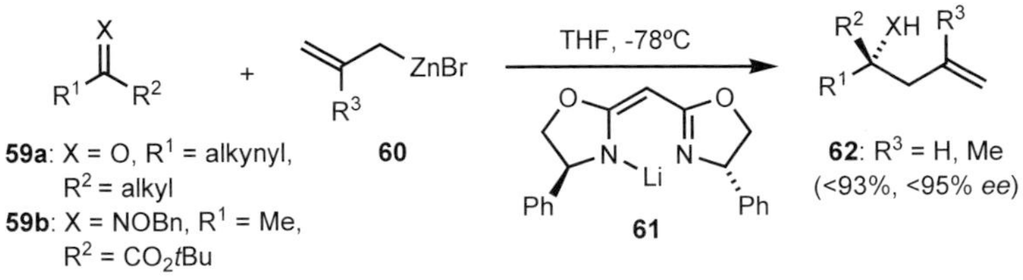
X
R1 R2
ZnBr
R3
THF, -78°C
Li
Ph
Ph
R2 XH R3
R1
59a: X = O, R1 = alkynyl, R2 = alkyl
59b: X = NOBn, R1 = Me, R2 = CO2tBu
60
61
62: R3 = H, Me (<93%, <95% ee)

Scheme 8.21 Enantioselective addition of allylzinc reagents to alkynyl ketones and α-ketoester oximes.

8.3.3 Alkynylation Processes

The modulated enantioselective addition of alkynylmetal derivatives to ketones has been widely studied using stoichiometric amounts of different amino alcohols. The first example, depicted in Scheme 8.22, is the addition of a large excess of lithium trimethylsilylacetylide **8c** to substituted cyclic ketones **63** in the presence of the chiral diamine **64** (2 equivalent). Under these conditions, the results were good and practically independent of the nature of the substituents. However, when the reaction was performed using the corresponding unsubstituted system (R or R' = H), the enantioselectivity dropped sharply [88]. These results were rationalized according to a transition state in which the alcoholate derived from compound **64** is able to chelate simultaneously the carbonyl compound (through the lithium atom) and the lithium alkynyl derivative through the nitrogen atom.

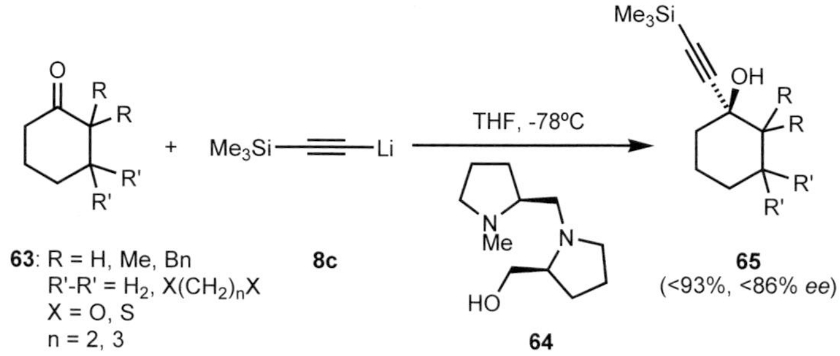

63: R = H, Me, Bn
R'-R' = H₂, X(CH₂)ₙX
X = O, S
n = 2, 3

Scheme 8.22 Enantioselective addition of alkynyllithium reagents to substituted cyclohexanones.

The second example reported is the asymmetric key step in the synthesis of a potent nonnucleoside inhibitor of HIV-1 reverse transcriptase (efavirenz). The key bond-forming reaction is the enantioselective addition of an acetylide **67** to the functionalized ketone **66** using a large excess of ephedrine derivative **68** [89]. After intensive work, the best conditions yielded an excellent *ee* for the corresponding tertiary alcohol **69** (Scheme 8.23).

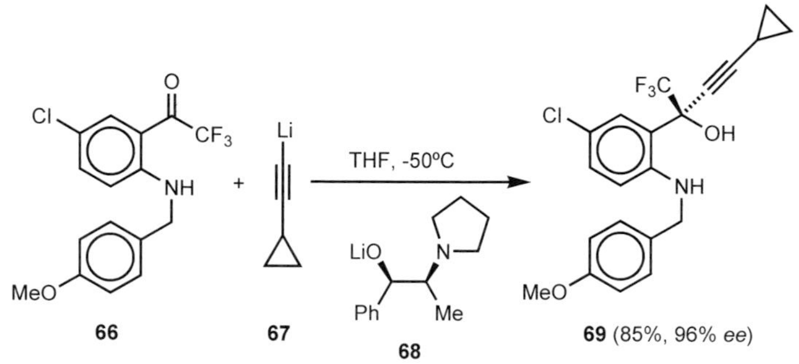

Scheme 8.23 Enantioselective addition of an alkynyllithium reagent to a functionalized phenone.

For the reaction outlined in Scheme 8.23 a wide study has been carried out, including the influence of the stoichiometry of the reagents, ⁶Li, ¹H and ¹³C-NMR studies, together with reactIR and semiempirical computational calculations, to elucidate each possible intermediate and a tentative transition state [90]. The first significant fact is that samples generated at a low temperature (below –70°C), warmed to room temperature, subsequently cooled and then reacted with phenone **66** at low temperature provided higher enantioselectivities than those whose temperature was kept all the time at −78°C. This fact, together with the presence of a positive nonlinear effect, was interpreted as proof of the existence of different aggregates with several ratios between the ligand and the acetylide derivative, the chemical equilibrium between them being very important. These studies have concluded that a dimer bearing two lithium ephedrine alkoxides and two lithium acetylide derivatives is the main aggregate (see structure **70** in Scheme 8.24). Moreover, according to MNDO calculations the dimeric structure **70** is the most stable. In addition, the substitution of a THF ligand on the dimer by the phenone is a moderately exothermic process, and this coordination takes place minimizing the steric interaction between the methyl moiety of the ligand and the phenone (**71**). In other words, the alkyl substituent of the phenone faces the chiral ligand, and this argument agrees with the calculated relative activation enthalphies for different phenones and the two possible coordination models.

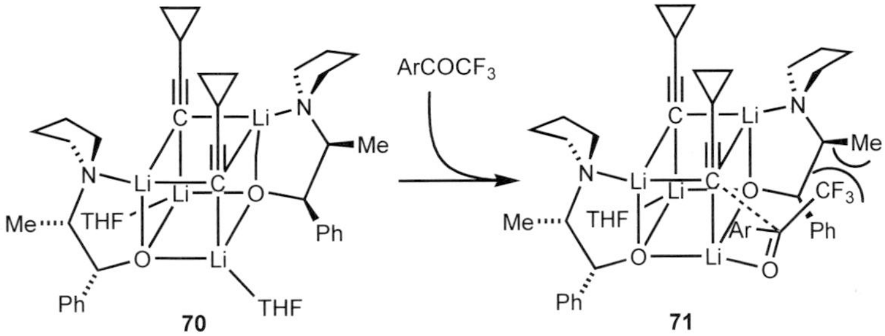

Scheme 8.24 Postulated aggregation and transition step for the alkynylation of functionalized phenones.

Another important fact is that the reaction needs two equivalents of the organolithium reagent if it is to provide both good chemical yield and high enantioselectivity. Although this suggests the presence of an active hydrogen at the nitrogen of the phenone **66**, IR spectroscopic studies at −90°C showed that the excess of organolithium derivative did not deprotonate the NH moiety. Thus, it seems that the hydrogen bond is very important to orient the aromatic fragment, so that a maximum of steric interaction with the ephedrine substituent and enantioselectivity are achieved, the 1,2-addition being faster than the corresponding deprotonation process.

An alternative protocol was further introduced in order to reduce the amount of chiral ligand. Thus, the reaction performed using an alkoxyzinc ephedrine derivative, instead of chiral lithium ephedrine alkoxide **68**, permitted the use of a substoichiometric amount of the chiral ligand. Under these new conditions, the starting ketone used **66** could be replaced by the corresponding NH_2 derivative, so avoiding the process of protection and final deprotection needed in the whole synthesis of efavirez. The best alkoxyzinc derivative was the corresponding 2,2,2-trifluoroethoxy one which was able to increase the enantioselectivity up to 98% [91].

The inhibitor efavirenz, as well as other related systems, can also be prepared through a 1,2-addition of alkynyllithium derivatives **73** to different *N*-acyl ketimines **72a** (Scheme 8.25). When the reaction was performed using a stoichiometric amount of quinine **74**, the protecting group R² of the distal nitrogen atom was crucial in order to achieve good enantioselectivities, the 9-anthranylmethyl derivative showing the best results (97% *ee*). It must be pointed out that the correlation between the enantioselectivity and the temperature showed a bell-type contour line, with a maximum at −25°C, which may reflect a change in the aggregation state of the lithium alkoxide acetylide complex [92], as in the case mentioned above.

The careen-derived ligand **75** has been proposed as an alternative modulator for this type of addition [93]. Under these new conditions the distal nitrogen in compound **72b** did not need to be protected and it was not necessary to use an excess of the expensive acetylide **72** to deprotonate both the ligand **75** and the substrate. This deprotonation could be carried out with the less expensive lithium bis(trimethylsilyl)amide and the *ee* of compound **76** was as high as 99%.

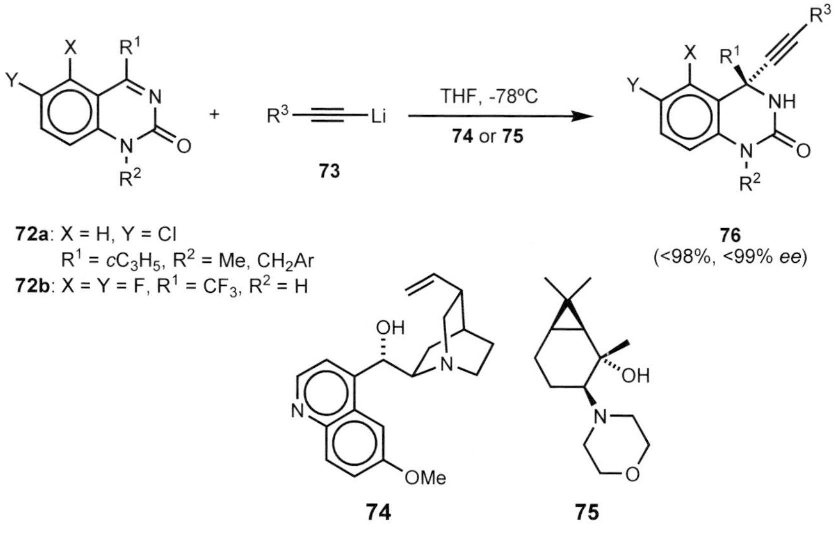

72a: X = H, Y = Cl
   R¹ = cC₃H₅, R² = Me, CH₂Ar
72b: X = Y = F, R¹ = CF₃, R² = H

Scheme 8.25 Enantioselective alkynylation of imine derivatives.

8.4
Enantioselective Additions by Promoted Processes

The preparation of chiral tertiary alcohols and the related amines presented in the previous section necessarily employs at least one equivalent of usually expensive and difficult to prepare chiral ligand. In order to reduce the amount of chiral ligands needed, the usual strategy is to consider the use of an organometallic reagent with lower nucleophilic character. However, under these new conditions, and in order to prevent the reaction failing, the chiral catalytic system must govern not only the topological course of the reaction (modulation) but also the chemical reaction itself (promotion). This new role of the catalyst may be accounted for either by activation of the organometallic reagent or, most classically, by activation of the carbonyl compound. Indeed, some catalysts are able to activate both the nucleophile and the carbonyl compound simultaneously [94].

8.4.1
Alkylation Processes

One of the ideal candidates for this type of reaction is organozinc reagents [95] since their nucleophilic character is very low. In fact, it is well known that they do not add to aldehydes in noncoordinating solvents. This behavior has permitted the development of a plethora of chiral promoters [96] such as amino alcohols, diols and disulfonamides, these last two types of compounds usually acting in combination with titanium [77] tetraisopropoxide, so that the catalyzed enantioselective addition of organozinc reagents to aldehydes has been achieved with excellent enantioselectivities. However, the situation is more complicated when the electrophilic counterpart of the reaction is a ketone since the addition process never takes place, even using the former promoters or at high temperatures, yielding only either the starting unchanged ketone or products arising from the reduction of the carbonyl group [97].

The first example of catalyzed enantioselective alkylation of simple ketones **5** was reported by our group in 1998 (Scheme 8.26), using the chiral isoborneolsulfonamide derivative **78** [98]. The addition of different dialkylzinc reagents **77** (even diethylzinc, in which the process of β-elimination is relatively favored) to simple ketones yielded the expected tertiary alcohols **79** with good-to-modest yields and enantioselectivities [99]. In fact, to date the only known moiety able to catalyze the addition of diethylzinc to ketones is an isoborneolsulfonamide. In order to improve these results a new ligand was prepared, in which the hydroxy group of ligand **78** was changed for a sulfonylamino group [100]. However, this change led to failure of the reaction.

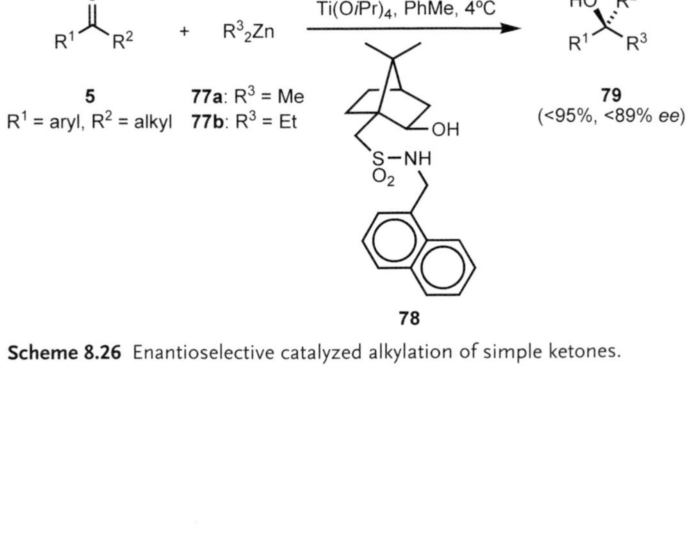

Scheme 8.26 Enantioselective catalyzed alkylation of simple ketones.

To overcome these difficulties, a simple mechanistic study was started with the hope that it would enlighten the further design of new ligands. The first fact was the presence of a small positive nonlinear effect when the reaction was performed using stoichiometric amounts of titanium and chiral ligand. This result is usually attributed to the presence of a bimetallic species in the catalytic cycle. However, when the amount of ligand **78** is reduced to 80% of the stoichiometric amount the aforementioned nonlinear effect disappeared, which could indicate that in the bimetallic species there is only one chiral ligand **80**. The enantioselectivity was practically independent of chemical yield (no autocatalysis effect was detected) and strongly dependent on the size of the ligand and the ketone [101]. All these facts were similar to those previously described for the known alkylation of aldehydes [102]. As a consequence, and assuming that the catalytic species has a pentacoordinated positively charged titanium center [103], a similar catalytic cycle and species involved were postulated for this new reaction (Scheme 8.27).

The alkylating agent is the corresponding alkyltitanium triisopropoxide **81**, obtained, in turn, by ligand exchange between dialkylzinc and titanium tetraisopropoxide. However, when the reaction was performed using freshly distilled methyltitanium triisopropoxide the corresponding tertiary alcohol was obtained as a racemic mixture. This fact might be interpreted by considering that the rates of addition of the nucleophile through the catalyzed or the uncatalyzed pathway are similar and therefore the concentration of the alkyltitanium derivative must be low to avoid the uncatalyzed process. The chiral ligand reacts with titanium tetraisopropoxide (2 equivalent) to yield the aggregate **80**, which through different ligand exchanges yields the new species **83**, which in a following step is able to coordinate the ketone **5** giving the properly called catalytic species **86**. The enantioselective addition takes place on this species to form the corresponding tertiary alcoholate complex **85**. After that, different ligand exchanges liberate (through the intermediate **82**) the tertiary alcoholate **84** renewing the starting aggregate **83**.

The assumption of the dinuclear catalytic species **86** in Scheme 8.27 calls our attention to the possibility of improving the ligand by eliminating the floppy isopropoxy bridges in **86**. The new ligand would be able to bind to two titanium atoms at the same time. In this way, by playing with the length and angles of the covalent linker, we could approach both titanium atoms and place them at the ideal separation for the addition. We started preparing different chiral bisisoborneolsulfonamide ligands derived from xylilendiamine and other aromatic amines [104]. Among them, ligand **87** gave the best results, the *ee* being up to 92%. Although the difference was not very high, this showed that the direction was correct. The following step was to put closer both isoborneolsulfonamide moieties through ethylenediamine derivatives. From this new set of ligands, the *exo*-diol (HOCSAC, **88**) derived from the corresponding 1,2-*trans*-bis(isoborneolsulfonamido)cyclohexane emerged as the best [105]. HOCSAC **88** is an excellent promoter for the catalytic enantioselective addition of dialkylzinc reagents to ketones (*ee* higher than 99%), even for dialkyl ketones. The reaction with this ligand is very fast and the conditions are the mildest reported so far. These facts can be attributed to the double activation process: on one hand a titanium center serves as a Lewis acid to

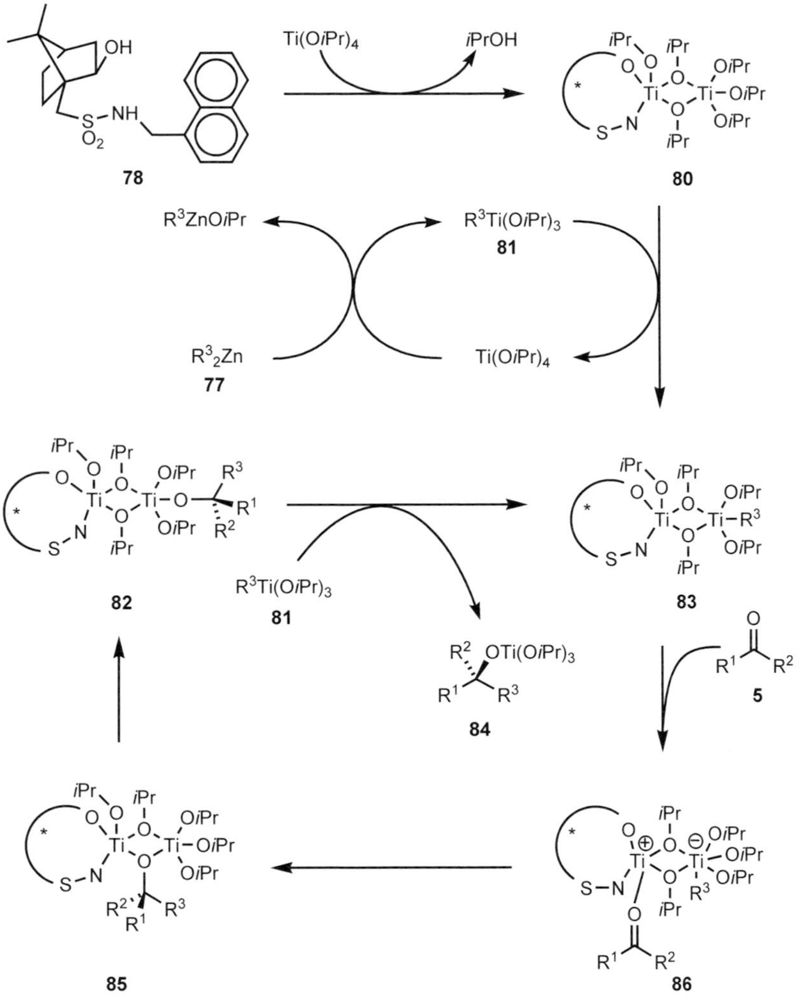

Scheme 8.27 Postulated catalytic cycles for the alkylation of simple ketones.

activate the ketone and on the other hand another titanium center activates the nucleophile by forming the corresponding alkyltitanium derivative. The HOC-SAC ligand has been used not only for the successful alkylation of simple ketones [105, 106] but also for cyclic α,β-unsaturated ketones [107], in all cases the enantioselectivity being very high (up to 99%) and homogeneous.

The aforementioned extraordinary enantioselectivity, independent of the nature of the ketone, permitted the use of HOCSAC as chiral catalytic ligand for the key step in the synthesis of (–)-frontalin [108], giving 89% of the *ee* yielded by the addition of dimethylzinc to a functionalized α,β-unsaturated ketone.

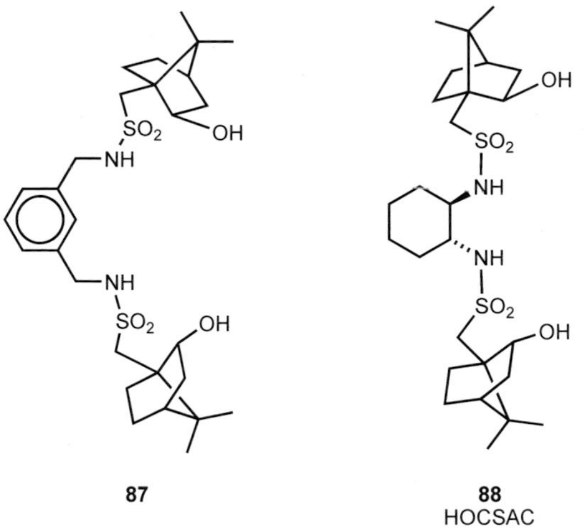

Structures 87–88

Ligand **90** has been used in the enantioselective alkylation of highly reactive ketones, such as α-ketoesters **89** (Scheme 8.28), the idea behind this ligand being the previously mentioned formation of a bimetallic complex (double activation). In this case, both iminic and hydroxy groups chelate a titanium atom which is the acid center, while the amine moieties serve to chelate the zinc reagent, which is actually the alkylic nucleophile. The *ee* never reached 80% [109] and the fact that only α-ketoesters are adequate ketone substrates makes this reaction rather limited for synthetic applications.

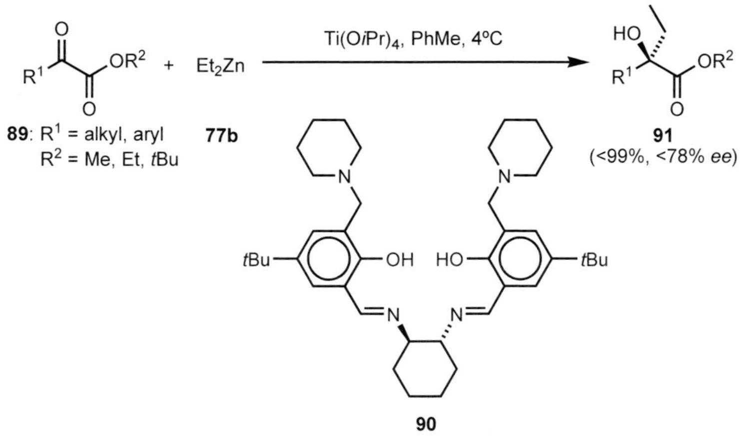

Scheme 8.28 Enantioselective catalyzed alkylation of α-ketoesters.

Ligand **92**, as in previous cases, has been designed to be able to chelate two zinc atoms at the same time and has been applied to the enantioselective alkylation of highly reactive α-ketoesters **89** [110]. In this case, the reaction is not only limited to the use of α-ketoesters but also to dimethylzinc reagents, since other dialkylzinc reagents failed. The best enantioselectivity (up to 96%) was obtained in the presence of substoichiometric amounts of isopropanol.

Finally, the enantioselective trifluoromethylation of different phenones with trifluoromethyltrimethylsilane promoted by *N*-benzylcinchonium fluoride derivative **93** at low temperature has been reported [111]. Although the chemical yield was excellent, the enantioselectivity for the tertiary alcohols was never higher than 51%.

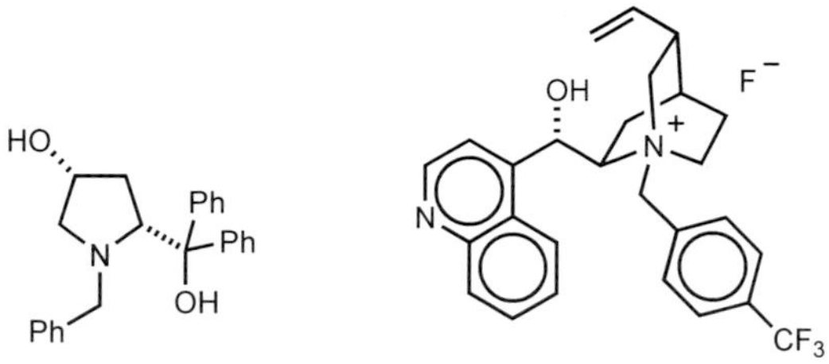

92   **93**

Structures 92–93

8.4.2
Allylation Processes

Another paradigm, through which readers can appreciate the evolution of catalytic systems, is the allylation of ketones. In the first example, depicted in Scheme 8.29, the allylation of acetophenone **5c** was achieved using an excess of tetraallyltin **14b** as the nucleophilic source and in the presence of an excess of BINOL **94** and methanol [112]. In this way, the homoallylic tertiary alcohol **56a** was obtained with moderate *ee* but the reaction represented a breakthrough in this field. In order to obtain good enantioselectivity the mixture of BINOL 94 and tetraallyltin must be heated to 45°C before addition of the ketone. This behavior seems to indicate that the first step in the process is the formation of a BINOL–allyltin complex that is more reactive than the initial allyltin derivative and therefore responsible for the addition.

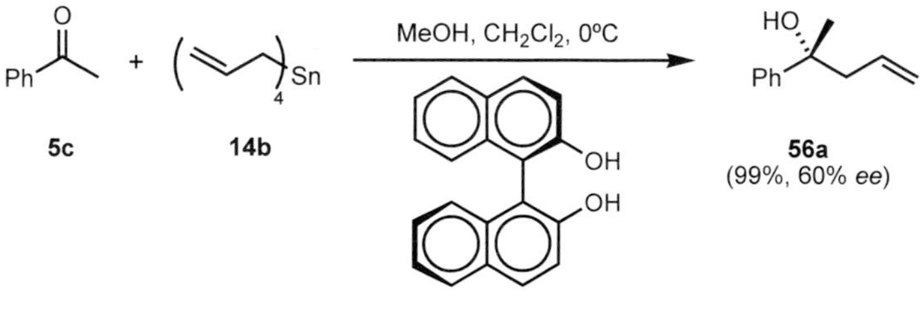

5c   **14b**

56a
(99%, 60% *ee*)

94

Scheme 8.29 Enantioselective allylation of acetophenone using tin derivatives.

Since the tin–sulphur bond is stronger than the corresponding tin–oxygen one, it was suggested that 2-hydroxy-2′-sulfanyl-1,1′-binaphthyl (the BINOL related system with one SH group) might improve the previous results by activation of the tin nucleophile. In fact, the allylation of different phenones using this chiral monothiol could be performed at room temperature and using substoichiometric amounts of the ligand without any further additive, the *ee* being up to 92% [113]. A mechanistic study showed the great impact of impurities on the enantioselectivity and as result of this study a tetrametallic species was postulated as the catalytic system.

Instead of activating the allyltin nucleophile, the activation of the ketone can also be performed by the use of a Lewis acid. Thus, the allylation of different ketones was carried out using substoichiometric amounts of a 1:1 mixture of BINOL **94** and diisopropoxytitanium dichloride [114] which leads to the system **53**. The *ee* reached 80% and was independent of the chemical yield, which ruled out the possibility of any competition of the tin alkoxide with the titanium center as Lewis acid. Finally it is worth noting that the electronic properties of the substituent had a negligible impact on the enantioselectivity. The results obtained using Lewis acid **53** could be improved by the addition of 20% of isopropanol, the *ee* being up to 96% [115]. This implies a double activation process: On one hand, the chiral Lewis acid **53** activates the ketone and on the other hand isopropanol activates the nucleophile.

The former reaction can also be performed using the less acidic titanium tetraisopropoxide instead of the aforementioned dichloride derivative. However, in this last case it is necessary to use achiral trityl diamines. The role of the achiral diamine is not very clear but it has been proposed that it has a function in the formation of bimetallic species. The diamine spacer can chelate two chiral titanium complexes at the same time to form a new bimetallic acid, and this has the possibility of simultaneously coordinating both titanium atoms to the carbonyl group of the ketone, activating it strongly for nucleophilic attack [116].

Another example of an allylation process is depicted in Scheme 8.30. In this case, the reaction of tin(II) catecholate **96** is able to carry out the oxidative addition to different allylbromides **95** catalyzed by copper(I) salts to yield the expected allyltin(IV) derivatives. The enantioselective allylation of benzyl pyruvate **89a** using stoichiometric amounts of diisopropyl tartrate **97** yielded the corresponding α-hydroxy acid derivative **98** with an *ee* up to 98% [117]. In its turn, the final substituted hydroxy acid has been used as starting material in the preparation of frontalin through a ring-closing metathesis [117b].

Not only have allyltin derivatives been used in enantioselective promoted allylation of ketones but also allylsilane derivatives can be used. The allylation of acetophenone **5c** using allyltrimethoxysilane **14c** can be performed using the Lewis-acid complex formed by copper(I) chloride and the chiral phosphane **99** and in the presence of ammonium fluoride salts for the activation of the nucleophile (Scheme 8.31). Although the results are modest [118], the use of other nucleophilic sources is remarkable.

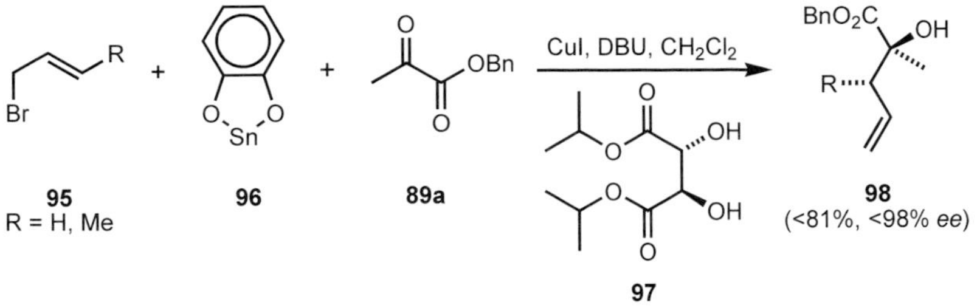

Scheme 8.30 Enantioselective allylation of benzyl pyruvate.

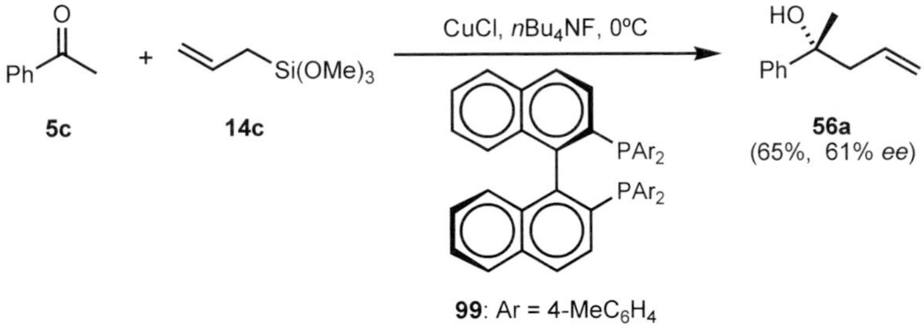

Scheme 8.31 Enantioselective allylation of acetophenone using silane derivatives.

Following the former idea, the use of allylsilyl trichloride has been reported as a nucleophilic source in the allylation of ketone-derived benzoylhydrazones [119], an excess of pseudoephedrine being used as catalyst. The reaction goes through the exchange between two chlorine atoms and the chiral ligand to yield the corresponding chiral allylsilyl chloride, which is able to coordinate the hydrazone through both the imine nitrogen atom and the acyl oxygen atom. The addition process takes place on this ate-type silyl complex, yielding the corresponding homoallylic alcohol with an *ee* of up to 97%.

8.4.3
Arylation Processes

The catalytic enantioselective addition of a large excess of pure diphenylzinc **77c** to ketones **5** has been promoted by the amino alcohol derivative **100** in the presence of methanol, yielding the corresponding tertiary alcohol **101** with reasonable enantioselectivity (Scheme 8.32). A simple mechanistic study showed that the catalytic cycle and the active species are probably similar to those described for the corresponding reaction with aldehydes, changing an alkyl moiety for a methoxide on the zinc atom of the chiral Lewis acid, the chiral system activating the ketone and the zinc reagent at the same time. The success of this reaction was attributed on the one hand to the increase in the Lewis-acid character of the zinc atom due to the presence of a methoxide moiety and, on the other hand, to the higher reactivity of diphenylzinc, which does not have hydrogen atoms in the α-position, limiting or making impossible the formation of zinc hydride and therefore the undesirable reduction processes [120].

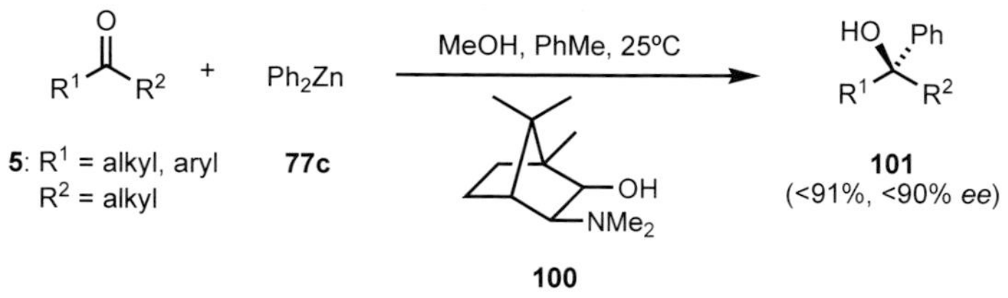
MeOH, PhMe, 25°C
Ph2Zn
5: R1 = alkyl, aryl
R2 = alkyl
77c
OH
NMe2
100
HO Ph
101
(<91%, <90% ee)

Scheme 8.32 Enantioselective arylation of ketones.

The previously mentioned HOCSAC ligand **88** [121, 122] has been used in the enantioselective addition of pure diphenylzinc **77c** to substituted phenones in the presence of titanium tetraisopropoxide [77] yielding the expected tertiary alcohols **101** with up to 96% *ee*. It must be pointed out that this ligand allowed the use of functionalized arylzinc reagents obtained in their turn by transmetallation from functionalized arylboronic acid and diethylzinc [121]. In this way, different functionalized diaryl alkanols were obtained with excellent chemical yields and *ee*s as high as 93%.

8.4.4
Alkenylation Processes

Following our previous idea of *in situ* generation of the zinc reagent, the alkenylation of ketones has been carried out using substoichiometric amounts of HOCSAC **88** as chiral ligand [123]. The *E*-alkenylzirconium derivative obtained by hydrozirconiation of terminal alkynes **102** was transmetallated using dimethylzinc **77a** to give the corresponding *E*-alkenylzinc reagent, which was trapped by reaction with different ketones **5** yielding, after hydrolysis, the corresponding chiral tertiary allylic alcohols **103** (Scheme 8.33).

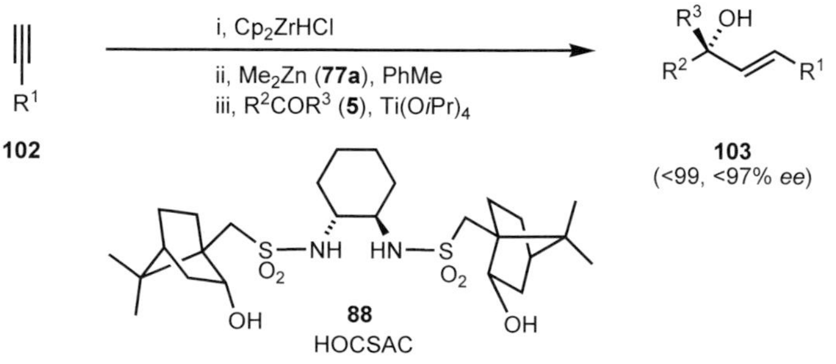
i, Cp2ZrHCl
ii, Me2Zn (77a), PhMe
iii, R2COR3 (5), Ti(OiPr)4
102
103
(<99, <97% ee)
S—NH HN—S
O2 O2
OH
88
HOCSAC
OH

Scheme 8.33 Enantioselective alkenylation of ketones.

8.4.5
Alkynylation Processes

The alkynylation of simple ketones has been carried out using phenylacetylene **102a**, excess of dimethylzinc **77a** and the chiral salen derivative **104** [124]. In this case, the success of the reaction has been attributed, on the one hand, to the *in situ* formed alkynylzinc derivative, which does not have a hydrogen atom at both the α and β positions, preventing the reduction process. On the other hand, the double activation takes place on both reagents, favoring the reaction. The chiral salen **104** is deprotonated to form the corresponding diphenolate–zinc system, which is acidic enough to chelate the ketone effectively. At the same time, one oxygen of the chiral ligand can chelate the alkynylzinc reagent, activating it and yielding the postulated catalytic active species (Scheme 8.34).

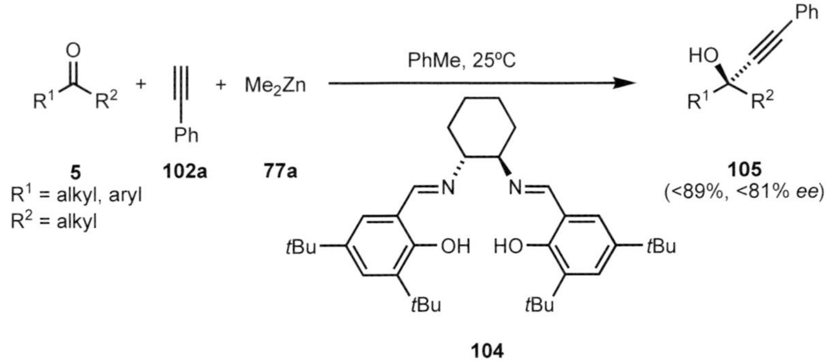

Scheme 8.34 Enantioselective alkynylation of ketones.

The Lewis acid obtained by combination of substoichiometric equal amounts of chiral ligand **78** and copper(II) triflate, instead of using ligand **104**, has been proposed as an alternative for the process depicted in Scheme 8.34 [125]. Under these new conditions, the results were perceptibly higher (*ee* up to 97%).

A simpler procedure for the alkynylation of ketones has been developed (Scheme 8.35). The reaction was performed using highly reactive α-ketoesters **89** and acetylenes **102** as the nucleophilic source (and simultaneously as solvent) yielding in general excellent results. In this case, the corresponding zinc acetylide is formed as a reactive intermediate in the course of the reaction. However, it is still unclear what kind of activation takes place: on the nucleophile, on the electrophile or both. The above protocol has been extended to cyclic N-acyl ketimines of type **72**, also giving excellent results [126].

8.4.6
Miscellaneous Processes

The only example in this section is the enantioselective acylation of cyclic enones **108** (Scheme 8.36). The reaction was catalyzed by substoichiometric amounts of

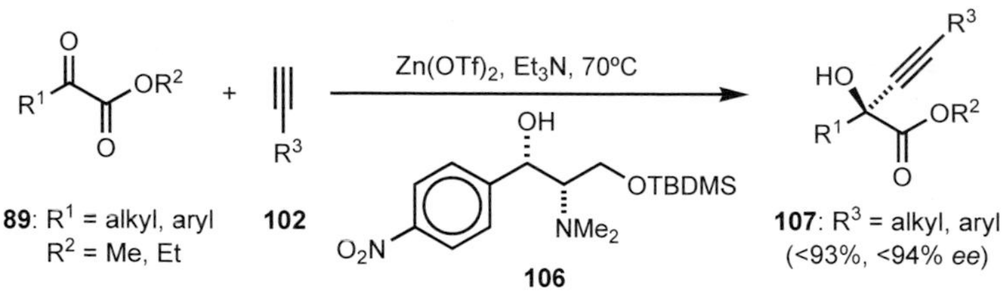

Scheme 8.35 Enantioselective alkynylation of α-ketoesters.

palladium and the chiral binaphthyl derivative **110** [127]. Although the best enantioselectivity obtained for the hydroxy ketones was not very good, it must be pointed out that the use of functionalized nucleophiles normally decreases the enantioselectivity. Moreover, the difference around the carbonyl moiety in the starting enone is very small, making more difficult the achievement of good results.

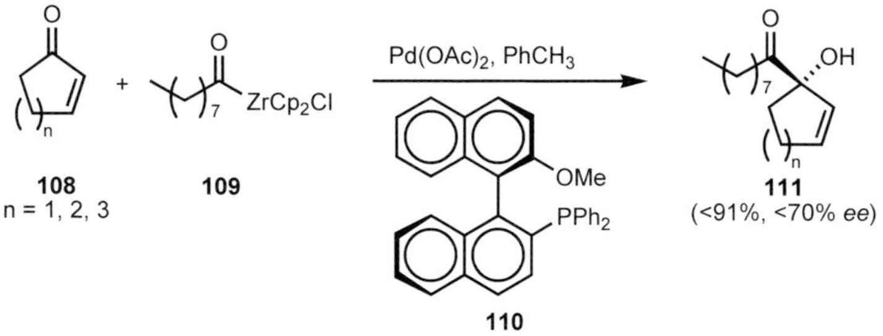

Scheme 8.36 Enantioselective acylation of cycloalkenones.

References

1 a) K. Mori, *Acc. Chem. Res.* **2000**, *33*, 102–110; b) E. Brenna, C. Fuganti, S. Serra, *Tetrahedron: Asymmetry* **2003**, *14*, 1–42.

2 a) G. Helmchen, R. W. Hoffmann, J. Mulzer, E. Schaumann (eds.) **1996**, *Stereoselective Synthesis (Houben-Weyl)*, Vols. 1–10, Thieme, Stuttgart; b) E. N. Jacobsen, A. Pfaltz, H. Yamamoto (eds.) **1999**, *Comprehensive Asymmetric Catalysis*, Vols. 1–3, Springer Verlag, Berlin; c) E. N. Jacobsen, A. Pfaltz, H. Yamamoto (eds.) **2004**, *Comprehensive Asymmetric Catalysis, Supplement 1*, Springer Verlag, Berlin.

3 For reviews on the enantioselective construction of quaternary stereocenters, see: a) K. Fuji, *Chem. Rev.* **1993**, *93*, 2037–2066; b) E. J. Corey, A. Guzmán-Pérez, *Angew. Chem.* **1998**, *110*, 402–415; *Angew. Chem. Int. Ed.* **1998**, *37*, 388–401; c) J. Christoffers, A. Mann, *Angew. Chem.* **2001**, *113*, 4725–4732; *Angew. Chem. Int. Ed.* **2001**, *40*,

4591–4597; d) J. Christoffers, A. Baro, *Angew. Chem.* **2003**, *115*, 1726–1728; *Angew. Chem. Int. Ed.* **2003**, *42*, 1688–1690.

4 For a review, see: D. J. Ramón, M. Yus, *Curr. Org. Chem.* **2004**, *8*, 149–183.

5 M. Yus, D. J. Ramón, *Recent Res. Devel. Org. Chem.* **2002**, *6*, 297–378.

6 For reviews, see: a) M. North, *Synlett* **1993**, 807–820; b) F. Effenberger, *Angew. Chem.* **1994**, *104*, 1609–1619; *Angew. Chem. Int. Ed. Engl.* **1994**, *33*, 1555–1564; c) R. J. H. Gregory, *Chem. Rev.* **1999**, *99*, 3649–3682; d) H. Gröger, *Chem. Eur. J.* **2001**, *7*, 5246–5251; e) M. North, *Tetrahedron: Asymmetry* **2003**, *14*, 147–176; f) J.-M. Brunel, I. P. Holmes, *Angew. Chem.* **2004**, *116*, 2810–2837; *Angew. Chem. Int. Ed.* **2004**, *43*, 2752–2778.

7 M. Bandini, A. Melloni, A. Umani-Ronchi, *Angew. Chem.* **2004**, *116*, 560–566; *Angew. Chem. Int. Ed.* **2004**, *43*, 550–556.

8 Y. Izumi, *Angew. Chem.* **1971**, *83*, 956–966; *Angew. Chem. Int. Ed. Engl.* **1971**, *10*, 871–881.

9 a) A. Bachki, F. Foubelo, M. Yus, *Tetrahedron: Asymmetry* **1995**, *6*, 1907–1910; b) A. Bachki, F. Foubelo, M. Yus, *Tetrahedron: Asymmetry* **1996**, *7*, 2997–3008.

10 M. García-Valverde, R. Pedrosa, M. Vicente, *Tetrahedron: Asymmetry* **1995**, *6*, 1787–1794.

11 P. Vossmann, K. Hornig, R. Fröhlich, E.-U. Würthwein, *Synthesis* **2001**, 1415–1426.

12 H. Paulsen, C. Graeve, D. Hoppe, *Synthesis* **1996**, 141–144.

13 For reviews, see: a) D. Hoppe, *Angew. Chem.* **1997**, *109*, 2376–2410; *Angew. Chem. Int. Ed. Engl.* **1997**, *36*, 2282–2316; b) M. C. Whisler, S. MacNeil, V. Snieckus, P. Beak, *Angew. Chem.* **2004**, *116*, 2256–2276; *Angew. Chem. Int. Ed.* **2004**, *43*, 2206–2225.

14 H. Paulsen, C. Graeve, R. Fröhlich, D. Hoppe, *Synthesis* **1996**, 145–148.

15 M. Tiffeneau, J. Levy, E. Ditz, *Bull. Soc. Chim. Fr.* **1935**, 1855–1866.

16 For reviews, see: a) R. M. Devant, H.-E. Radunz **1996**, in *Stereoselective Synthesis (Houben-Weyl)*, Vol. 2, ed. G. Helmchen, R. W. Hoffmann, J. Mulzer, E. Schaumann, Thieme, Stuttgart, pp. 1151–1268; b) D. Hoppe **1996**, in *Stereoselective Synthesis (Houben-Weyl)*, Vol. 3, ed. G. Helmchen, R. W. Hoffmann, J. Mulzer, E. Schaumann, Thieme, Stuttgart, **1996**, pp. 1401–1409, 1541–1550.

17 a) G. Chelucci, F. Soccolini, *Tetrahedron: Asymmetry* **1992**, *3*, 1235–1238; b) M. Genov, K. Kostova, V. Dimitrov, *Tetrahedron: Asymmetry* **1997**, *8*, 1869–1876; c) S. Panev, V. Dimitrov, *Tetrahedron: Asymmetry* **2000**, *11*, 1517–1526; d) Q. Xu, G. Wang, X. Pan, A. S. C. Chan, *Tetrahedron: Asymmetry* **2001**, *12*, 381–385.

18 a) D. J. Ramón, M. Yus, *Tetrahedron: Asymmetry* **1997**, *8*, 2479–2496; b) I. Gómez, E. Alonso, D. J. Ramón, M. Yus, *Tetrahedron* **2000**, *56*, 4043–4052.

19 M. Yus, T. Soler, F. Foubelo, *Tetrahedron: Asymmetry* **2001**, *12*, 801–810.

20 For reviews, see: a) C. Nájera, M. Yus, *Trends Org. Chem.* **1991**, *2*, 155–181; b) C. Nájera, M. Yus, *Recent Res. Devel. Org. Chem.* **1997**, *1*, 67–96; c) C. Nájera, M. Yus, *Curr. Org. Chem.* **2003**, *7*, 867–926.

21 S. Lavaire, R. Plantier-Royon, C. Portella, *Tetrahedron: Asymmetry* **1998**, *9*, 213–226.

22 For reviews, see: a) M. Yus, F. Foubelo, *Rev. Heteroatom Chem.* **1997**, *17*, 73–107; b) F. Foubelo, M. Yus, *Trends Org. Chem.* **1998**, *7*, 1–26; c) M. Yus, F. Foubelo, *Targets in Heterocyclic Systems* **2002**, *6*, 136–171; d) M. Yus, *Pure Appl. Chem.* **2003**, *75*, 1453–1475.

23 For reviews on different aspects of this methodology, see: a) M. Yus, *Chem. Soc. Rev.* **1996**, *25*, 155–161; b) D. J. Ramón, M. Yus, *Eur. J. Org. Chem.* **2000**, 225–237; c) M. Yus, *Synlett* **2001**, 1197–1205; d) D. J. Ramón, M. Yus, *Rev. Cubana Quim.* **2002**, *14*, 76–115; e) M. Yus, D. J. Ramón, *Latv. J. Chem.* **2002**, 79–92; f) M. Yus **2004**, in *The Chemistry of Organolithium Compounds*, ed. Z. Rapoport, I. Marek, J. Wiley & Sons, Chichester, **2004**, pp. 647–747.

24 T. Soler, A. Bachki, L. R. Falvello, F. Foubelo, M. Yus, *Tetrahedron: Asymmetry* **2000**, *11*, 493–517.

25 a) M. Carda, F. González, S. Rodríguez, J. A. Marco, *Tetrahedron: Asymmetry* **1992**, *3*, 1511–1514; b) M. Carda, F. González, S. Rodríguez, J. A. Marco, *Tetrahedron: Asymmetry* **1993**, *4*, 1799–1802; c) J. A. Marco, M. Carda, F. González, S. Rodríguez, E. Castillo, J. Murga, *J. Org. Chem.* **1998**, *63*, 698–707.

26 V. S. Safont, V. Moliner, M. Oliva, R. Castillo, J. Andrés, F. González, M. Carda, *J. Org. Chem.* **1996**, *61*, 467–3475.

27 M. Carda, P. Casabó, F. González, S. Rodríguez, L. R. Domingo, J. A. Marco, *Tetrahedron: Asymmetry* **1997**, *8*, 559–577.

28 M. Carda, E. Castillo, S. Rodríguez, J. Murga, J. A. Marco, *Tetrahedron: Asymmetry* **1998**, *9*, 1117–1120.

29 M. Carda, E. Castillo, S. Rodríguez, F. González, J. A. Marco, *Tetrahedron: Asymmetry* **2001**, *12*, 1417–1429;

Corrigendum: *Tetrahedron: Asymmetry* **2001**, *12*, 3061.

30 a) M. Carda, E. Castillo, S. Rodríguez, J. A. Marco, *Tetrahedron Lett.* **2000**, *41*, 5511–5513; b) M. Carda, S. Rodríguez, E. Castillo, A. Bellido, S. Díaz-Oltra, J. A. Marco, *Tetrahedron* **2003**, *59*, 857–864; c) J. A. Marco, M. Carda, S. Rodríguez, E. Castillo, M. N. Kneeteman, *Tetrahedron* **2003**, *59*, 4085–4101.

31 M. Carda, E. Castillo, S. Rodríguez, E. Falomir, J. A. Marco, *Tetrahedron Lett.* **1998**, *39*, 8895–8896.

32 H. Chikashita, Y. Nakamura, H. Uemura, K. Itoh, *Chem. Lett.* **1992**, 439–442.

33 A. Ojida, N. Matsunaga, T. Kaku, A. Tasaka, *Tetrahedron: Asymmetry* **2004**, *15*, 1555–1559.

34 a) A. Barbero, F. J. Pulido, J. A. Rincón, P. Cuadrado, D. Galisteo, H. Martínez-García, *Angew. Chem.* **2001**, *113*, 2159–2161; *Angew. Chem. Int. Ed.* **2001**, *40*, 2101–2103; b) A. Barbero, F. J. Pulido, J. A. Rincón, *J. Am. Chem. Soc.* **2003**, *125*, 12049–12056.

35 J. Nokami, K. Nomiyama, S. M. Shafi, K. Kataoka, *Org. Lett.* **2004**, *6*, 1261–1264.

36 a) D. J. Cram, F. A. A. Elhafez, *J. Am. Chem. Soc.* **1952**, *74*, 5828–5835; b) D. J. Cram, J. D. Knight, *J. Am. Chem. Soc.* **1952**, *74*, 5835–585838.

37 a) M. Chérest, H. Felkin, N. Prudent, *Tetrahedron Lett.* **1968**, 2199–2204; b) N. T. Anh, O. Eisenstein, *Tetrahedron Lett.* **1976**, 155–158; c) A. Mengel, O. Reiser, *Chem. Rev.* **1999**, *99*, 1191–1223.

38 O. Takahashi, K. Saito, Y. Kohno, H. Suezawa, S. Ishihara, M. Nishio, *New J. Chem.* **2004**, 335–360.

39 W. C. Still, J. H. McDonald, *Tetrahedron Lett.* **1980**, *21*, 1031–1034.

40 P. V. Ramachandran, H. Liu, M. V. R. Reddy, H. C. Brown, *Org. Lett.* **2003**, *5*, 3755–3757.

41 R. D. Pace, G. W. Kabalka, *J. Org. Chem.* **1995**, *60*, 4838–4899.

42 J. L. García Ruano, A. Tito, R. Culebras, *Tetrahedron* **1996**, *52*, 2177–2186.

43 a) S. V. Ley, L. R. Cox, G. Meek, K.-H. Metten, C. Piqué, J. M. Worrall, *J. Chem. Soc., Perkin Trans. 1* **1997**, 3299–3313; b) S. V. Ley, L. R. Cox, *J. Chem. Soc., Perkin Trans. 1* **1997**, 3315–3325.

44 K. Suzuki, K. Inomata, Y. Endo, *Heterocycles* **2003**, *60*, 2743–2748.

45 a) M. Sakamoto, T. Iwamoto, N. Nono, M. Ando, W. Arai, T. Mino, T. Fujita, *Chem. Commun.* **2004**, 1002–1003; b) M. Sakamoto, S. Kobaru, T. Mino, T. Fujita, *Chem. Commun.* **2004**, 1002–1003.

46 V. Prelog, *Bull. Soc. Chim. Fr.* **1956**, 987–995.

47 K. Mikami, M. Shimizu, H.-C. Zhang, B. E. Maryanoff, *Tetrahedron* **2001**, *57*, 2917–2951.

48 A. Ichikawa, S. Hiradate, A. Sugio, S. Kuwahara, M. Watanabe, N. Harada, *Tetrahedron: Asymmetry* **2000**, *11*, 2669–2675.

49 D. Basavaiah, S. Pandiaraju, M. Bakthadoss, K. Muthukumaran, *Tetrahedron: Asymmetry* **1996**, *7*, 997–1000.

50 a) Y. Tamai, T. Hattori, M. Date, H. Takayama, Y. Kamikubo, Y. Minato, S. Miyano, *J. Chem. Soc., Perkin Trans. 1* **1999**, 1141–1142; b) Y. Tamai, T. Hattori, M. Date, S. Koike, Y. Kamikubo, M. Akiyama, K. Seino, H. Takayama, T. Oyama, S. Miyano, *J. Chem. Soc., Perkin Trans. 1* **1999**, 1685–1694.

51 T. Fujisawa, T. Watai, T. Sugiyama, Y. Ukaji, *Chem. Lett.* **1989**, 2045–2048.

52 A. B. Charette, A. F. Benslimane, C. Mellon, *Tetrahedron Lett.* **1995**, *47*, 8557–8560.

53 a) T. Fujisawa, M. Watanabe, T. Sato, *Chem. Lett.* **1984**, 2055–2058; b) T. Fujisawa, M. Funabora, Y. Ukaji, T. Sato, *Chem. Lett.* **1988**, 59–62.

54 T. Fujisawa, I. Takemura, Y. Ukaji, *Tetrahedron Lett.* **1990**, *38*, 5479–5482.

55 J. L. García Ruano, M. M. Rodríguez-Fernández, M. C. Maestro, *Tetrahedron* **2004**, *60*, 5701–5710.

56 P. Maglioli, O. De Lucchi, G. Delogu, G. Valle, *Tetrahedron: Asymmetry* **1992**, *3*, 365–366.

57 K. M. Akhoon, D. C. Myles, *J. Org. Chem.* **1997**, *62*, 6041–6045.

58 P. A. Rose, S. R. Abrams, A. C. Shaw, *Tetrahedron: Asymmetry* **1992**, *3*, 443–450.

59 a) F. Martínez-Ramos, M. E. Vargas-Díaz, L. Chacón-García, J. Tamariz, P. Joseph-Nathan, L. G. Zepeda, *Tetrahedron: Asymmetry* **2001**, *12*, 3095–3103; b) M. E. Vargas-Díaz, L. Chacón-García, P. Velázquez, J. Tamariz, P. Joseph-Nathan, L. G. Zepeda, *Tetrahedron: Asymmetry* **2003**, *14*, 3225–3232; c) L. Chacón-García, S. Lagunas-Rivera, S. Pérez-Estrada, M. E. Vargas-Díaz, P. Joseph-Nathan, J. Tamariz, L. G. Zepeda, *Tetrahedron Lett.* **2004**, *45*, 2141–2145.

60 A. I. Meyers, M. A. Hanagan, L. M. Trefonas, R. J. Baker, *Tetrahedron* **1983**, *39*, 1991–1999.

61 a) C. M. Schuch, R. A. Pilli, *Tetrahedron: Asymmetry* **2000**, *11*, 753–764; b) C. M. Schuch, R. A. Pilli, *Tetrahedron: Asymmetry* **2002**, *13*, 1973–1980.

62 E. Alonso, D. J. Ramón, M. Yus, *Tetrahedron* **1997**, *53*, 2641–2652.

63 N. Sewald, L. C. Seymour, K. Burger, S. N. Osipov, A. F. Kolomiets, A. V. Fokin, *Tetrahedron: Asymmetry* **1994**, *5*, 1051–1060.

64 a) J. A. Marco, M. Carda, J. Murga, S. Rodríguez, E. Falomir, M. Oliva, *Tetrahedron: Asymmetry* **1998**, *9*, 1679–1701; b) R. Portolés, J. Murga, E. Falomir, M. Carda, S. Uriel, J. A. Marco, *Synlett* **2002**, 711–714.

65 N. G. v. Keyserlink, J. Martens, *Eur. J. Org. Chem.* **2002**, 301–308.

66 a) D. H. Hua, N. Lagneau, H. Wang, J. Chen, *Tetrahedron: Asymmetry* **1995**, *6*, 349–352; b) J. P. McMahon, J. A. Ellman, *Org. Lett.* **2004**, *6*, 1645–1647.

67 a) C. Cimarelli, G. Palmieri, E. Volpini, *Tetrahedron: Asymmetry* **2002**, *13*, 2011–2018; b) C. Cimarelli, G. Palmieri, E. Volpini, *J. Org. Chem.* **2003**, *68*, 1200–1206.

68 A. Job, C. F. Janek, W. Bettray, R. Peters, D. Enders, *Tetrahedron* **2002**, *58*, 2253–2329.

69 D. Enders, U. Reinhold, *Tetrahedron: Asymmetry* **1997**, *8*, 1895–1946.

70 V. Breuil-Desvergnes, P. Compain, J.-M. Vatèle, J. Goré, *Tetrahedron Lett.* **1999**, *40*, 5009–5012.

71 H. L. Cohen, G. F. Wright, *J. Org. Chem.* **1953**, *18*, 432–446.

72 T. D. Inch, G. J. Lewis, G. L. Sainbury, L. Gordon, D. J. Sellers, *Tetrahedron Lett.* **1969**, 3657–3660.

73 A. I. Meyers, M. E. Ford, *Tetrahedron Lett.* **1974**, 1341–1344.

74 a) B. Weber, D. Seebach, *Angew. Chem.* **1992**, *104*, 96–97; *Angew. Chem. Int. Ed. Engl.* **1992**, *31*, 84–86; b) B. Weber, D. Seebach, *Tetrahedron* **1994**, *50*, 6117–6128.

75 G. Zadel, E. Breitmaier, *GIT. Fachz. Lab.* **1993**, *3*, 212–214.

76 For reviews on this reaction, see: a) O. G. Kulinkovich, A. de Meijere, *Chem. Rev.* **2000**, *100*, 2789–2834; b) F. Sato, H. Urabe, S. Okamoto, *Chem. Rev.* **2000**, *100*, 2835–2886; c) O. G. Kulinkovich, *Pure Appl. Chem.* **2000**, *72*, 1715–1719; d) O. G. Kulinkovich, *Chem. Rev.* **2003**, *103*, 2597–2632.

77 For a review on titanium-catalyzed enantioselective reactions, see: D. J. Ramón, M. Yus, *Recent Res. Devel. Org. Chem.* **1998**, *2*, 489–523.

78 E. J. Corey, S. A. Rao, M. C. Noe, *J. Am. Chem. Soc.* **1994**, *116*, 9345–9346.

79 G. Boireau, D. Abenhaim, J. Bourdais, E. Henry-Basch, *Tetrahedron Lett.* **1976**, 4781–4782.

80 L. H. Gade, P. Renner, H. Memmler, F. Fecher, C. H. Galka, M. Laubender, S. Radojevic, M. McPartlin, J. W. Lauther, *Chem. Eur. J.* **2001**, *7*, 2563–2580.

81 A. B. Charette, A. Gagnon, *Tetrahedron: Asymmetry* **1999**, *10*, 1961–1968.

82 E. Risberg, P. Somfai, *Tetrahedron: Asymmetry* **2002**, *13*, 1957–1959.

83 P. K. Jadhav, K. S. Bhat, T. Perumal, H. C. Brown, *J. Org. Chem.* **1986**, *51*, 432–439.

84 M. Riediker, R. O. Duthaler, *Angew. Chem.* **1989**, *101*, 488–490; *Angew. Chem. Int. Ed. Engl.* **1989**, *28*, 494–495.

85 T.-P. Loh, J.-R. Zhou, X.-R. Li, *Tetrahedron Lett.* **1999**, *40*, 9333–9336.

86 M. Nakamura, A. Hirai, M. Sogi, E. Nakamura, *J. Am. Chem. Soc.* **1998**, *120*, 5846–5847.

87 S. Hanessian, R.-Y. Yang, *Tetrahedron Lett.* **1996**, *37*, 8997–9000.

88 K. Scharpwinkel, S. Matull, H. J. Schäfer, *Tetrahedron: Asymmetry* **1996**, *7*, 2497–2500.

89 A. S. Thompson, E. G. Corley, M. F. Huntington, E. J. J. Grabowski, *Tetrahedron Lett.* **1995**, *36*, 8937–8940.

90 A. S. Thompson, E. G. Corley, M. F. Huntington, E. J. J. Grabowski, J. F. Remenar, D. B. Collum, *J. Am. Chem. Soc.* **1998**, *120*, 2028–2038.

91 L. Tan, C.-y. Chen, R. D. Tillyer, E. J. J. Grabowski, P. J. Reider, *Angew. Chem.* **1999**, *111*, 724–727; *Angew. Chem. Int. Ed.* **1999**, *38*, 711–712.

92 M. A. Huffman, N. Yasuda, A. E. DeCamp, E. J. J. Grabowski, *J. Org. Chem.* **1995**, *60*, 1590–1594.

93 G. S. Kauffman, G. D. Harris, R. L. Dorow, B. R. P. Stone, R. L. Parsons, J. A. Pesti, N. A. Magnus, J. M. Fortunak, P. N. Confalone, W. A. Nugent, *Org. Lett.* **2000**, *2*, 3119–3121.

94 a) H. Gröger, *Chem. Eur. J.* **2001**, *7*, 5246–5251; b) G. J. Rowlands, *Tetrahedron* **2001**, *57*, 1865–1882; c) S. Woodward, *Tetrahedron* **2002**, *58*, 1017–1050; d) M. Shibasaki, M. Kanai, K. Funabashi, *Chem. Commun.* **2002**, 1989–1999.

95 D. J. Ramón, M. Yus, *Angew. Chem.* **2004**, *116*, 286–289; *Angew. Chem. Int. Ed.* **2004**, *43*, 284–287.

96 L. Pu, H.-B. Yu, *Chem. Rev.* **2001**, *101*, 757–824.

97 a) J. Boersma **1982**, in *Comprehensive Organometallic Chemistry*, Vol. 2, ed. G. Wilkinson, Pergamon, Oxford, pp. 823–862; b) R. Noyori, S. Suga, K. Kawai, S. Okada, M. Kitamura, N. Oguni, T. Kanedo, Y. Matsuda *J. Organomet. Chem.* **1990**, *382*, 19–37; c) M. Watanabe, K. Soai, *J. Chem. Soc., Perkin Trans. 1* **1994**, 3125–3128; d) P. Knochel, **1995**, in *Encyclopedia of Reagents for Organic Synthesis*, Vol. 3, ed. L. A. Paquette, John Wiley & Sons, Chichester, pp. 1861–1866.

98 D. J. Ramón, M. Yus, *Tetrahedron: Asymmetry* **1997**, *8*, 2479–2496.

99 D. J. Ramón, M. Yus, *Tetrahedron Lett.* **1998**, *39*, 1239–1242.

100 O. Prieto, D. J. Ramón, M. Yus, *Tetrahedron: Asymmetry* **2000**, *11*, 1629–1644.

101 D. J. Ramón, M. Yus, *Tetrahedron* **1998**, *54*, 5651–5666.

102 For studies on the mechanism of the enantioselective addition of dialkylzinc to aldehydes in the presence of titanium alkoxides, see for the TADDol ligand: a) D. Seebach, D. A. Plattner, A. K. Beck, Y. M. Wang, D. Hunziker, W. Petter, *Helv. Chim. Acta* **1992**, *75*, 2171–2209; b) Y. N. Ito, X. Ariza, A. K. Beck, A. Bohá?, C. Ganter, R. E. Gawley, F. N. M. Kühnle, J. Tuleja, Y. M. Wang, D. Seebach, *Helv. Chim. Acta* **1994**, *77*, 2071–2110; c) B. Weber, D. Seebach, *Tetrahedron* **1994**, *50*, 7473–7484. For the BINOL ligand: d) J. Balsells, T. J. Davis, P. Carroll, P. J. Walsh, *J. Am. Chem. Soc.* **2002**, *124*, 10336–10348. For 1,2-hydroxysulfonamide ligands: e) K.-H. Wu, H.-M. Gau, *Organometallics* **2004**, *23*, 580–588.

103 D. J. Ramón, G. Guillena, D. Seebach, *Helv. Chim. Acta* **1996**, *79*, 875–894.

104 M. Yus, D. J. Ramón, O. Prieto, *Tetrahedron: Asymmetry* **2003**, *14*, 1103–1114.

105 M. Yus, D. J. Ramón, O. Prieto, *Tetrahedron: Asymmetry* **2002**, *13*, 2291–2293.

106 a) C. García, L. K. LaRochelle, P. J. Walsh, *J. Am. Chem. Soc.* **2002**, *124*, 10970–10971; b) T. Mukaiyama, T. Shintou, K. Fukumoto, *J. Am. Chem. Soc.* **2003**, *125*, 10538–10539.

107 S.-J. Jeon, P. J. Walsh, *J. Am. Chem. Soc.* **2003**, *125*, 9544–9545.

108 M. Yus, D. J. Ramón, O. Prieto, *Eur. J. Org. Chem.* **2003**, 2745–2748.

109 a) E. F. DiMauro, M. C. Kozlowski, *J. Am. Chem. Soc.* **2002**, *124*, 12668–12669; b) E. F. DiMauro, M. C. Kozlowski, *Org. Lett.* **2002**, *4*, 3781–3784.

110 K. Funabashi, M. Jachmann, M. Kanai, M. Shibasaki, *Angew. Chem.* **2003**, *115*, 5647–5650; *Angew. Chem. Int. Ed.* **2003**, *42*, 5489–5492.

111 K. Iseki, T. Nagai, Y. Kobayashi, *Tetrahedron Lett.* **1994**, *35*, 3137–3138.

112 M. Yasuda, N. Kitahara, T. Fujibayashi, A. Baba, *Chem. Lett.* **1998**, 743–744.

113 a) A. Cunningham, S. Woodward, *Synlett* **2002**, 43–44; Corrigendum:

Synlett **2004**, 914; b) A. Cunningham,
V. Mokal-Parekh, C. Wilson,
S. Woodward, *Org. Biomol. Chem.* **2004**,
2, 741–748.

114 S. Casolari, D. D'Addario, E. Tagliavini,
Org. Lett. **1999**, *1*, 1061–1063.

115 K. M. Waltz, J. Gavenonis, P. J. Walsh,
Angew. Chem. **2002**, *114*, 3849–3852;
Angew. Chem. Int. Ed. **2002**, *41*,
3697–3699.

116 a) H. Hanawa, S. Kii, K. Maruoka,
Adv. Synth. Catal. **2001**, *343*, 51–56;
b) S. Kii, K. Maruoka, *Chirality* **2003**, *15*,
68–70.

117 a) K. Yamada, T. Tozawa, M. Nishida,
T. Mukaiyama, *Bull. Chem. Soc. Jpn.*
1997, *70*, 2301–2308; b) M. Scholl,
R. H. Grubbs, *Tetrahedron Lett.* **1999**,
40, 1425–1428.

118 S. Yamasaki, K. Fujii, R. Wada,
M. Kanai, M. Shibasaki, *J. Am. Chem.
Soc.* **2002**, *124*, 6536–6537.

119 R. Berger, K. Duff, J. L. Leighton, *J. Am.
Chem. Soc.* **2004**, *126*, 5686–5687.

120 P. I. Dosa, G. C. Fu, *J. Am. Chem. Soc.*
1998, *120*, 445–446.

121 O. Prieto, D. J. Ramón, M. Yus,
Tetrahedron: Asymmetry **2003**, *14*,
1955–1957.

122 For further examples of this reaction,
see: C. García, P. J. Walsh, *Org. Lett.*
2003, *5*, 3641–3644.

123 H. Li, P. J. Walsh, *J. Am. Chem. Soc.*
2004, *126*, 6538–6539.

124 a) P. G. Cozzi, *Angew. Chem.* **2003**, *115*,
3001–3004; *Angew. Chem. Int. Ed.* **2003**,
42, 2895–2898; b) for the use of a related
BINOL-salen ligand, see: B. Saito,
T. Katsuki, *Synlett* **2004**, 1557–1560.

125 G. Lu, X. Li, X. Jia, W. L. Chan, A. S. C.
Chan, *Angew. Chem.* **2003**, *115*,
5211–5212; *Angew. Chem. Int. Ed.* **2003**,
42, 5057–5058.

126 a) B. Jiang, X. Tang, *Org. Lett.* **2002**,
4, 3451–3453; b) B. Jiang, Y.-G. Si,
Angew. Chem. **2004**, *116*, 218–220; *Angew.
Chem. Int. Ed.* **2004**, *43*, 216–218.

127 Y. Hanzawa, N. Tabuchi, K. Saito,
S. Noguchi, T. Taguchi, *Angew. Chem.*
1999, *111*, 2552–2555; *Angew. Chem. Int.
Ed.* **1999**, *38*, 2395–2398.

9
Asymmetric Allylic Alkylation
Manfred Braun

9.1
Introduction

Among the manifold substitution patterns at a stereogenic quaternary carbon atom in both targets and intermediates of organic syntheses [1], those containing an allylic moiety are of particular interest. This is mainly because the carbon–carbon double bond that is introduced by means of the allylation process offers a large variety of synthetic transformations [2]. From a retrosynthetic point of view [3], asymmetric allylation can be accomplished by the concepts outlined in Scheme 9.1: either a nucleophile is combined with electrophilic allylation synthons **3a** or **3b** (paths a and b) or a combination with inverse polarity is chosen, so that an electrophilic carbon atom and nucleophilic allylation synthons **7a** or **7b** (paths c and d) link together.

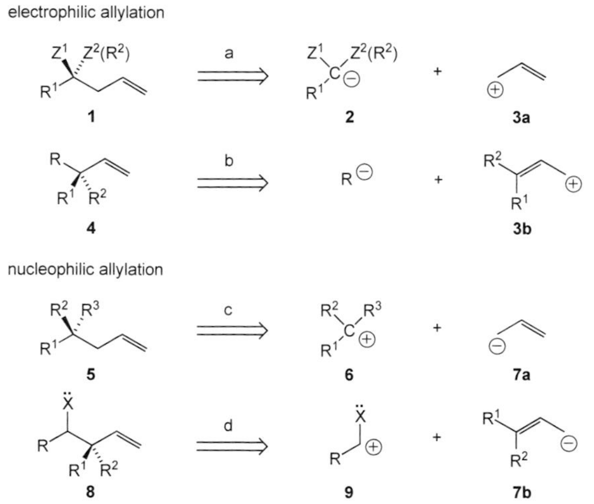

electrophilic allylation

nucleophilic allylation

Scheme 9.1 Retrosynthesis of electrophilic (paths a and b) and nucleophilic (paths c and d) allylic alkylation. R, R^1, R^2, R^3 = alkyl, aryl; Z^1, Z^2 = electron withdrawing groups; $\ddot{X}$ = electron donating group.

Quaternary Stereocenters: Challenges and Solutions for Organic Synthesis. Edited by Jens Christoffers, Angelika Baro
Copyright © 2005 WILEY-VCH Verlag GmbH & Co. KGaA, Weinheim
ISBN: 3-527-31107-6

An electrophilic allylation according to path a obviously requires the substrate **1** to carry one or two electron withdrawing groups as substituents Z^1 and Z^2. When this concept is aimed to lead to nonracemic products, enolates **2** have proven themselves to be the most suitable substrates: they readily react with allyl halides, the synthetic equivalents of the electrophilic synthon **3a**, without further mediation to yield the products **1** with a quaternary carbon center (Section 9.2.1).

A more recently developed, promising approach directed towards the synthesis of **1** is based on a palladium-catalyzed allylation that uses in situ generated η^3 transition metal complexes as electrophiles (Section 9.2.2). The particular advantage of this approach lies in the fact that enantioselectivity can be provided by chiral ligands attached to the transition metal.

According to the same polarity pattern, products that contain a stereogenic quaternary carbon center in the allyl moiety become accessible when a nucleophile R^- is combined with an electrophilic synthon **3b** that reacts in its γ-position (path b). Obviously the substituents R^1 and R^2 at the carbon–carbon double bond have to be nonidentical if the formation of a stereogenic carbon center in the product **4** is required. This relatively new approach to the synthesis of stereoselective variants will be discussed in Section 9.2.3.

A completely different route to products with a quaternary carbon center **5** is based on the nucleophilic allylation of a cationic substrate **6**. Here, allyl silanes turned out to be suitable equivalents of the nucleophilic allyl synthon **7a** (path c). Enantioselective variants have been developed and will be discussed in Section 9.3.1. In addition, nucleophilic allylation opens the way to a different substitution pattern, represented by the target **8**, which contains the stereogenic carbon center in the allylic position. It results from a combination of a cationic synthon **9** with the ambident nucleophile **7b** provided that the latter is attacked at the γ-carbon atom (path d). Stereoselective variants of this approach are based on allyl silanes and boronates serving as equivalents of synthon **7b** (Section 9.3.2). Aside from the polar combinations outlined in the retrosynthetic Scheme 9.1, radical reactions and rearrangements have been used occasionally in order to obtain compounds with stereogenic quaternary carbon centers (Section 9.4).

9.2
Electrophilic Allylic Alkylation

9.2.1
Direct Allylation of Enolates

The reaction of an α-disubstituted enolate with an electrophilic allylic reagent is an obvious idea in order to generate quaternary carbon centers carrying an allyl residue [Eq. (1)]. The reactivity of the enolate has to be high enough to overcome the increase in steric hindrance that arises when an enolate **10** is converted into an α-trisubstituted carbonyl compound **11**. This is best achieved by using "preformed enolates", which are generated and transformed under aprotic conditions [4–10].

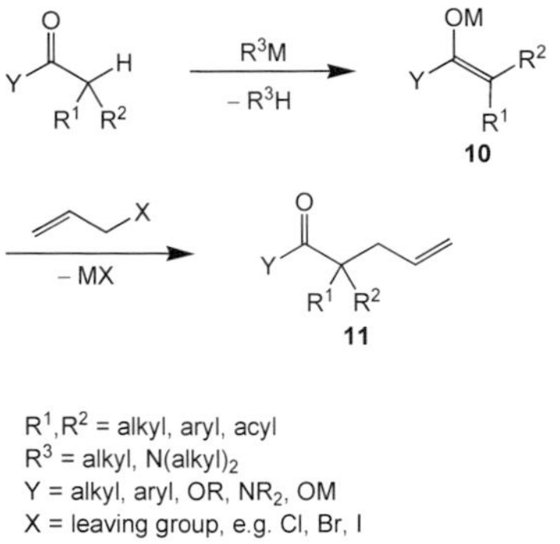

$$R^1, R^2 = \text{alkyl, aryl, acyl}$$
$$R^3 = \text{alkyl, N(alkyl)}_2$$
$$Y = \text{alkyl, aryl, OR, NR}_2, \text{OM}$$
$$X = \text{leaving group, e.g. Cl, Br, I}$$

The reaction outlined in Eq. (1) leads to a stereogenic quaternary carbon center only when the substituents R^1 and R^2 in the α-position are nonidentical. Numerous alkylations have been described, for compounds including the enolates of ketones, carboxylic esters, amides, and dianions of carboxylic acids. However, in most cases, either identical residues ($R^1 = R^2$) were applied so that no stereogenic center was formed or, if the substituents R^1 and R^2 differed, the reaction occurred without stereocontrol, resulting in racemic products **11** [4, 10].

If the allylation of an enolate is aimed at stereoselectivity, control of the enolate geometry is evidently a prerequisite. This means that only those enolates that are single diastereomers with respect to the carbon–carbon double bond [either (*E*) or (*Z*) enolates **10**] may lead to the exclusive or predominant formation of one stereoisomer of the product **11**. Various reliable methods for the control of enolate geometry have been elaborated in the course of the development of "preformed enolates" [4–10]. However, almost all these methods are restricted to α-monosubstituted enolates, and are not suitable for α-disubstituted ones, which are evidently the suitable intermediates for electrophilic allylation reactions leading to quaternary centers. The problem has been solved by using enolates with a "fixed geometry." This is provided in most cases by incorporating the carbonyl compound into a five- to eight-membered ring or by taking advantage of a chelating effect. Examples **12** and **13** represent typical enolates with a "fixed geometry" (Scheme 9.2).

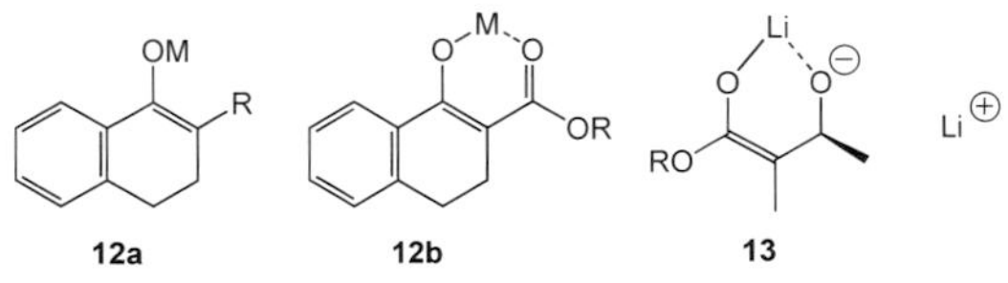

Scheme 9.2 Examples of α-disubstituted enolates with "fixed geometry" (M = Li, Na, K).

Considering the stereochemical outcome of the allylic alkylation, either diastereomeric products **11** may arise or enantiomeric ones. This depends on whether the enolate **10** is either chiral or achiral. In the latter case, the products of the allylation step are enantiomers, and stereocontrol can be achieved by the use of chiral ligands that are either attached at the enolate metal or coordinate to the metal of an electrophilic allyl complex. In chiral enolates **10** leading to diastereomeric allylation products **11**, either the residue R^1 or R^2 or the ipso substituent Y may contain a stereogenic unit.

In a seminal investigation, the allylation of doubly deprotonated β-hydroxycarboxylic esters was studied (Scheme 9.3) [11]. Thus, the ester **14**, which is a pure enantiomer with respect to the stereogenic carbinol center (C-3) but an epimeric mixture at C-2, was treated with lithium isopropylcyclohexyl amide (2 equivalents). The dilithiated species thus formed is assumed to have the rigid cyclic structure **13** due to chelation of the lithium by the two oxygen atoms. Evidently, the stereocenter at C-2 is destroyed by the formation of the enolate, which, as a consequence, is a single enantiomerically pure compound. Its treatment with allyl bromide leads to the formation of the diastereomeric esters **15a** and **15b** in a ratio of 94:6. Subsequent oxidation of the secondary alcohols **15a,b** delivered the ketone **16** in 91% *ee*. The stereochemical outcome of the reaction is a *lk*-topicity [12], which means that, in the (*S*)-enolate **13**, the electrophilic allyl bromide approaches from the *Si* face to the enolate double bond. This result is plausibly explained by the assumption that the electrophile approaches the "rigid enolate" from the sterically less hindered bottom side, *anti* to the methyl residue, as shown in the model **17** [11].

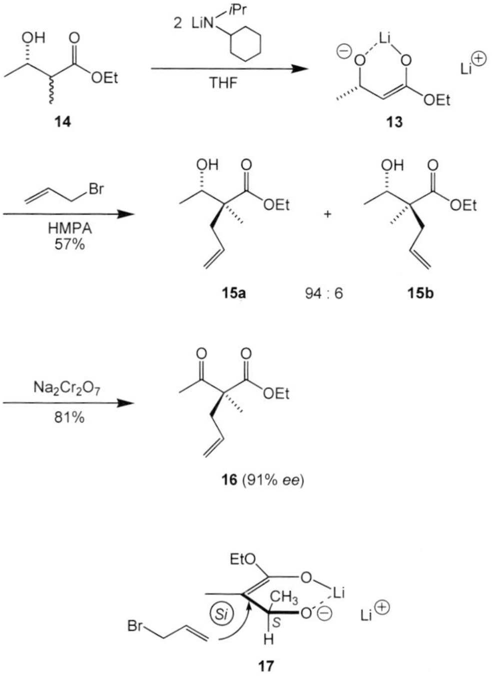

Scheme 9.3 Allylation of epimeric hydroxycarboxylic ester **14**, oxidation to ketoester **16** and *lk*-topicity in the model **17**.

An acetal derived from C_2 symmetric 1,2-cyclohexanediol serves as a temporary chiral auxiliary group in a different route to β-ketoesters containing a stereogenic quaternary center. The acetals **18**, which are diastereomeric mixtures with regard to the α-substitutent, were treated with an excess of lithium diisopropylamide, allyl bromide and HMPA. Thus, enol ethers **19** were obtained in a highly diastereoselective manner (*dr* >97.5:2.5). Evidently, enolate formation and concomitant opening of the dioxolane ring occur upon double deprotonation. Cleavage of the enol ether moiety permits removal of the chiral auxiliary group and liberates the β-ketoesters **20**. A rationale for the predominant *Si*-face attack on the intermediate enolate **21**, wherein the alkoxide lithium is chelated by two oxygen atoms, is shown in Scheme 9.4. It seems to be plausible that the *Re*-face of the enolate moiety might be shielded by the cyclohexanediol group so that the electrophilic allyl bromide approaches from the *Si*-face predominantly [13].

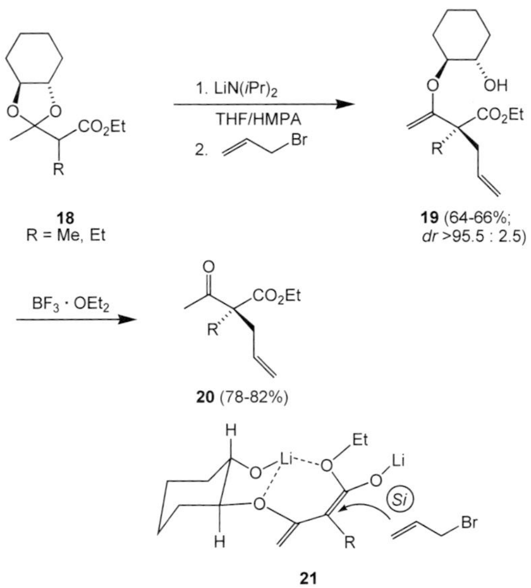

Scheme 9.4 Deprotonation and allylation of cyclohexane-1,2-diol-derived esters **18**. Model **21** of the approach of the allylation reagent to the intermediate enolate.

A rare example of a preformed enolate with a "fixed geometry" but without metal complexation is the intermediate **24** postulated in the allylation of the chiral lactam **22**, which is deprotonated with sodium diisopropylamide (Scheme 9.5). The sodium enolate moiety in **24** thus generated is assumed to orient itself perpendicular to the lactam π-system. The electrophile, allyl iodide, then approaches predominantly from the less hindered *Re*-face. Thus, the lactam **23**, bearing a quaternary carbon center in the carboxylic skeleton, is formed in a diastereomeric ratio of 20.5:1. The *Z*-configuration of the sodium enolate has been shown to exceed 49:1 by NOE measurements of the *N,O*-ketene actal [14].

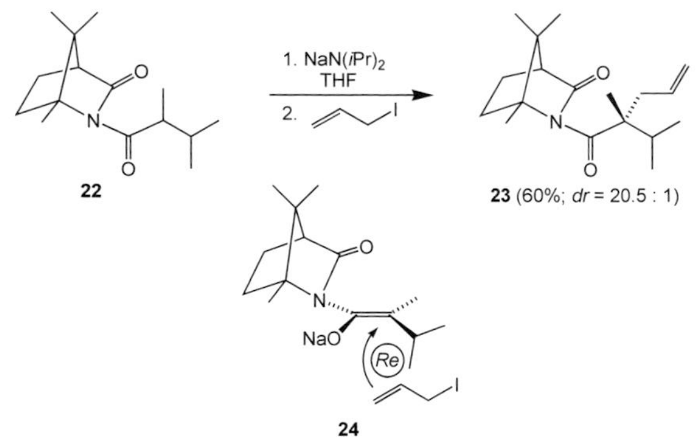

Scheme 9.5 Allylation of the chiral lactam **22** and model for the electrophilic attack on the non-chelated enolate **24**.

When five- or six-membered lactones or ketones are used as substrates for electrophilic allylations, the corresponding enolates have *per se* a "fixed geometry". In a diastereoselective variant, a one-pot procedure allowed conversion of the lactone **25** into the allylation product **28**, which is formed in a diastereomeric ratio of 10:1 (Scheme 9.6). In the key step, the enolate **27** was generated from sulfone **26** by reductive metallation and subsequently attacked by the electrophile, allyl iodide, from the less hindered *Si*-face, so that a *trans*-configuration of the allyl residue and the vicinal methyl group results [15].

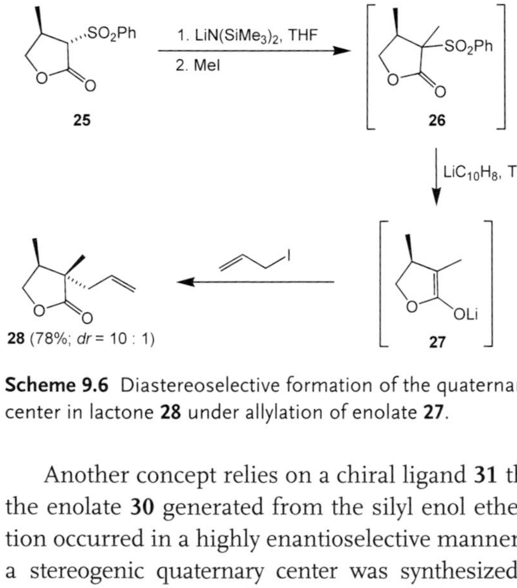

Scheme 9.6 Diastereoselective formation of the quaternary carbon center in lactone **28** under allylation of enolate **27**.

Another concept relies on a chiral ligand **31** that chelates the lithium atom in the enolate **30** generated from the silyl enol ether **29** (Scheme 9.7). Thus, allylation occurred in a highly enantioselective manner so that the tetralone **32** bearing a stereogenic quaternary center was synthesized in 97% *ee* [16]. In a different approach, an indanone-derived enolate has been reacted with a chiral allyl sulfonium salt. However, only marginal *ee* was obtained by this protocol [17].

Stereocontrol in electrophilic allylation reactions can also be exhibited in planar chiral arene complexes, as illustrated (Scheme 9.8) by the ruthenium complex **33**,

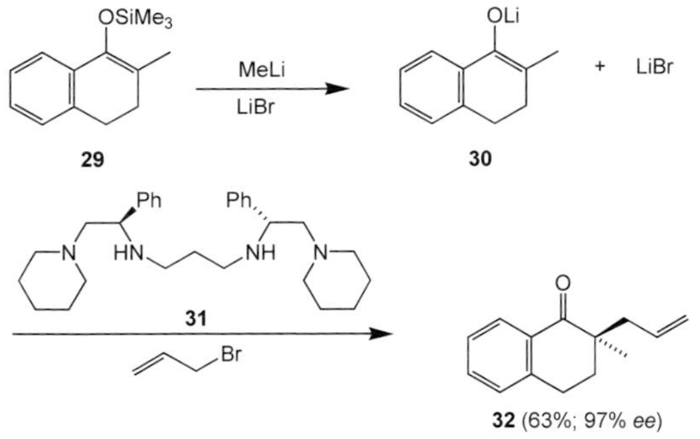

Scheme 9.7 Enantioselective allylation of lithium enolate **30** in the presence of chiral ligand **31**.

which is used as a diastereomeric mixture (1.5:1.0). When treated with sodium hydride, deprotonation occurs and the stereogenic center in the benzylic position vanishes. Various electrophilic reagents (allyl bromide, methylallyl bromide and dimethylallyl bromide) attack the intermediate **34** from the face opposite to the η^5-bound transition metal. As a result, the products **35** were obtained as single diastereomers. Subsequent decomplexation liberated the tetrahydroquinoline **36**. The method, which has been extended to the corresponding indole system, delivered racemic product **36** starting from racemic metal complex **33**. It could, in principle, be applied to enantiomerically pure starting materials [18].

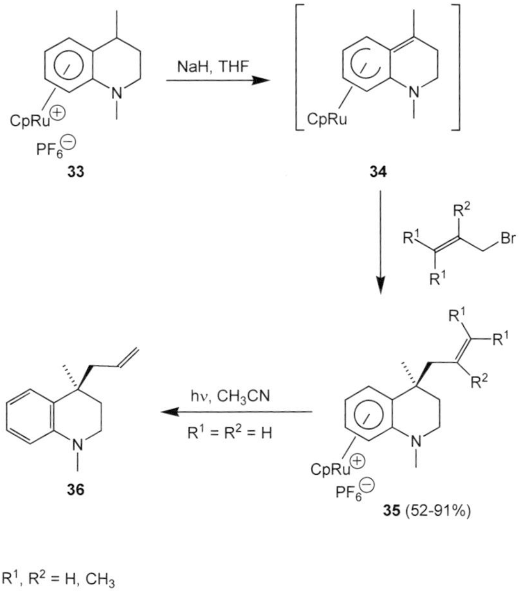

$R^1, R^2 = H, CH_3$

Scheme 9.8 Diastereoselective allylation of planar chiral ruthenium complex **33**.

In all the procedures presented in this section, the chiral auxiliary, either covalently bound or used as an additive, had to be applied in stoichiometric amounts. Remarkable progress came from the substitution reaction of palladium allyl complexes. Such complexes function as intermediates and originate from chiral palladium catalysts.

9.2.2
Palladium-catalyzed Allylation

The reaction of carbon nucleophiles with allylpalladium complexes has developed into a particularly efficient method for carbon–carbon bond formation. This is mainly due to the fact that the generation of the allylic complex is accomplished *in situ* and requires only catalytic amounts of the transition metal. Thus, allylic substrates, mostly acetates or carbonates, are treated with suitable palladium(0) reagents to generate π-allyl palladium complexes **37** (Scheme 9.9), which are stabilized by phosphines. The transition metal adopts the oxidation stage +2 in the complexes **37** that exhibit a distinct electrophilic reactivity. As a consequence, they readily react with nucleophiles so that allylic alkylation products **38** are obtained and the transition metal catalyst is liberated adopting oxidation stage 0 thereby [19].

It turned out that stabilized, "soft" carbon nucleophiles were almost exclusively employed in this procedure, typically the enolates of malonic esters and β-ketoesters or deprotonated sulfones. The extension of this method to preformed nonstabilized enolates has only occasionally been developed [20–22]. Enantioselective variants of this reaction are mainly based on chiral phosphine ligands at the palladium atom [23].

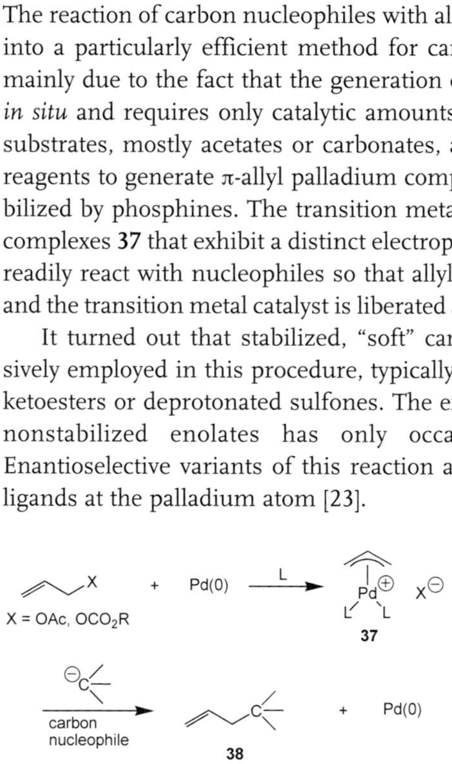

Scheme 9.9 General scheme of palladium-catalyzed allylic substitution.

Transition metal-catalyzed allylic substitution has only occasionally been used to generate stereogenic quaternary carbon centers. This is because neither symmetrical malonates nor mono- or unsubstituted allylic substrates, which represent a kind of standard combination, lead to that structural feature. Seminal work aimed at building a quaternary carbon center by means of an enantioselective palladium-catalyzed allylation is based on the chiral ferrocenyl ligand **39** [Eq. (2)]. Thus, acetylcyclohexanone was first deprotonated by means of sodium hydride to give an enolate that shows the type of "rigid geometry" of compound **12**. The allylation product **40** was obtained in a remarkable *ee* of 81%. It is assumed that an attractive interaction between the terminal hydroxy groups of the ligand and the prochiral enolate plays a key role with respect to enantioselectivity [24].

(2)

Chiral sulfoxides or sulfonamides were also used in order to mediate the same type of enantioselective allylation, as shown in Eq. (3). However, enantioselectivity was only moderate, so that the allylation product **40** was obtained in 50% *ee* [25].

(3)

Phosphine ligands that are based on the C_2 symmetric 1,2-diaminocyclohexane proved to be more efficient for this purpose, as shown in Scheme 9.10. When α-methyl tetralone **41** was submitted to the allylation protocol, the alkene **32** was obtained in 88% *ee* under optimum conditions. The protocol applied for this purpose includes a deprotonation of the ketone **41** and subsequent transmetallation with trimethyltin chloride. Thus, the nucleophilic species is obviously a tin enolate that reacts with the allyl palladium complex generated *in situ* from allyl acetate and carrying C_2 symmetric phosphine ligand **42**. It directs the allylation of the tetralone-derived enolate to occur predominantly form the *Si*-face so that (*R*)-**32** forms as the main enantiomer. The "mnemonic" shown in Scheme 9.10 offers some plausibility for the stereochemical outcome of the reaction, and considerable steric hindrance becomes evident in the disfavored topicity. The drawing indicates the C_2 symmetry of the chiral ligand **42** [21].

The method has been extended to various substituted allyl acetates and allyl carbonates. In addition to ketone **41**, ketene thioacetal **43** has been submitted to the protocol and yielded the allylation product **44** in 82% *ee* [Eq. (4)] [21].

(4)

An enhancement of enantioselectivity was observed when the chiral ligand **42** was replaced by the ferrocene derivative **45** that is based on the same C_2 symmetric

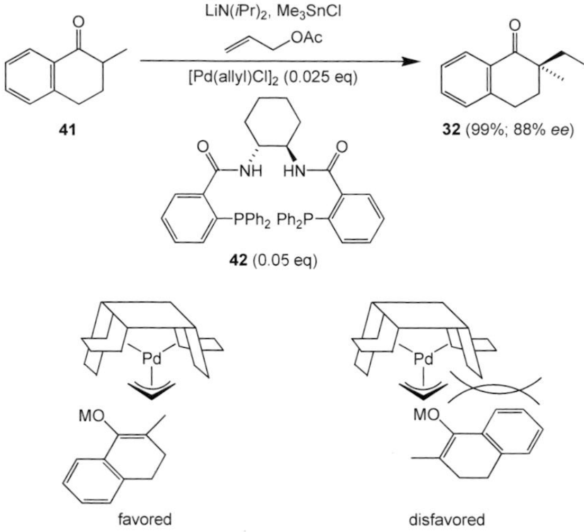

Scheme 9.10 Palladium-catalyzed allylation of α-methyl tetralone **41** mediated by the C_2 symmetric ligand **42**. Rationale for the stereochemical outcome.

"core unit", 1,2-diaminocyclohexane, and combines central and planar chirality. Thus, allylation product **32** was obtained from the lithium enolate of methyl tetralone **41** in 95% *ee*, as shown in Eq. (5). Allyl carbonates and 2-methylallyl carbonates served as precursors of the palladium complexes generated *in situ* [26].

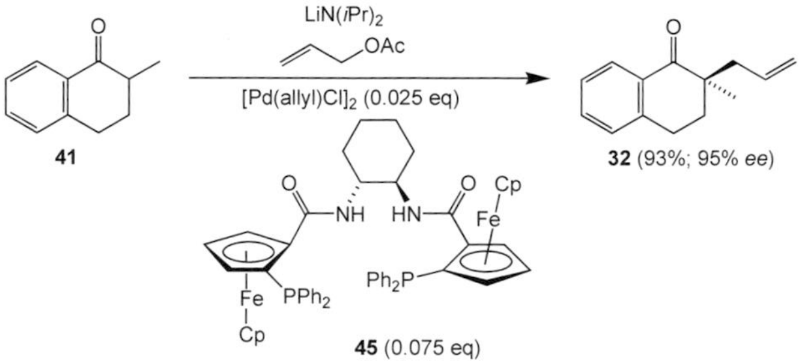

$$\tag{5}$$

Palladium-catalyzed allylations of β-diketones have also been mediated by the well-known BINAP ligand [27]. Thus, cyclic ketones **46** as well as open-chain β-dicarbonyl compounds **48** are first converted into the corresponding potassium enolates, which deliver the allylation products **47** and **49**, respectively, when treated with allyl or cinnamyl acetate in the presence of a palladium-(*R*)-BINAP catalyst, as shown in Eqs. (6, 7). The prerequisite "rigidity" of the enolate is given in the former case by the five- to eight-membered ring. In the latter case, a (*Z*)-configuration of the enolate is due to chelation of the metal in a six-membered ring. Enantiomeric excess of the allylation products **47** and **49** of up to 89% were reached [28].

$$(6)$$

46
n = 1 - 4; R = H, Ph

47 (87-95%; 77-89% ee)

(*R*)-BINAP

$$(7)$$

48
R = H, OMe

49 (90-96%; 80-83% ee)

(*R*)-BINAP

Quite a different type of ligand, the diaminophosphine oxide **52** featuring a stereogenic phosphorus atom, serves to mediate the enantioselective allylation of β-ketoesters **50** and **51**, as shown in Scheme 9.11. In this procedure, cinnamyl acetate and some derivatives thereof served as precursors of the allyl palladium complexes. It turned out that *N,O*-bis(trimethylsilyl)acetamide (BSA) was crucial in order to bring about catalytic activity. It is assumed and supported by NMR studies that silylation occurs at the phosphorus and the secondary amine nitrogen atom giving the "reactive" ligand **53**. The transition metal is than coordinated to the trivalent phosphorus of two molecules of **53** in a monodentate fashion. The complex **54** thus formed is loaded with the substituted allyl acetate thereby generating the suitable reactive electrophilic allylation species **55**. Enantioselectivities ranging from 72 to 93% are reached depending on the particular substitution pattern of the ester, the size of the ring and the cinnamic acetate component [29].

In all of the palladium-catalyzed allylation reactions discussed so far, enantioselective construction of the quaternary carbon center was provided by chiral ligands bound to the transition metal. In a different approach, the chiral nonracemic allyl acetate **56** served as the precursor for the formation of a palladium allyl complex that is, again, a chiral nonracemic species and is attacked by the nucleophilic enolate with inversion of the configuration, as shown in Scheme 9.12. High regio- and diastereoselectivities were obtained with achiral phosphine ligands such as carboxylic acid **57**. The new carbon–carbon bond forms between the *Re*-face of the allyl palladium complex (exclusively) and the *Re*-face of the enolate (predominantly). Thus, ketoesters **58** and **59** are produced in a diastereomeric ratio of 93:7 (Scheme 9.12). The protocol has been successfully applied in a synthesis of (–)-acetomycin **60** [30].

Although it is a fairly new concept, enantioselective palladium-catalyzed allylic substitution has developed very rapidly into an efficient tool for the formation of stereogenic quaternary centers carrying an allyl residue. The advantages of this

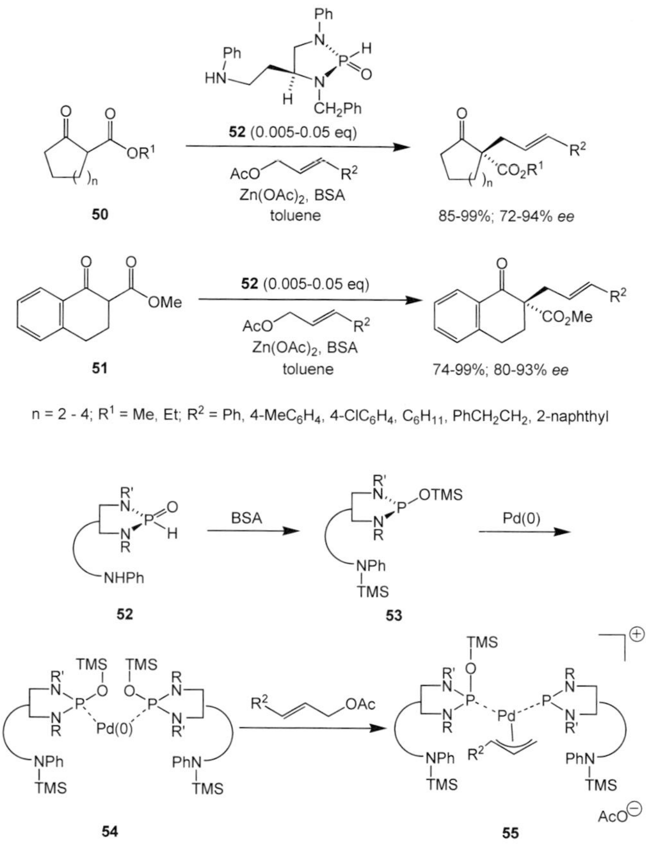

Scheme 9.11 Enantioselective allylation of ketoesters **50** and **51** mediated by chiral diaminophosphine oxide **52**. *In situ* generation of the active catalytic species.

method clearly lie in the highly electrophilic character of the allyl complexes and, more importantly, the fact that only catalytic amounts of the chiral auxiliary are required. Almost all of the protocols rely on cyclic ketone or ketoester enolates, thus underlining the importance of the "rigid enolate" as a prerequisite to high enantioselectivity.

9.2.3
γ-Attack on Electrophilic Allylic Substrates

When a carbon nucleophile is allowed to react with a substituted allylic electrophile, the problem of regioselectivity arises. If the electrophilic allylic substrate, which carries different residues R^1 and R^2 at the double bond, can be caused to react in its γ-position exclusively, the formation of a quaternary carbon center results, as shown in Scheme 9.1, path b. Various organocopper reagents have been shown to

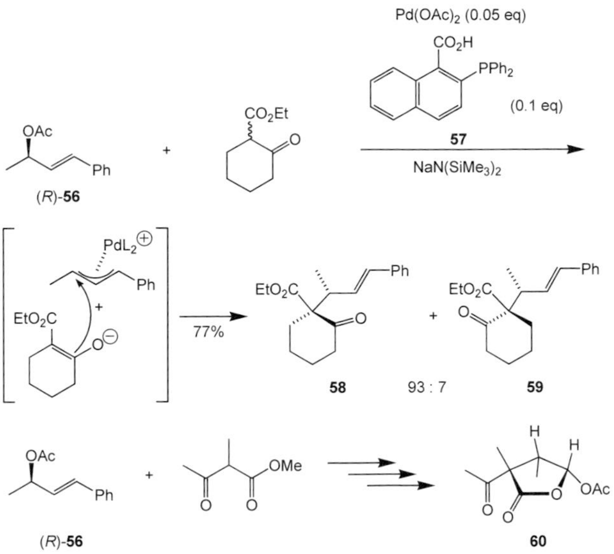

Scheme 9.12 Stereoselective formation of ketoester **58** starting from chiral nonracemic allyl acetate **56**. Synthesis of (−)-acetomycin **60**.

react selectively at the γ-carbon atom of electrophiles under allylic rearrangement [31]. Even if the γ-position carries two substituents, thus causing significant steric hindrance, predominant γ-attack is possible, as shown in Eq. (8) [32].

$$\text{(8)}$$

However, this concept has only been used to obtain chiral compounds with a stereogenic center in an enantioselective manner only recently (Scheme 9.13). Thus, a series of chiral ligands have been used to form copper complexes that, in turn, catalyze the addition of various dialkyl zinc compounds to the allylic substrate **61**, which contains a trisubstituted double bond [33–35]. Under these conditions, high regioselectivities are obtained so that the ratio of regioisomers usually surpasses 98:2 in favor of the γ-substitution products.

The enantioselectivity strongly depends on the particular ligands, of which compounds **63** and **64** are representative examples. So far, optimum enantioselectivities (up to 92%) have been obtained with the axially chiral ligand **64**, as shown by the conversion of phosphate **61** into allylation product **62**, a representative example that illustrates this concept [35]. Copper carbene complexes carrying ligands related to **64** also permit the use of Grignard reagents instead of dialkyl zinc compounds, although this is at the expense of lower enantioselectivities [36].

When an organometallic reagent is reacted with an allylic substrate in an S_N2' manner according to path b in Scheme 9.1, the configuration of the resulting stereogenic quaternary center clearly depends on the configuration of the

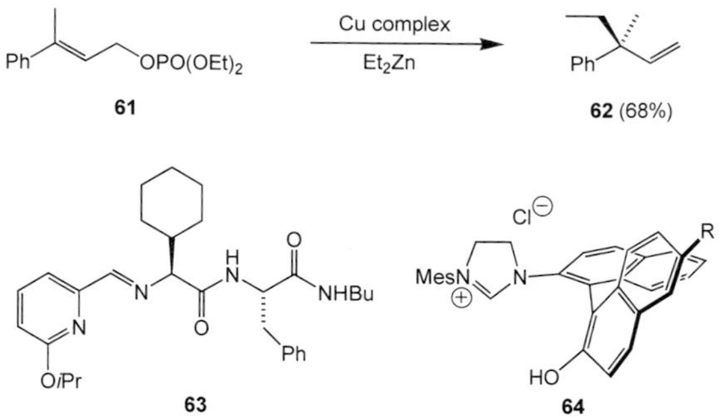

Scheme 9.13 Enantioselective γ-allylation mediated by chiral copper catalysts with ligands **63** and **64**.

trisubstituted double bond. This correlation has been elegantly shown by using the homochiral esters (*E*)- and (*Z*)-**65** [Eqs. (9, 10)]. When treated with dipentyl zinc in the presence of CuCN·2LiCl, enantiomeric products (*S*)-**66** and (*R*)-**66** result from the diastereomers (*E*)- and (*Z*)-**65**, respectively. The stereochemical outcome is an *anti*-substitution at the allylic substrate, occurring with high γ-selectivity [37]. Both enantiomers of **66** are obtained in 94% *ee*, thus demonstrating the efficiency of the chirality transfer.

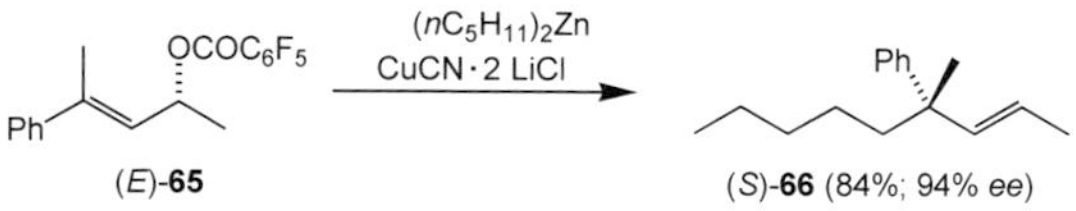

$$(9)$$

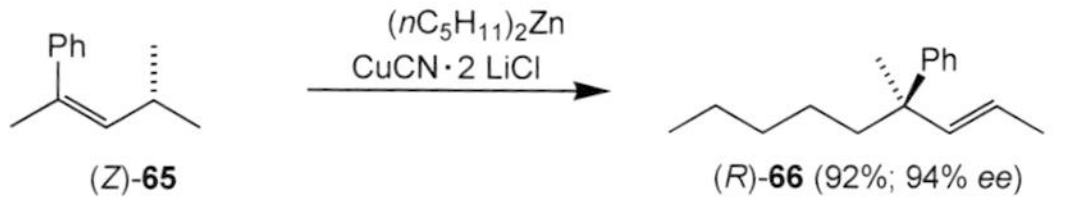

$$(10)$$

9.3
Nucleophilic Allylic Alkylation

9.3.1
Allylation of Trisubstituted Electrophilic Carbon Centers

The construction of quaternary carbon centers by combining a tertiary carbenium ion or, more generally, a trisubstituted electrophilic carbon atom with a nucleophilic allyl reagent, as outlined in Scheme 9.1, path b, is an obvious idea. This type of reaction has been observed in the course of geminal dialkylation of aryl alkyl ketones [38]. A stereoselective variant, however, was not disclosed until more recently. Thus, racemic silyl ethers **67** derived from methyl indanone and tetralone were treated with allyl trimethylsilane in the presence of the chiral titanium complex **68**. As shown in Scheme 9.14 [39], the allylation products **69** resulted, along with Me₃SiOSiMe₃, which is formed as a byproduct in stoichiometric amounts. The alkenes **69** were obtained in high enantiomeric excesses (93 to >98% *ee*) – a remarkable result inasmuch as racemic starting materials **67** have been used. The fact that the chemical yield of the products **69** is 94–96% clearly shows the conversion of both enantiomers of the starting material. Obviously, this reaction cannot be a simple kinetic resolution, but a dynamic kinetic asymmetric transformation [40].

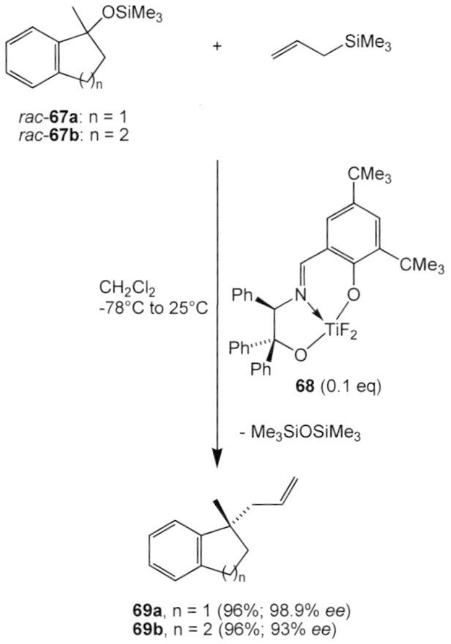

Scheme 9.14 Enantioselective allylation of racemic silyl ethers **67** catalyzed by titanium complex **68**.

The conversion of both enantiomers of the racemic silyl ethers **67** into the alkenes (*S*)-**69** with high enantiomeric excesses is rationalized as a dynamic kinetic asymmetric transformation as outlined in Scheme 9.15 for the indanyl series (**67a** to **69a**). A prerequisite for that kind of dynamic process is an *in situ* racemization that can occur either at the stage of the starting material or of an intermediate. Thus, it is assumed that the chiral Lewis acid **68** and both enantiomers of the silyl ether **67a** form two diastereomeric contact ion pairs, (*R*)-**67a·68** and (*S*)-**67a·68**. It seems to be plausible that they equilibrate rapidly via a planar, achiral carbenium ion **70**. Finally, one of the ion pairs, (*S*)-**67a·68** is postulated to react significantly faster with allyltrimethyl silane than its diastereomer (*R*)-**67a·68**. Based on the stereochemistry observed in Sakurai reactions, it is assumed that the allylic nucleophile attacks the indanyl moiety from the face that is not blocked by the titanium. Evidence for a substantial positive partial charge in the postulated intermediates **67a·68** comes from ^{13}C-NMR measurements [39].

This approach seems to be particularly appropriate for the construction of quaternary carbon centers carrying an allyl moiety, inasmuch as substrates that are trisubstituted and carry leaving groups in an activated position are suitable precursors. The allylic substituent is therefore attached in a highly enantioselective manner.

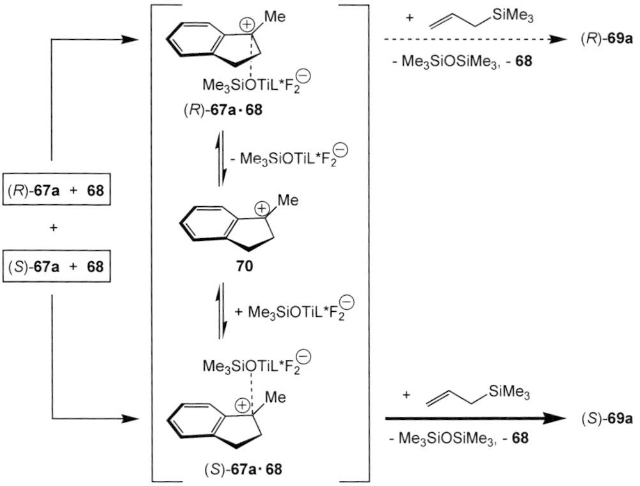

Scheme 9.15 Dynamic kinetic asymmetric allylic alkylation of racemic silyl ether *rac*-**67a**, mediated by titanium complex **68**.

9.3.2
γ-Addition of Allylic Nucleophiles to Aldehydes

The reaction of an activated aldehyde with a nucleophilic allylic compound that reacts in its γ-position is another approach that leads to a stereogenic quaternary center, provided that the carbon–carbon double bond carries different substituents R^1 and R^2, as shown in Scheme 9.1, path d. However, this concept, too, has not been applied frequently. In a diastereoselective approach, allyl borations of aldehydes were mediated by scandium triflate [Eq. (11)]. Thus, α-methylene-γ-lactones **72** were obtained from allylboronates **71** in diastereomeric ratios higher than 20:1. The method requires the allylboronate to consist of a single diastereomer with respect to the trisubstituted double bond. The thermal reaction is also feasible, but occurs in lower diastereoselectivity [41].

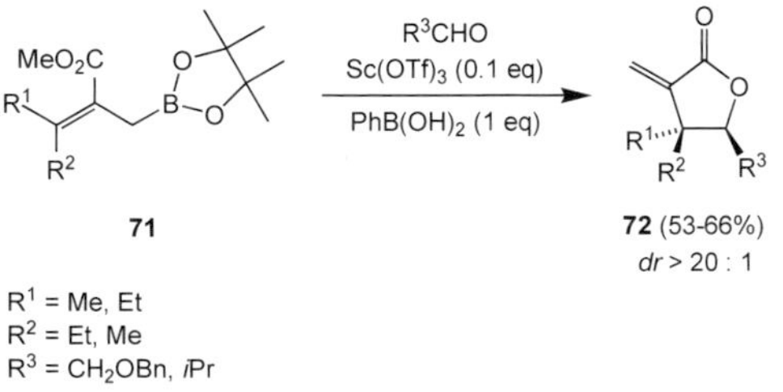

$$R^1 = Me, Et$$
$$R^2 = Et, Me$$
$$R^3 = CH_2OBn, iPr$$

(11)

71

72 (53-66%)
dr > 20 : 1

A diastereoselective and enantioselective approach that follows this concept has been realized by using a novel chiral bisphosphoramide (*R,R*)-**73**. It allows the addition of the isomeric trichlorosilanes (*E*)- and (*Z*)-**74** to benzaldehyde to be directed in a highly diastereoselective and enantioselective manner. Thus, the *anti*-diastereomer of the homoallylic alcohol **75** results from (*E*)-**74** whereas *syn*-**75** is predominantly obtained from (*Z*)-**74**, as shown in Scheme 9.16. A stereochemical correlation between double-bond geometry and configuration of the products opens a way to both stereoisomers of **75** [42].

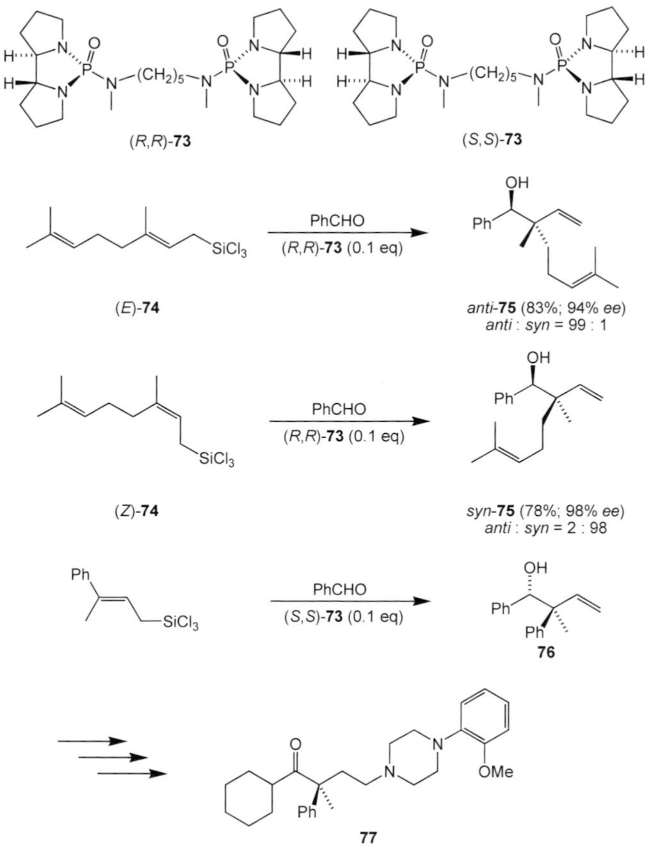

(*R*,*R*)-**73**   (*S*,*S*)-**73**

(*E*)-**74**   PhCHO, (*R*,*R*)-**73** (0.1 eq)   *anti*-**75** (83%; 94% *ee*) *anti* : *syn* = 99 : 1

(*Z*)-**74**   PhCHO, (*R*,*R*)-**73** (0.1 eq)   *syn*-**75** (78%; 98% *ee*) *anti* : *syn* = 2 : 98

Ph—SiCl₃   PhCHO, (*S*,*S*)-**73** (0.1 eq)   **76**

77

Scheme 9.16 Diastereoselective and enantioselective allylation mediated by chiral catalysts (*R*,*R*)- and (*S*,*S*)-**73**.

The homoallylic alcohol **76** is accessible from allyltrichlorosilane and benzaldehyde by means of the enantiomeric catalyst (*S*,*S*)-**73**. Alcohol **76** has been converted into the serotonine agonist **77** [43].

9.4
Miscellaneous Methods

As well as the intermolecular polarity-mediated reactions, the various Cope and Claisen-type rearrangements open an alternative route to quaternary carbon centers. Although this concept has been repeatedly used for the purpose, the targets obtained were either achiral owing to the substitution pattern at the quaternary carbon or, if they were chiral, the rearrangement took place only in a diastereoselective but not enantioselective way [44].

In a typical example, amino acids containing quaternary carbon centers are accessible by ester enolate Claisen rearrangements. Thus, the deprotonation of *N*-protected α-aminoester (*E*)-**78** with lithium diisopropylamide followed by trans-metallation with zinc chloride smoothly led to the formation of carboxylic acid **79**, which was protected as its methyl ester **80** by treatment with diazomethane. A diastereomeric ratio of 96:4 in favor of ester **80** was obtained (Scheme 9.17). The stereochemical outcome is easily rationalized by postulating the Claisen rearrangement to occur via the chair-like transition state **81** [45].

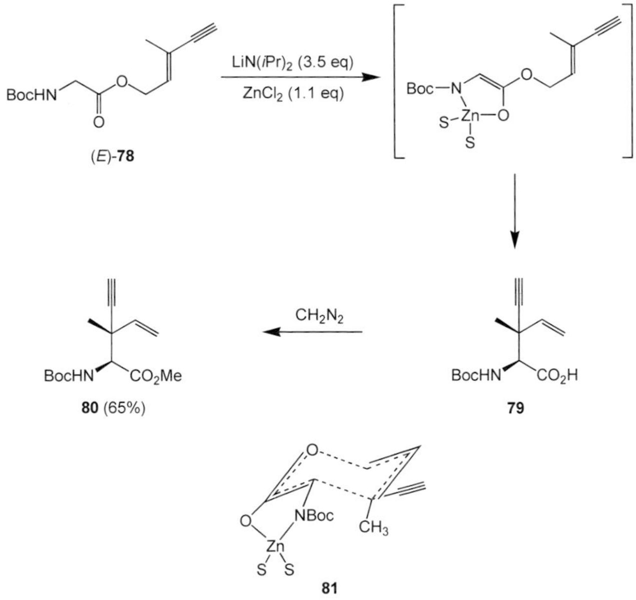

Scheme 9.17 Claisen rearrangement of *N*-protected α-amino esters **78**. Transition state model **81** (S = solvent).

When (*Z*)-**78** was chosen as the starting material the same diastereomer **80** was obtained, although with distinctly lower diastereoselectivity (73:27). A boat-like transition state is postulated in order to explain this result. The protocol allows the sterically demanding α-amino acids to be obtained in a diastereoselective manner [45]. An enantioselective variant has also been disclosed; it is based on the use of quinine as a chiral additive and leads to α-amino acids related to **79** in up to 93% *ee* [46].

An auxiliary-based concept was realized in the aza-Claisen rearrangement of N,O-acetal **82**. Remarkable diastereoselectivity was obtained, as shown by the ratio of products in Eq. (12), but the chemical yield was rather poor (13%) [47].

$$(12)$$

A suitable substituted enantiomerically pure secondary alcohol can serve as a starting material for compounds with quaternary carbon centers if a chirality transfer is possible. In an illustrative example, shown in Eq. (13), the oxy-Cope rearrangement of deprotonated carbinol **83**, delivered aldehyde **84** in 83% *ee* [48].

$$(13)$$

Enantioselective Claisen–Ireland rearrangements have been efficiently mediated by the chiral diazaborolidine **85**. Representative examples of this concept are given in Eqs. (14) [49] and (15) [50]. Thus, polyenes **86** and **87**, both serving as intermediates in the synthesis of diterpenoids, became accessible in over 99 and 98% *ee*, respectively. The chiral reagent **85** is required in stoichiometric amounts.

$$(14)$$

$$(15)$$

Finally, an enantioselective allylation has been reported that relies on radical-mediated coupling, as shown in Eq. (16). When iodolactones **88** were treated with allyltributylstannane in the presence of trimethyl aluminum and the chiral ligand **89**, allylation products **90** were obtained in enantiomeric excesses up to 91%. The amount of the chiral Lewis acid that was generated from **89** and trimethyl aluminum varied from 0.1 to 1.0 equivalent [51].

R = Me, CH$_2$OMe, CH$_2$OEt, CH$_2$OBn

(16)

9.5
Outlook

Among the various methods for the synthesis of quaternary carbon centers, asymmetric allylic substitution can be considered as a relatively new concept. Indeed, most of the methods that permit the enantioselective creation of carbon centers bearing allyl moieties have been reported since the mid-1990s. As success in the methodology will undoubtedly continue, one can envisage that more and more procedures for asymmetric allylic alkylations aimed at the synthesis of targets with quaternary carbon centers will be developed in the near future.

References

1 (a) E. J. Corey, A. Guzman-Perez, *Angew. Chem.* **1998**, *110*, 402; *Angew. Chem. Int. Ed.* **1998**, *37*, 388; (b) J. Christoffers, A. Mann, *Angew. Chem.* **2001**, *113*, 4725; *Angew. Chem. Int. Ed.* **2001**, *40*, 4591.

2 H. Söll, in *Houben Weyl, Methoden der Organischen Chemie*, Vol. V, 1b, Thieme, Stuttgart, **1972**, pp. 946–1179.

3 D. Seebach, *Angew. Chem.* **1979**, *91*, 259; *Angew. Chem. Int. Ed. Engl.* **1979**, *18*, 239.

4 C. H. Heathcock **1992**, in *Modern Synthetic Methods 1992*, ed. R. Scheffold, VHCA, VCH, Basel, Weinheim, pp. 1–102.

5 C. H. Heathcock **1993**, in *Comprehensive Organic Synthesis*, Vol. 2, ed. B. Trost, Pergamon, Oxford, Ch. 1.6.

6 D. Seebach, *Angew. Chem.* **1988**, *100*, 1685; *Angew. Chem. Int. Ed. Engl.* **1988**, *27*, 1624.

7 C. Lambert, P. v. R. Schleyer **1993**, in *Houben Weyl, Methoden der Organischen Chemie*, Vol. E 19d, ed. M. Hanack, Thieme, Stuttgart, pp. 75–82.

8 L. M. Jackman, J. Bortiatynski **1992**, in *Advances in Carbanion Chemistry*, Vol. 1, ed.

V. Snieckus, JAI Press, Greenwich, CN, pp. 45–87.

9 P. G. Williard **1991**, in *Comprehensive Organic Synthesis*, Vol. 1, ed. B. M. Trost, Pergamon, Oxford, Ch. 1.1.

10 D. A. Evans **1984**, in *Asymmetric Synthesis*, Vol. 3, Part B, ed. J. D. Morrison, Academic Press, New York, Ch. 1.

11 G. Fráter, U. Müller, W. Günther, *Tetrahedron* **1984**, *40*, 1269.

12 For a definition of the descriptors "*lk*" and "*ul*", see: D. Seebach, V. Prelog, *Angew. Chem.* **1982**, *94*, 696; *Angew. Chem. Int. Ed. Engl.* **1982**, *21*, 654.

13 M. Tanaka, M. Oba, K. Tamai, H. Suemune, *J. Org. Chem.* **2001**, *66*, 2667.

14 R. K. Boeckman, Jr., D. J. Boehmler, R. A. Musselman, *Org. Lett.* **2001**, *3*, 3777.

15 P. A. Evans, L. J. Kennedy, *Org. Lett.* **2000**, *2*, 2213.

16 Y. Yamashita, K. Odashima, K. Koga, *Tetrahedron Lett.* **1999**, *40*, 2803.

17 K. Umemura, H. Matsuyama, M. Kobayashi, N. Kamigata, *Bull. Chem. Soc. Jpn.* **1989**, *62*, 3026.

18 F. C. Pigge, S. Fang, N. P. Rath, *Tetrahedron Lett.* **1999**, *40*, 2251.

19 For reviews, see: (a) B. M. Trost, *Tetrahedron* **1977**, *33*, 2615; (b) J. Tsuji **1980**, *Organic Synthesis with Palladium Compounds*, Springer, New York; (c) B. M. Trost, *Acc. Chem. Res.* **1980**, *13*, 385; (d) B. M. Trost, *Pure Appl. Chem.* **1981**, *53*, 2357; (e) J. Tsuji, *Pure Appl. Chem.* **1982**, *54*, 197; (f) S. A. Godleski **1991**, in *Comprehensive Organic Synthesis*, Vol. 4, ed. B. M. Trost, Pergamon, Oxford, pp. 585–661.

20 (a) J. C. Fiaud, J.-L. Malleron, *J. Chem. Soc., Chem. Commun.* **1981**, 1159; (b) B. Åkermark, A. Jutand, *J. Organomet. Chem.* **1981**, *217*, C41; (c) E. Negishi, H. Matsushita, S. Chatterjee, R. A. John, *J. Org. Chem.* **1982**, *47*, 3188; (d) B. M. Trost, E. Keinan, *Tetrahedron Lett* **1980**, *21*, 2591; (e) B. M. Trost, C. R. Self, *J. Org. Chem.* **1984**, *49*, 468; (f) U. Kazmaier, F. L. Zumpe, *Angew. Chem.* **1999**, *111*, 1572; *Angew. Chem. Int. Ed.* **1999**, *38*, 1468; (g) U. Kazmaier, *Current Org. Chem.* **2003**, *7*, 317.

21 B. M. Trost, G. M. Schroeder, *J. Am. Chem. Soc.* **1999**, *121*, 6759. Very recently, enantioselective palladium-catalyzed allylic alkylations of ketones have been performed using allyl enol carbonates: D. C. Behenna, B. M. Stoltz, *J. Am. Chem. Soc.* **2004**, *126*, 15044; B. M. Trost, J. Xu, *J. Am. Chem. Soc.* **2005**, *127*, 2846.

22 M. Braun, F. Laicher, T. Meier, *Angew. Chem.* **2000**, *112*, 3637; *Angew. Chem. Int. Ed.* **2000**, *39*, 3494.

23 For reviews, see: (a) O. Reiser, *Angew. Chem.* **1993**, *105*, 576; *Angew. Chem. Int. Ed. Engl.* **1993**, *32*, 547; (b) J. M. J. Williams, *Synlett* **1996**, 705; (c) B. M. Trost, D. L. Van Vranken, *Chem. Rev.* **1996**, *96*, 395; (d) G. Helmchen, *J. Organomet. Chem.* **1999**, *576*, 203; (e) G. Helmchen, A. Pfaltz, *Acc. Chem. Res.* **2000**, *33*, 336.

24 T. Hayashi, K. Kanehira, T. Hagihara, M. Kumada, *J. Org. Chem.* **1988**, *53*, 113.

25 (a) K. Hiroi, A. Hidaka, R. Sezaki, Y. Imamura, *Chem. Pharm. Bull.* **1997**, *45*, 769; (b) K. Hiroi, Y. Suzuki, *Heterocycles* **1997**, *46*, 77.

26 S.-L. You, X.-L. Hou, L.-X. Dai, X.-Z. Zhu, *Org. Lett.* **2001**, *3*, 149.

27 (a) A. Miyashita, A. Yasuda, H. Takaya, K. Toriumi, T. Ito, T. Souchi, R. Noyori, *J. Am. Chem. Soc.* **1980**, *102*, 7932; (b) H. Takaya, S. Akutagawa, R. Noyori, *Org. Synth.* **1988**, *67*, 20.

28 R. Kuwano, K. Uchida, Y. Ito, *Org. Lett.* **2003**, *5*, 2177.

29 T. Nemoto, T. Matsumoto, T. Masuda, T. Hitomi, K. Hatano, Y. Hamada, *J. Am. Chem. Soc.* **2004**, *126*, 3690.

30 J. Uenishi, M. Kawatsura, D. Ikeda, N. Muraoka, *Eur. J. Org. Chem.* **2003**, 3909.

31 For reviews, see: (a) G. H. Posner, *Org. React.* **1972**, *19*, 1; (b) G. H. Posner, *Org. React.* **1975**, *22*, 253; (c) B. H. Lipshutz, S. Sengupta, *Org. React.* **1992**, *41*, 135; (d) N. Krause **1996**, *Metallorganische Chemie*, Spektrum Akademischer Verlag, Heidelberg, Ch. 8.3.

32 Y. Yamamoto, S. Yamamoto, H. Yatagai, K. Maruyama, *J. Am. Chem. Soc.* **1980**, *102*, 2318.

33 C. A. Luchaco-Cullis, H. Mizutani, K. E. Murphy, A. H. Hoveyda, *Angew. Chem.* **2001**, *113*, 1504; *Angew. Chem. Int. Ed.* **2001**, *40*, 1456.

34 M. A. Kacprzynski, A. H. Hoveyda, *J. Am. Chem. Soc.* **2004**, *126*, 10676.

35 A. O. Larsen, W. Leu, C. N. Oberhuber, J. E. Campbell, A. H. Hoveyda, *J. Am. Chem. Soc.* **2004**, *126*, 11130.

36 S. Tominaga, Y. Oi, T. Kato, D. K. An, S. Okamoto, *Tetrahedron Lett.* **2004**, *45*, 5585.

37 N. Harrington-Frost, H. Leuser, M. I. Calaza, F. F. Kneisel, P. Knochel, *Org. Lett.* **2003**, *5*, 2111.

38 M. T. Reetz, *Top. Curr. Chem.* **1982**, *106*, 1.

39 M. Braun, W. Kotter, *Angew. Chem.* **2004**, *116*, 520; *Angew. Chem. Int. Ed.* **2004**, *43*, 514.

40 Cf.: (a) R. S. Ward, *Tetrahedron: Asymmetry* **1995**, *6*, 1475; (b) R. Noyori, M. Tokunaga, M. Kitamura, *Bull. Chem. Soc. Jpn.* **1995**, *68*, 36; (c) H. Stecher, K. Faber, *Synthesis* **1997**, 1; (d) E. J. Ebbers, G. J. A. Ariaans, J. P. M. Houbiers, A. Bruggink, B. Zwanenburg, *Tetrahedron* **1997**, *53*,

9417; (e) U. T. Strauss, U. Felfer, K. Faber, *Tetrahedron: Asymmetry* **1999**, *10*, 107; (f) S. Caddick, K. Jenkins, *Chem. Soc. Rev.* **1996**, *25*, 447; (g) F. F. Huerta, A. B. E. Minidis, J.-E. Bäckvall, *Chem. Soc. Rev.* **2001**, *30*, 321.

41 J. W. J. Kennedy, D. G. Hall, *J. Org. Chem.* **2004**, *69*, 4412.

42 S. E. Denmark, J. Fu, *J. Am. Chem. Soc.* **2001**, *123*, 9488.

43 S. E. Denmark, J. Fu, *Org. Lett.* **2002**, *4*, 1951.

44 See, for example: (a) A. Srikrishna, P. Praveen Kumar, *Tetrahedron Lett.* **1995**, *36*, 6313; (b) A. Srikrishna, M. Srinivasa Rao, *Tetrahedron Lett.* **2002**, *43*, 151; (c) A. Srikrishna, M. Srinivasa Rao, *Synlett* **2002**, 340. For reviews, see: (d) F. E.

Ziegler, *Chem. Rev.* **1988**, *88*, 1423; (e) S. Pereira, M. Srebnik, *Aldrichimica Acta* **1993**, *26*, 17.

45 U. Kazmaier, *J. Org. Chem.* **1996**, *61*, 3694.

46 A. Krebs, U. Kazmaier, *Tetrahedron Lett.* **1996**, *37*, 7945.

47 M. J. Kurth, E. G. Brown, *Synthesis* **1988**, 362.

48 E. Lee, I.-J. Shin, T.-S. Kim, *J. Am. Chem. Soc.* **1990**, *112*, 260.

49 E. J. Corey, B. E. Roberts, B. R. Dixon, *J. Am. Chem. Soc.* **1995**, *117*, 193.

50 E. J. Corey, R. S. Kania, *J. Am. Chem. Soc.* **1996**, *118*, 1229.

51 M. Murakata, T. Jono, Y. Mizuno, O. Hoshino, *J. Am. Chem. Soc.* **1997**, *119*, 11713.

10
Phase-Transfer Catalysis
Takashi Ooi and Keiji Maruoka

10.1
Introduction

Bond-forming and cleavage reactions by phase-transfer catalysis (PTC) have long been regarded as a practical method for organic synthesis because of their simple experimental operations, mild reaction conditions, inexpensive and environmentally benign reagents and solvents, and their potential for being carried out on a large scale. Nowadays, they appear to be a prime synthetic tool, and are appreciated in various fields of organic chemistry. In addition, since phase-transfer catalysis offers yet more advantages such as increasing yield, reducing reaction time and/or temperature, and eliminating solvent, it has found widespread industrial applications. Accordingly, it is reasonably accepted that phase-transfer catalysis has become a standard method in organic synthesis [1].

On the other hand, the development of asymmetric phase-transfer catalysis, based on the use of structurally well-defined chiral, nonracemic catalysts, has progressed rather slowly. However, the enormous efforts that have been made have certainly resulted in notable achievements in this field, making it feasible to perform various bond-formation reactions under phase-transfer-catalyzed conditions [2]. The main purpose of this chapter is to visualize the range of processes available for the construction of quaternary carbon centers with synthetically useful levels of stereoselectivity by using chiral phase-transfer catalysts. The reaction patterns involved in this event are itemized starting from alkylation, in which carbon–heteroatom bond formation reactions are included taking a broad view of what constitutes a quaternary stereocenter.

Quaternary Stereocenters: Challenges and Solutions for Organic Synthesis. Edited by Jens Christoffers, Angelika Baro
Copyright © 2005 WILEY-VCH Verlag GmbH & Co. KGaA, Weinheim
ISBN: 3-527-31107-6

10.2 Carbon–Carbon Bond Formation Through PTC

10.2.1 Alkylation

10.2.1.1 Ketones Possessing Acidic α-Protons

In 1984, the Merck research group reported that the methylation of phenylindanone derivative **1a** under liquid (toluene)–liquid (50% NaOH aqueous solution) phase-transfer conditions with cinchonine-derived quaternary ammonium salt **2a** as catalyst gave the corresponding alkylated product **3a** in excellent yield with high enantiomeric excess (Scheme 10.1) [3]. They made careful and systematic studies of this reaction, proposing the intermediacy of a tight ion pair **4** fixed by an electrostatic effect and hydrogen bonding as well as π–π stacking interactions to account for the result. This is the first successful example not only of a phase-transfer-catalyzed asymmetric alkylation but also of a highly enantioselective construction of quaternary stereocenters under phase-transfer conditions. The effectiveness of the catalysis was also demonstrated in the reaction of α-*n*-propyl analogue **1b** with 1,3-dichloro-2-butene as included in Scheme 10.1 [4].

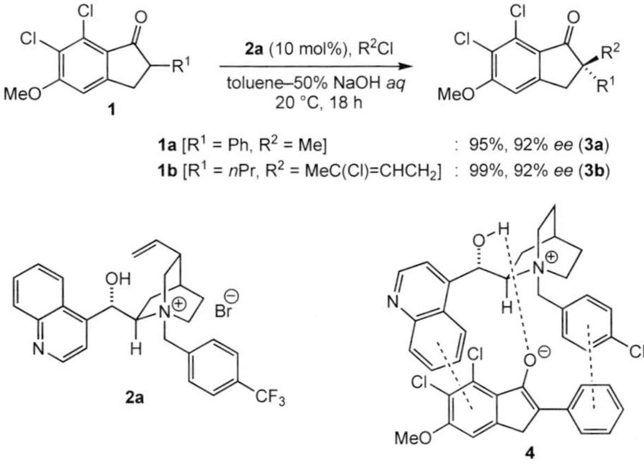
Cl
O
Cl
R1
MeO
1
2a (10 mol%), R2Cl
toluene–50% NaOH aq
20 °C, 18 h
R2
R1
1a [R1 = Ph, R2 = Me] : 95%, 92% ee (3a)
1b [R1 = nPr, R2 = MeC(Cl)=CHCH2] : 99%, 92% ee (3b)
OH
N
H
Br
2a
CF3
O
H
CF3
4

Scheme 10.1 Asymmetric phase-transfer-catalyzed alkylation of indanone derivatives.

This alkylation strategy was successfully applied to the asymmetric cyanomethylation of oxindole **5** with catalyst **2b** having a 3,4-dichlorophenylmethyl group on nitrogen, allowing a practical stereoselective synthesis of (–)-esermethole **7**, a penultimate intermediate to a clinically useful anticholinesterase agent (–)-physostigmine (Scheme 10.2) [5]. In this study, the generality of the alkylation was demonstrated with other alkyl halides and, in particular, the allylation product can also be converted to (–)-physostigmine via ozonolysis and reductive amination.

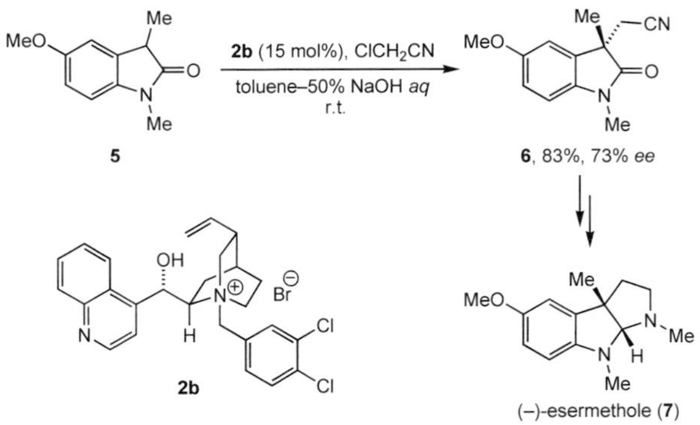

Scheme 10.2 Asymmetric cyanomethylation of oxindole.

The asymmetric alkylation of isomeric tetralone derivative **8** with 1,5-dibromopentane was also examined under similar phase-transfer conditions with cinchonidine-derived catalyst **9a**. The reaction must be carried out under completely oxygen-free conditions to avoid rapid oxidation at the α-position. The involvement of a similar ion pair to that of the indanone alkylation (**4** in Scheme 10.1) was suggested by the configuration of product **10**, which was proven to be *R* by the synthesis of (–)-Wy-16,225 (**11**), a potent analgesic agent (Scheme 10.3) [6].

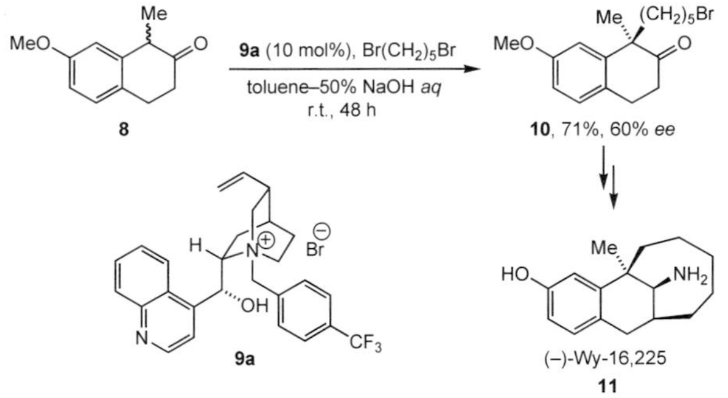

Scheme 10.3 Asymmetric alkylation of isomeric tetralone aimed at the synthesis of (–)-Wy-16,225.

Owing to the distinctive properties of organofluorine compounds, their preparation in an optically pure form, particularly by way of a practical synthetic protocol, has continued to be an important problem [7]. In this connection, Shioiri, Arai and coworkers studied the asymmetric alkylation of α-fluorotetralone **12** under phase-transfer conditions using a chiral quaternary ammonium bromide of type **2** [8]. Screening of the catalyst structure in the benzylation of **12** by varying the arylmethyl substituent on nitrogen eventually revealed that **2c**, with the 2,3,4,5,6-pentamethylphenylmethyl group, was the catalyst of choice. Thus, the reaction of **12** with benzyl bromide in the presence of **2c** (10 mol%) and KOH in toluene at –10°C for 24 h afforded the desired product **13a** in 71% yield with 80% *ee*.

This system tolerated several benzylic bromides with substituents of different steric and electronic properties as shown in Scheme 10.4.

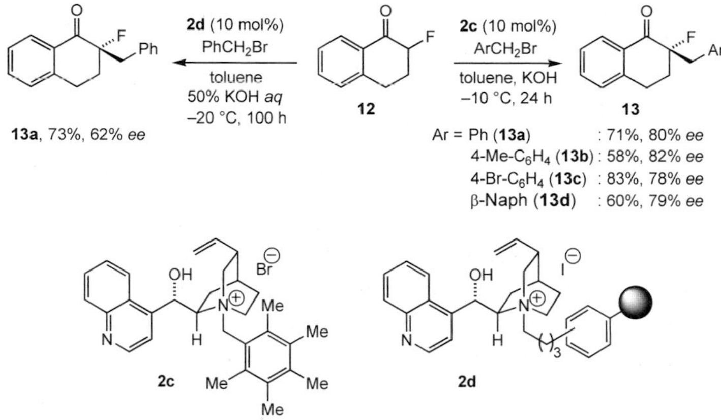

Scheme 10.4 Asymmetric alkylation of α-fluorotetralone.

The cinchoninium iodide linked to polystyrene **2d** was also found to be effective for the benzylation of **12** in toluene–50% aqueous KOH, giving **13a** in 73% yield and 62% *ee* after stirring at –20°C for 100 h (Scheme 10.4) [9]. Although the enantioselectivity was lower, the immobilization of the catalyst offered several practical advantages including ease of catalyst recycling and product purification.

10.2.1.2 Glycinate Schiff Base: Synthesis of α,α-Dialkyl-α-amino Acids

Nonproteinogenic, chiral α,α-dialkyl-α-amino acids possessing stereochemically stable quaternary carbon centers have been significant synthetic targets, not only because they are often effective enzyme inhibitors but also because they are indispensable for the elucidation of enzymatic mechanisms. Accordingly, numerous studies have been conducted to develop truly efficient methods for their preparation [10], and phase-transfer catalysis has made unique contributions.

Chiral 3,6-dihydro-2*H*-1,4-oxazin-2-ones **16** act as very reactive chiral cyclic alanine equivalents and can be diastereoselectively alkylated under solid–liquid phase-transfer conditions with K_2CO_3 as base. Hydrolysis of the resulting oxazinones **17** allows the preparation of enantiomerically enriched (*S*)-α-methyl-α-amino acids **18** (Scheme 10.5) [11].

In 1992, O'Donnell and co-workers succeeded in obtaining optically active α-methyl-α-amino acid derivatives **20** catalytically through the phase-transfer alkylation of *p*-chlorobenzaldehyde imine of alanine *tert*-butyl ester **19a** with cinchonine-derived **2e** as catalyst as illustrated in Scheme 10.6 [12]. Examination of the effect of different base systems revealed the importance of using the mixed-solid base KOH/K_2CO_3. Although the enantioselectivities are moderate, this study is the first example of the preparation of optically active α,α-dialkyl-α-amino acids by chiral phase-transfer catalysis.

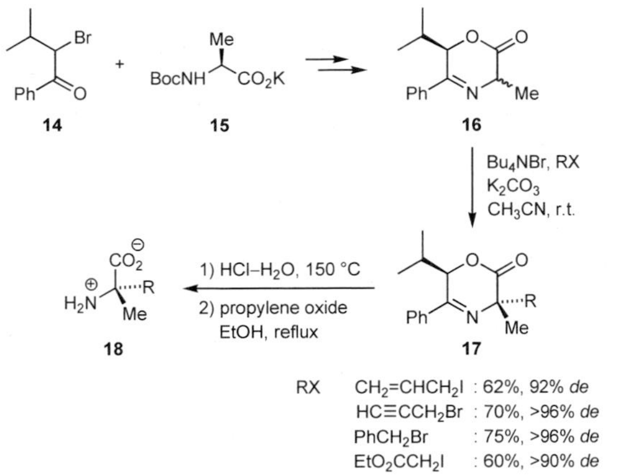

Scheme 10.5 Synthesis of optically active α,α-dialkyl-α-amino acids via diastereoselective alkylation.

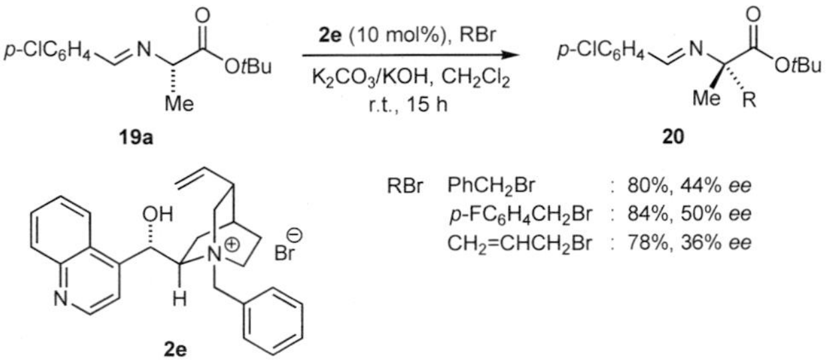

Scheme 10.6 The first catalytic system using **19a** as a key substrate.

In 1999, Lygo demonstrated that the use of *N*-anthracenylmethyldihydrocin-chonidinium chloride **9b** significantly improved the enantioselectivity of alkylations with substituted benzyl bromides, enhancing the utility of this approach to α,α-dialkyl-α-amino acids [13]. Here, the solid base KOH/K$_2$CO$_3$ must be freshly prepared before use if reproducible results are to be obtained. The lack of stereoselectivity in the reactions with other electrophiles was ascribed to competing, nonselective background alkylation (Scheme 10.7).

Jew, Park and co-workers made systematic investigations to develop a more efficient system for the asymmetric synthesis of α-alkylalanines by chiral phase-transfer catalysis [14]. Consequently, sterically more demanding 2-naphthyl aldimine *tert*-butyl ester **22** was identified as a suitable substrate, and its alkylation in the presence of the stronger base rubidium hydroxide and *O*(9)-allyl-*N*-2′,3′,4′-trifluorobenzyldihydrocinchonidinium bromide (**9c**) at lower reaction temperature (–35°C) led to the highest enantioselectivity (Scheme 10.8).

Takemoto and coworkers developed palladium-catalyzed asymmetric allylic alkylation of **19a** using allyl acetate and a chiral phase-transfer catalyst [15]. The proper choice of the achiral palladium ligand (PhO)$_3$P was crucial to achieving high enantioselectivity, but with no chiral phosphine ligand on palladium, the

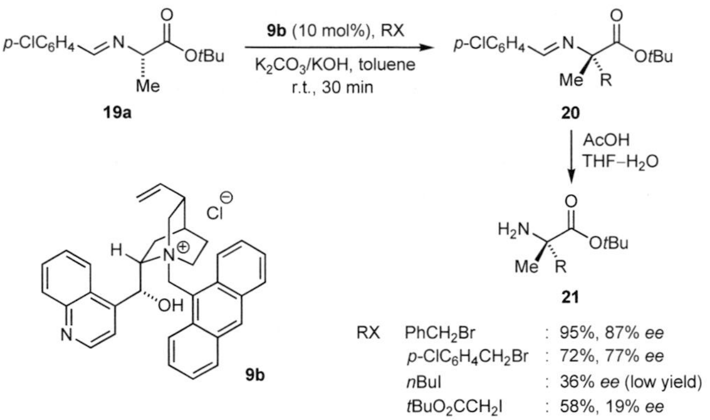

RX PhCH₂Br   : 95%, 87% *ee*
    p-ClC₆H₄CH₂Br   : 72%, 77% *ee*
    *n*BuI   : 36% *ee* (low yield)
    *t*BuO₂CCH₂I   : 58%, 19% *ee*

Scheme 10.7 Significant improvement of enantioselectivity with catalyst **9b**.

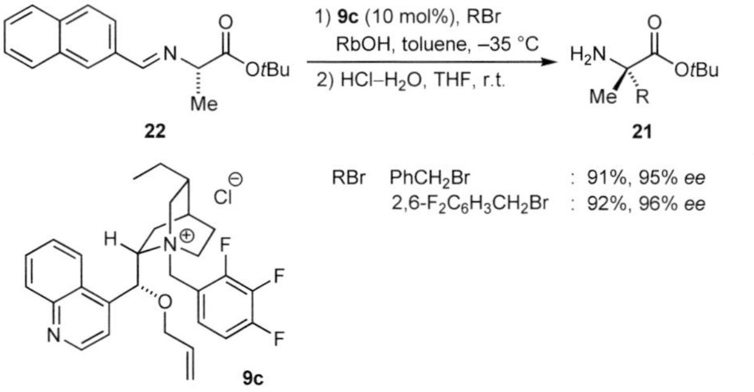

RBr PhCH₂Br   : 91%, 95% *ee*
     2,6-F₂C₆H₃CH₂Br   : 92%, 96% *ee*

Scheme 10.8 Optimal substrate, catalyst structure and reaction conditions for attaining high enantioselectivity.

desired allylation product **23** was obtained with 83% *ee* after hydrolysis of the imine moiety and subsequent benzoylation as depicted in Scheme 10.9.

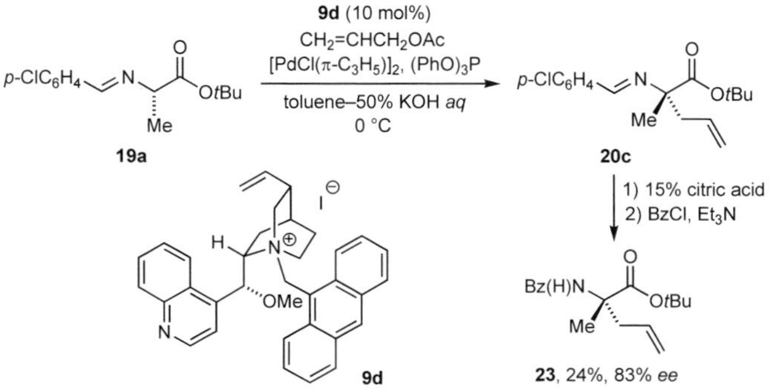

23, 24%, 83% *ee*

Scheme 10.9 The combined use of achiral palladium catalysis and chiral phase-transfer catalysis. (Bz = benzoyl)

Other types of chiral phase-transfer catalysis are also employable for the enantioselective alkylation of alanine-derived imines **24**. Enantiopure (4*R*,5*R*)- or (4*S*,5*S*)-2,2-dimethyl-α,α,α′,α′-tetraphenyl-1,3-dioxolane-4,5-dimethanol (**26**, TADDOL) and 2-hydroxy-2′-amino-1,1′-binaphthyl (**27**, NOBIN), upon *in situ* deprotonation with solid NaOH or NaH, act as chiral bases [16, 17]. Their chelating ability toward the sodium cation is crucial for making the sodium enolate soluble in toluene as well as for achieving the enantiofacial differentiation in the transition state (Scheme 10.10).

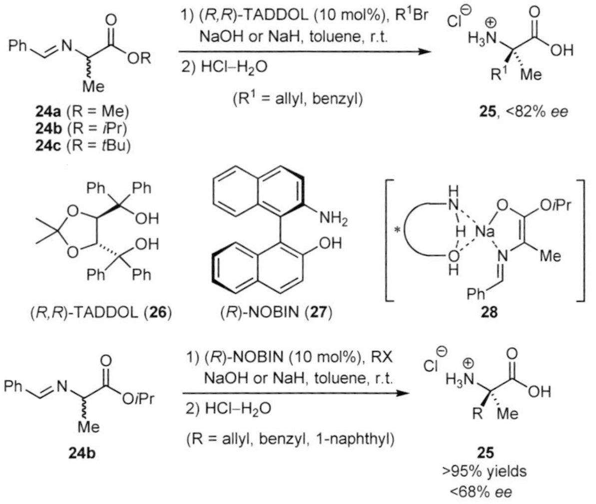

Scheme 10.10 Use of anions from TADDOL or NOBIN as chiral bases.

Copper(II)–salen complex **31** was found to function as a chiral phase-transfer catalyst. With 2 mol% of **31**, aldimine Schiff bases of various α-alkyl-α-amino acid methyl esters **29** can be alkylated enantioselectively (Scheme 10.11) [18]. Since hydrolysis of the alkylated imines occurred in the reaction mixture, a re-esterification step with methanol and acetyl chloride was introduced as part of the reaction.

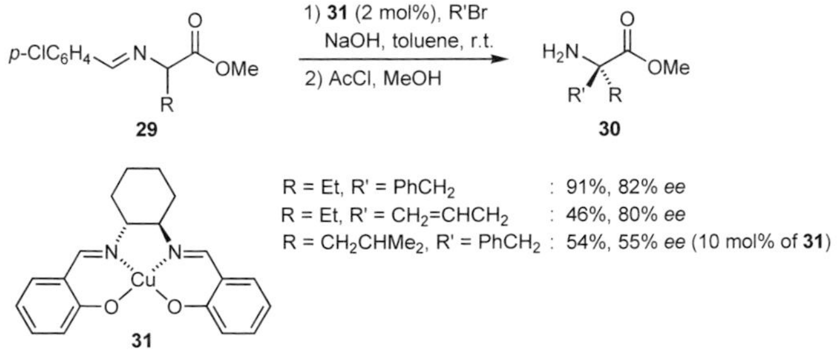

Scheme 10.11 A transition metal complex as chiral phase-transfer catalyst.

Since the aldimine Schiff base **34** can be readily prepared from glycine **32**, direct stereoselective introduction of two different side chains to **34** by appropriate chiral phase-transfer catalysis would provide an attractive yet powerful strategy for the asymmetric synthesis of structurally diverse α,α-dialkyl-α-amino acids. This possibility of a one-pot asymmetric double alkylation has been realized by using C_2-symmetric chiral quaternary ammonium bromide (*S,S*)-**36a** (Scheme 10.12).

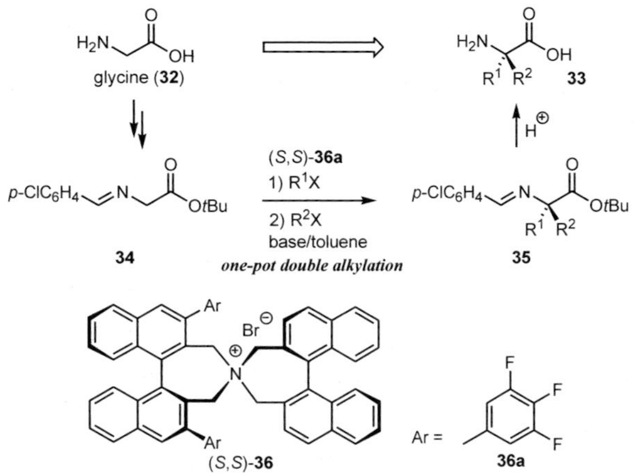

Scheme 10.12 One-pot double alkylation using C_2-symmetric chiral quaternary ammonium bromide (*S,S*)-**36a**.

Initial treatment of the toluene solution of **34** and (*S,S*)-**36a** (1 mol%) with allyl bromide (1 equivalent) and CsOH·H$_2$O (5 equivalent) at −10°C and the subsequent reaction with benzyl bromide (1.2 equivalent) at 0°C resulted in formation of the double alkylation product **37a** in 80% yield with 98% *ee* after hydrolysis. Notably, in the double alkylation of **34** by addition of the halides in the reverse order, the absolute configuration of the product **37a** was confirmed to be opposite, indicating the intervention of the expected chiral ammonium enolate **38** in the second alkylation stage (Scheme 10.13) [19].

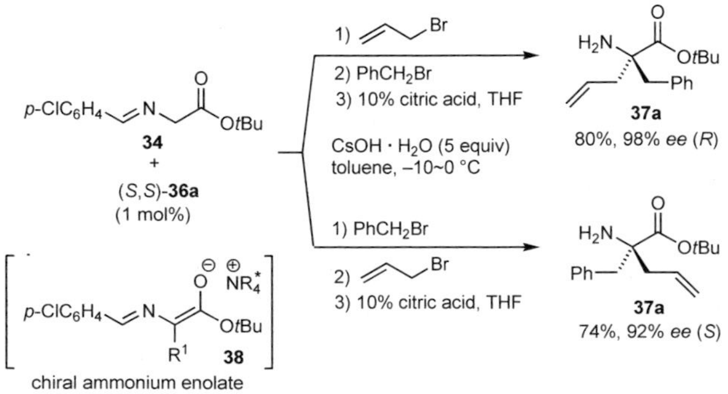

Scheme 10.13 Highly enantioselective, one-pot double alkylation.

Since the stereochemistry of the newly created quaternary carbon center was apparently determined in the second alkylation process, the core of this method should be applicable to the asymmetric alkylation of aldimine Schiff bases **19** derived from the corresponding α-amino acids. Indeed, *dl*-alanine-, phenylalanine- and leucine-derived imines **19a–c** can be alkylated smoothly under similar conditions, affording the desired noncoded amino acid esters **37a–c**, respectively, with excellent asymmetric induction as exemplified in Scheme 10.14 [19].

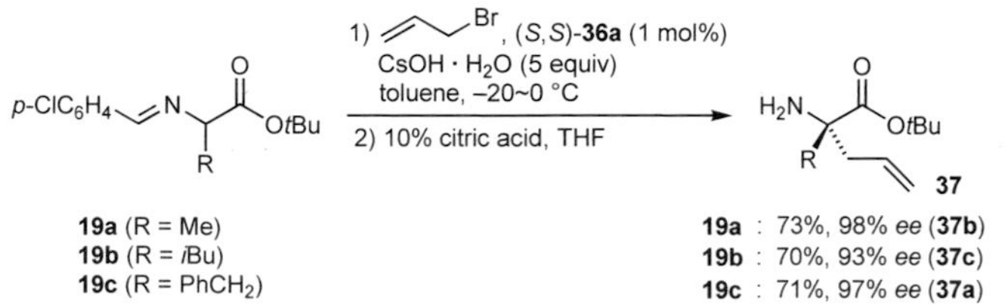

19a (R = Me)
19b (R = *i*Bu)
19c (R = PhCH₂)

19a : 73%, 98% *ee* (**37b**)
19b : 70%, 93% *ee* (**37c**)
19c : 71%, 97% *ee* (**37a**)

Scheme 10.14 The effectiveness of **36a** in the alkylation of aldimine Schiff bases derived from α-alkyl-α-amino acids.

This powerful quaternization method permitted the catalytic asymmetric synthesis of quaternary isoquinoline derivatives with **19a** as a substrate [20]. When **19a** was treated with α,α′-dibromo-*o*-xylene, CsOH · H₂O and (*S*,*S*)-**36a** (1 mol%) in toluene at 0°C, the transient monoalkylation product was rapidly produced, which was transformed into the desired **38** (64%, 88% *ee*) during the workup procedure. Catalytic asymmetric phase-transfer alkylation of **19a** with functionalized benzyl bromide **39** followed by sequential treatment with 1-N HCl and then excess NaHCO₃ furnished the corresponding dihydroisoquinoline derivative **40** in 87% with 94% *ee* (Scheme 10.15).

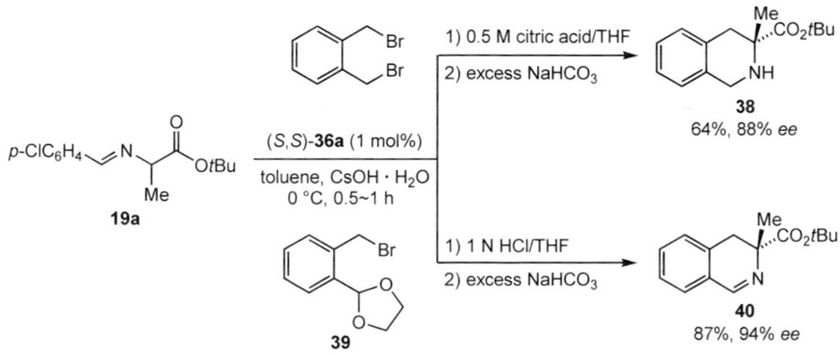

38
64%, 88% *ee*

19a

39

40
87%, 94% *ee*

Scheme 10.15 Catalytic asymmetric synthesis of quaternary isoquinoline derivatives.

The salient feature of chiral quaternary ammonium salts of type **36** was highlighted by the highly diastereoselective *N*-terminal functionalization of Schiff-base-activated peptides [21]. In particular, **36b**, with a sterically hindered aromatic substituent as Ar, allowed an asymmetric construction of noncoded α, α-dialkyl-α-amino acid residues at the peptide terminal as represented by the stereoselective alkylation of the dipeptide, L-Ala–L-Phe derivative **41**

(Scheme 10.16). It is important to note that (*S,S*)-**36b** is a matched catalyst for the alkylation of **41**.

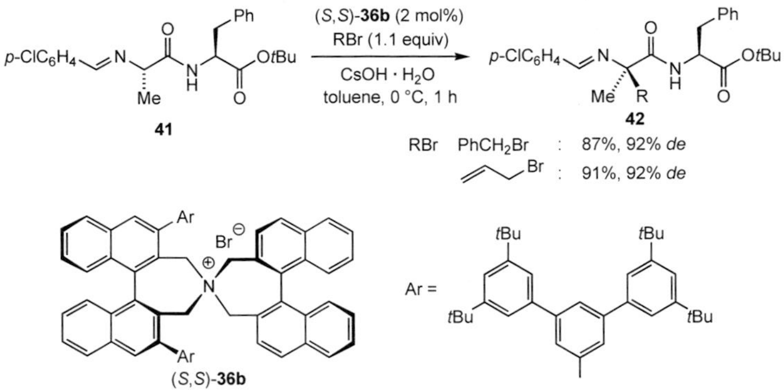

Scheme 10.16 Diastereoselective *N*-terminal alkylation of dipeptides.

10.2.1.3 β-Keto Esters

Manabe designed chiral phosphonium salts of type **45** that have a multiple hydrogen-bonding site [22]. Their utility as chiral phase-transfer catalysts has been demonstrated in the asymmetric alkylation of β-keto ester **43a**, in which a chiral quaternary carbon center is established. Although the reactivity and selectivity remained to be improved, this study provided a conceptual advance for the development of new chiral onium salts (Scheme 10.17).

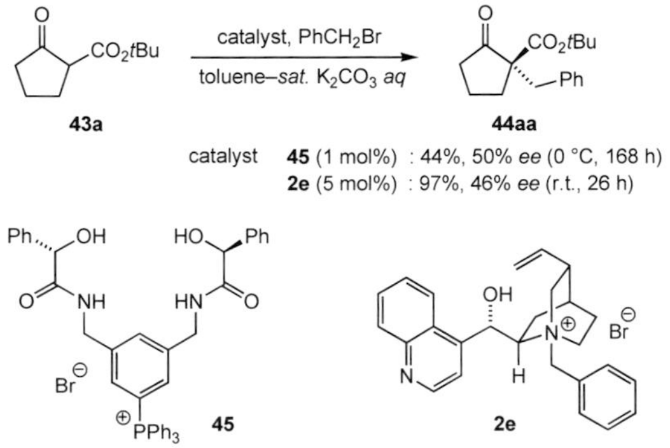

Scheme 10.17 Asymmetric alkylation of β-keto esters.

The cinchona-alkaloid derivative **2e** also catalyzed the benzylation of **43a** under similar conditions to give **44aa** in excellent chemical yield with 46% *ee* as included in Scheme 10.17 [23].

More efficient, highly enantioselective construction of a quaternary stereocenter from β-keto esters under phase-transfer conditions has been achieved using

N-spiro chiral quaternary ammonium bromide **36c** as catalyst [24]. This system has a broader generality in terms of the structure of β-keto esters and alkyl halides (Scheme 10.18). The resulting alkylation products **44** can be readily converted into the corresponding β-hydroxy esters **46** and β-amino esters **47**, respectively, as illustrated in Scheme 10.19.

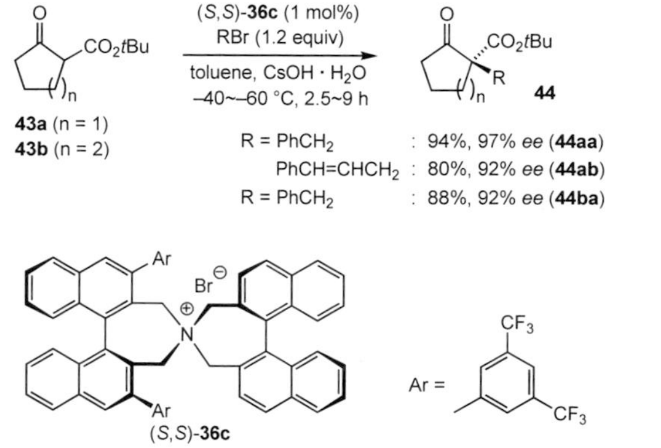
(S,S)-36c (1 mol%)
RBr (1.2 equiv)
toluene, CsOH · H2O
−40~−60 °C, 2.5~9 h
CO2tBu
43a (n = 1)
43b (n = 2)
44
R = PhCH2 : 94%, 97% ee (44aa)
PhCH=CHCH2 : 80%, 92% ee (44ab)
R = PhCH2 : 88%, 92% ee (44ba)
Ar
Br
N
(S,S)-36c
Ar =
CF3

Scheme 10.18 Advantage of **36** for construction of quaternary stereocenters from β-keto esters.

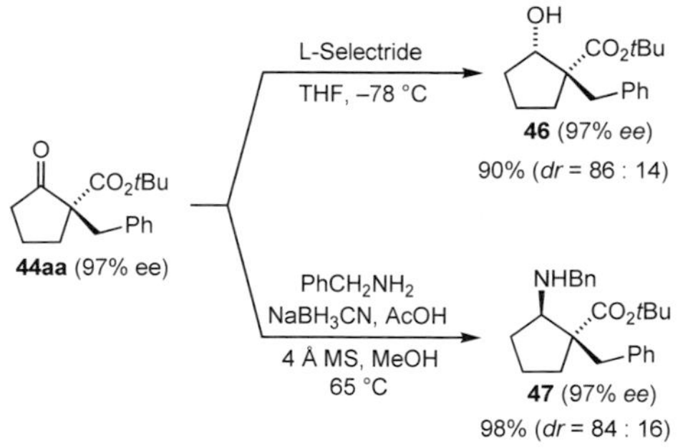
L-Selectride
THF, −78 °C
OH
CO2tBu
Ph
46 (97% ee)
90% (dr = 86 : 14)
44aa (97% ee)
PhCH2NH2
NaBH3CN, AcOH
4 Å MS, MeOH
65 °C
NHBn
47 (97% ee)
98% (dr = 84 : 16)

Scheme 10.19 Facile conversion of **44** into β-hydroxy and β-amino esters.

10.2.2
Michael Addition Reaction

The asymmetric Michael addition of a carbonyl substrate possessing an acidic methine proton to electron-deficient olefins, particularly α,β-unsaturated carbonyl compounds, represents a fundamental yet useful approach to the creation of a stereochemically defined, new quaternary carbon center [25]. The first successful, phase-transfer-catalyzed process was based on the use of well-designed chiral crown ethers **54** and **55** as catalyst [26]. In the presence of **54**, β-keto ester **48** added to methyl vinyl ketone (**49**, MVK) in moderate yield but with virtually complete stereochemical control. Further, crown ether **55** was shown to be effective in the reaction of methyl 2-phenylpropionate **51** with methyl acrylate **52**, affording the corresponding Michael adduct **53** in 80% yield with 83% *ee* (Scheme 10.20).

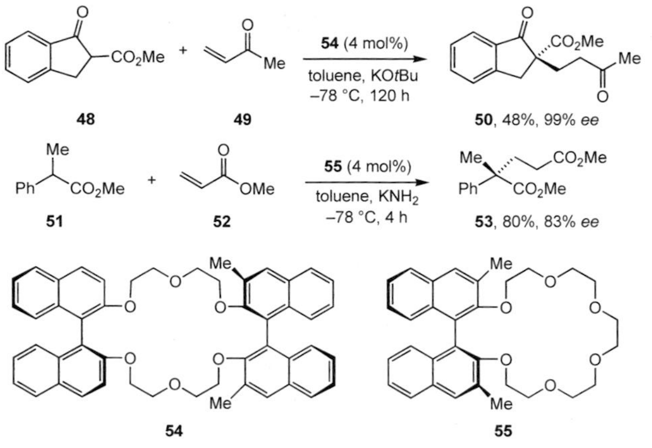

Scheme 10.20 Chiral crown ethers as phase-transfer catalysts.

As part of a research effort for the stereoselective functionalization of indanone derivative **1**, the Merck group reported the Michael addition of **1b** to MVK **49** catalyzed by their original catalyst **2a**, which proceeded smoothly in toluene–50% aqueous NaOH to give diketone **56b** in 95% yield with 80% *ee* (Scheme 10.21) [27]. This approach can be extended to the reaction of phenylindanone derivative **1a** with **49**, though the chemical yield and enantioselectivity were only moderate as indicated in Scheme 10.21 [28].

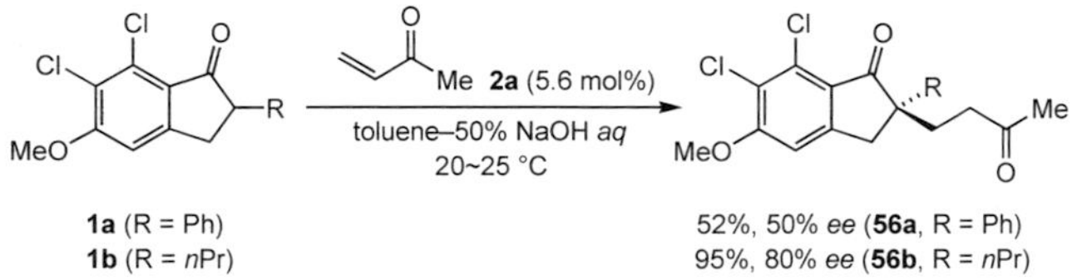

1a (R = Ph)
1b (R = *n*Pr)

52%, 50% *ee* (**56a**, R = Ph)
95%, 80% *ee* (**56b**, R = *n*Pr)

Scheme 10.21 Asymmetric Michael addition of an indanone derivative to methyl vinyl ketone.

This type of catalyst has further been applied to asymmetric Michael addition of tetralone derivatives to enones followed by a one-pot Robinson annulation. For example, the reaction of **8** with ethyl vinyl ketone **57** under the influence of dihydrocinchonidine-derived **9e** in toluene–60% aqueous KOH afforded the intermediary Michael adduct **58**. Subsequent addition of 18-crown-6 and continuous stirring at room temperature for 12 h furnished the cyclic enone **59** in 81% yield with 81% *ee*, which can be derivatized into tricyclic enone (+)-podocarp-8(14)-en-13-one **60**, a key intermediate for the synthesis of several diterpenes (Scheme 10.22) [6]. A similar Michael addition–Robinson annulation sequence has also been accomplished with α-phenylcyclohexanone **61** and **49** using chiral ammonium bromide **9a** (Scheme 10.23) [6, 28].

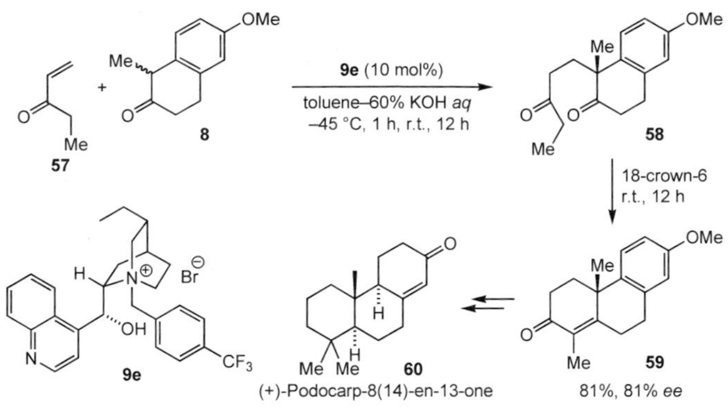

Scheme 10.22 Sequential Michael addition and Robinson annulation in one-pot.

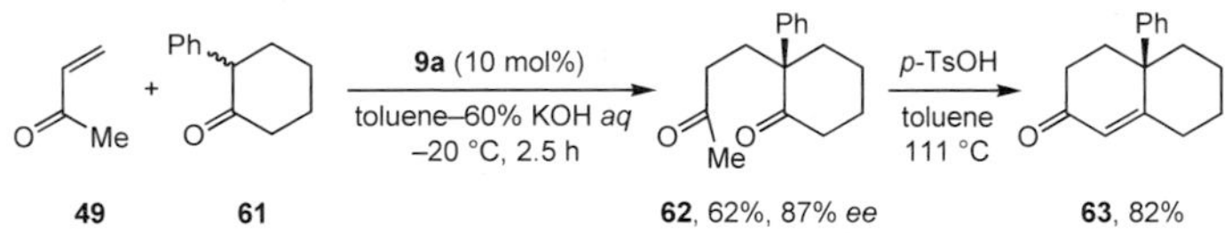

Scheme 10.23 Michael reaction–Robinson annulation between α-phenylcyclohexanone and MVK **49**.

Enantioselective Michael addition of β-keto esters to α,β-unsaturated carbonyl compounds is a useful way to synthesize chiral quaternary carbon centers, as illustrated by the example at the beginning of this section. A characteristic feature of C_2-symmetric chiral phase-transfer catalyst **36** in this type of transformation is that it allows the use of α,β-unsaturated aldehydes as Michael acceptors, leading to the construction of quaternary stereocenters with three different functionalities of carbonyl origin. For instance, treatment of 2-*tert*-butoxycarbonylcyclopentanone **43a** with acrolein **65** in the presence of (*S,S*)-**36c** (2 mol%) and K_2CO_3 (10 mol%) in cumene at –78°C followed by stirring at –40°C for 2 h resulted in clean formation of the corresponding Michael adduct, which was isolated as its acetal **66a** in 84% yield with 79% *ee*. It is of interest that the use of fluorenyl ester **64** as substrate improved the enantioselectivity to 90% *ee*. The addition of **64** to **49** was also feasible under similar conditions and the desired **67** was obtained quantitatively with 97% *ee* (Scheme 10.24) [24].

10.2.3
The Darzens Reaction

The Darzens reaction represents one of the most powerful methods for the synthesis of α,β-epoxy carbonyl and related compounds. Shioiri and Arai demonstrated that cyclic α-chloro ketones undergo the asymmetric Darzens condensation with various aldehydes under phase-transfer conditions using **2a** as catalyst and

LiOH as base in dibutyl ether at room temperature, furnishing enantiomerically enriched epoxy ketones with quaternary carbon centers (Scheme 10.25) [29].

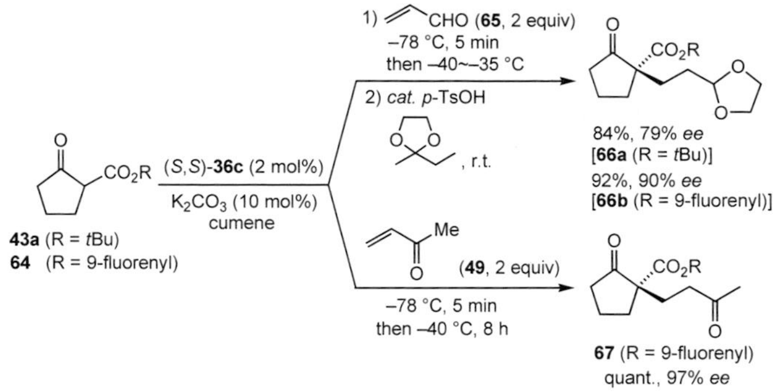

Scheme 10.24 Asymmetric Michael addition of β-keto esters to acrolein and MVK **49**.

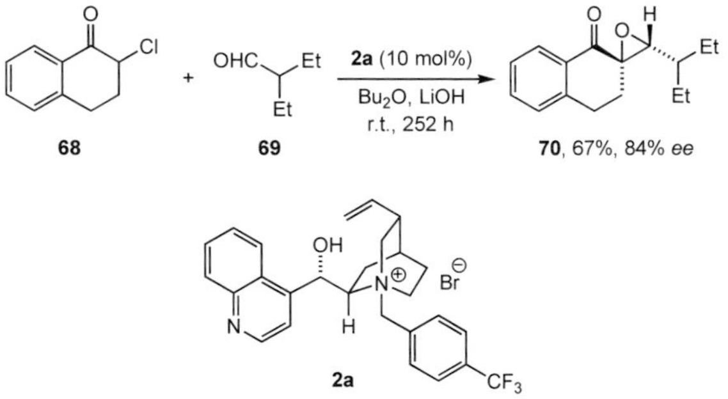

Scheme 10.25 The asymmetric Darzens reaction of α-chloro ketones with aldehydes.

They applied this approach to the preparation of optically active α,β-epoxy sulfones bearing quaternary stereocenters through the reaction of methyl ketone derivatives and chloromethyl phenyl sulfone **72** [30]. Thorough investigation revealed that the reaction was generally slow and the desired epoxides were obtained as a diastereomeric mixture, whose ratio was quite sensitive to the ketone substituent and catalyst structure. However, the enantioselectivity of the major isomer **73** reached 60% *ee* with quinine-derived ammonium bromide **75a** as catalyst in the condensation of acetophenone **71** and **72** (Scheme 10.26).

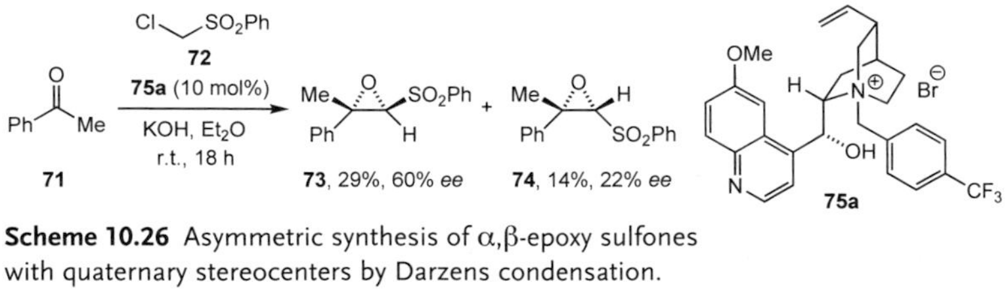

Scheme 10.26 Asymmetric synthesis of α,β-epoxy sulfones with quaternary stereocenters by Darzens condensation.

10.2.4
Cyclopropanation

The asymmetric cyclopropanation of α-bromocyclohexenone **76** with cyanomethyl phenyl sulfone **77** has been achieved under phase-transfer conditions by use of cinchona-alkaloid-derived catalyst, which constructs chiral quaternary carbons on the cyclopropane rings [31]. An extensive survey of the structure of quinine-based ammonium bromide as well as the reaction conditions revealed that the desired product **78** was obtained in 60% yield with 60% *ee* by performing the reaction in toluene at room temperature in the presence of **75b** with the 2,4-dimethylphenyl-methyl group and K_2CO_3. Cyanoacetate **79** was also found to be a quite effective carbon nucleophile. In this case, chlorobenzene was an ideal solvent and intro-duction of a benzyl group with an electron-withdrawing substituent on nitrogen had a beneficial effect on the enantioselectivity of **80** (Scheme 10.27).

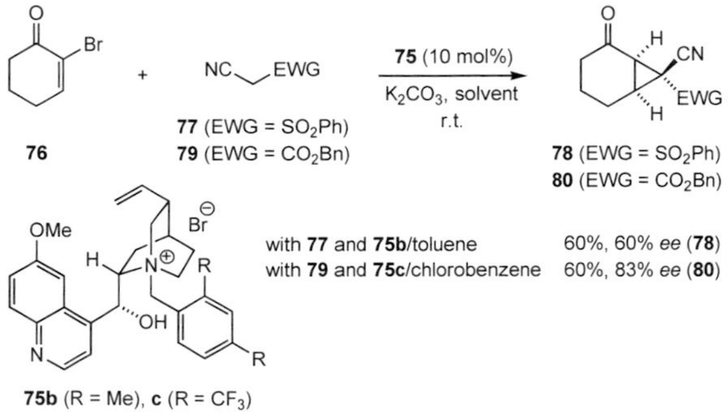

Scheme 10.27 Asymmetric cyclopropanation of α-bromocy-clohexenone under phase-transfer conditions.

10.3
Carbon–Heteroatom Bond Formation Through PTC

10.3.1
α-Hydroxylation

The catalytic enantioselective α-hydroxylation of tetralone and indanone deriva-tives **81a** and **81b** with molecular oxygen using chiral phase-transfer catalysts has been repeatedly examined since the first report by the Shioiri group [23, 32]. The cinchonine-derived **2a** and **2f** were employed as reliable catalysts, leading to the formation of α-hydroxy ketones with quaternary carbons **82** in more than 90% yields with good enantioselectivities. In addition, this oxidation process also demonstrated the effectiveness of chiral crown ether **83** (Scheme 10.28 and Tab. 10.1) [33].

Further, α,β-unsaturated ketones appeared to be good candidates for α-hydrox-ylation. For instance, (*E*)-2-ethylidene-1-tetralone **84** was oxidized to α-hydroxy ketone **85** under similar conditions in 73% yield with 55% *ee* (Scheme 10.29) [32a].

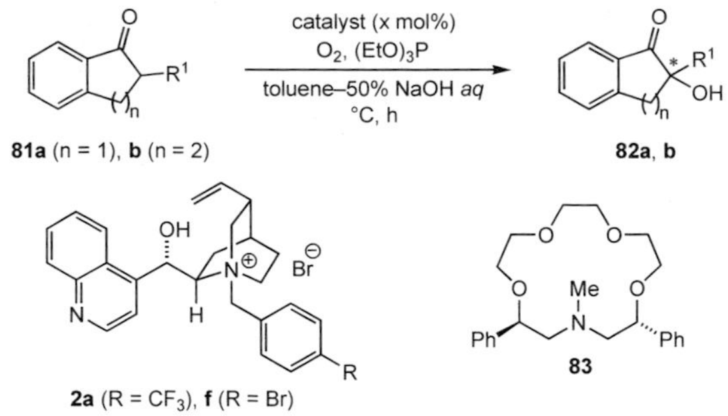

Scheme 10.28 Catalytic asymmetric α-hydroxylation of indanone and tetralone derivatives.

10.3.2
Epoxidation

The catalytic asymmetric epoxidation of electron-deficient olefins, particularly α,β-unsaturated ketones, has been the subject of numerous investigations and a number of useful methodologies have been elaborated [34]. Among these, the method utilizing chiral phase-transfer catalysis occupies a unique place featuring its practical advantages, and it allows highly enantioselective epoxidation of *trans*-α,β-unsaturated ketones [35, 36b]. On the other hand, however, enantiocontrol in the epoxidation of *cis* enones is still a difficult task, and successful examples are limited to the epoxidation of naphthoquinones and isoflavones. A typical reaction recipe for naphthoquinone epoxidation involves a treatment of 2-substituted naphthoquinone **86** with 30% H_2O_2 and LiOH in chloroform in the presence of chiral ammonium bromide, such as **75d**, and **88**, affording the corresponding epoxides **87** with quaternary carbon centers with good enantioselectivities [36]. Interestingly, use of the deaza derivative **88** as catalyst provided the enhanced enantioselectivity (Scheme 10.30) [23].

The asymmetric phase-transfer-catalyzed epoxidation of isoflavones **89** has been realized by Adam and coworkers using cinchonine-derived **2a** as catalyst and

Table 10.1 Catalytic asymmetric α-hydroxylation of indanone and tetralone derivatives **81a,b**.

Entry	Substrate (81, R^1)	Catalyst (x)	Conditions (°C, h)	Yield (%)	ee (%)	Config.	Product (82, R^1)
1	81a, Me	2a (5)	r.t., 12	94	73	S	82a, Me
2	81b, Me	2a (5)	r.t., 24	95	70	S	82b, Me
3	81b, Et	2a (5)	r.t., 24	98	72	S	82b, Et
4	81b, Et	2f (5)	r.t., 42	n.d.	54.4	S	82b, Et
5	81b, Et	83 (10)	−20–6, 24	95	67	R	82b, Et
6	81b, allyl	83 (10)	−20–6, 24	89	72	S	82b, allyl

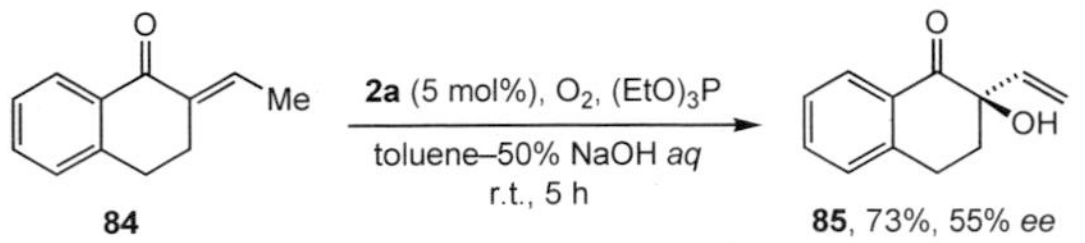

Scheme 10.29 Asymmetric α-hydroxylation of α,β-unsaturated ketones.

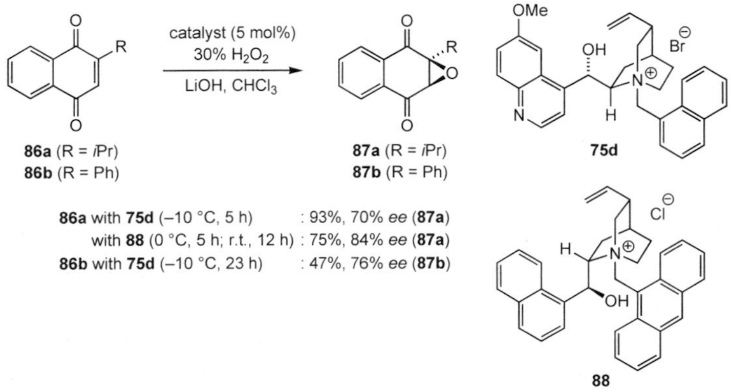

84 → **85**, 73%, 55% *ee*

2a (5 mol%), O₂, (EtO)₃P
toluene–50% NaOH *aq*
r.t., 5 h

Scheme 10.30 Asymmetric epoxidation of naphthoquinones.

86a (R = *i*Pr) **86b** (R = Ph)

catalyst (5 mol%)
30% H₂O₂
LiOH, CHCl₃

87a (R = *i*Pr) **87b** (R = Ph)

75d

86a with **75d** (–10 °C, 5 h)   : 93%, 70% *ee* (**87a**)
with **88** (0 °C, 5 h; r.t., 12 h) : 75%, 84% *ee* (**87a**)
86b with **75d** (–10 °C, 23 h)   : 47%, 76% *ee* (**87b**)

88

commercially available cumyl hydroperoxide **90** as oxidant. Upon reducing the catalyst loading to 1 mol%, isoflavone epoxide **91a** was obtained almost quantitatively with excellent enantioselectivity. The 2-methyl derivative **89b** afforded the corresponding epoxide **91b** possessing two consecutive quaternary stereogenic centers in 97% yield with 89% *ee* (Scheme 10.31) [37].

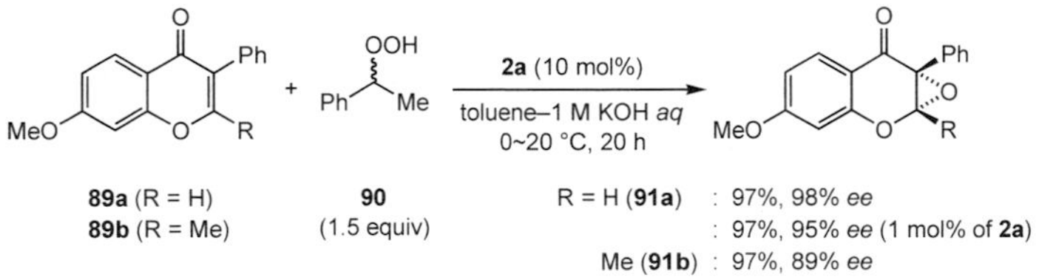

89a (R = H)   **90**   **91a**
89b (R = Me) (1.5 equiv)

2a (10 mol%)
toluene–1 M KOH *aq*
0~20 °C, 20 h

R = H (**91a**) : 97%, 98% *ee*
                 : 97%, 95% *ee* (1 mol% of **2a**)
Me (**91b**) : 97%, 89% *ee*

Scheme 10.31 Asymmetric epoxidation of isoflavones.

Aggarwal and coworkers introduced a novel process for asymmetric epoxidation of simple alkenes using Oxone as oxidant, which was catalyzed by chiral amines as shown in Scheme 10.32 [38]. Thorough mechanistic investigations provided compelling evidence to support the dual role of a protonated chiral secondary ammonium salt as the phase-transfer catalyst to bring the oxidant into solution and as the activator of Oxone through hydrogen bonding in the key oxidizing agent **96**. It was pointed out that the presence of a number of possible forms such as **96a–c** was likely to be a factor responsible for the moderate enantioselectivity.

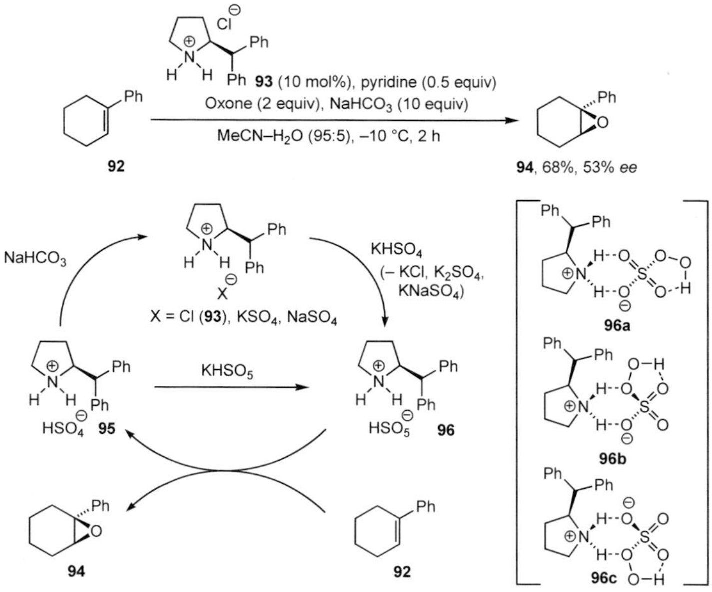

Scheme 10.32 Chiral amine-catalyzed asymmetric epoxidation of simple olefins using Oxone as oxidant.

10.3.3
α-Fluorination

In view of the importance of optically active organofluorine compounds in various fields of chemistry, catalytic enantioselective fluorination of carbonyl substrates has emerged as a long-awaited method [39] and the asymmetric electrophilic fluorination of β-keto ester **97** under phase-transfer conditions certainly belongs to this category. The combined use of appropriately modified catalyst **2g** and *N*-fluorobenzenesulfonimide **98** as a fluorinating agent in toluene with base (K₂CO₃) allowed the desired **99** to be isolated in 92% yield with 69% *ee* (Scheme 10.33) [40].

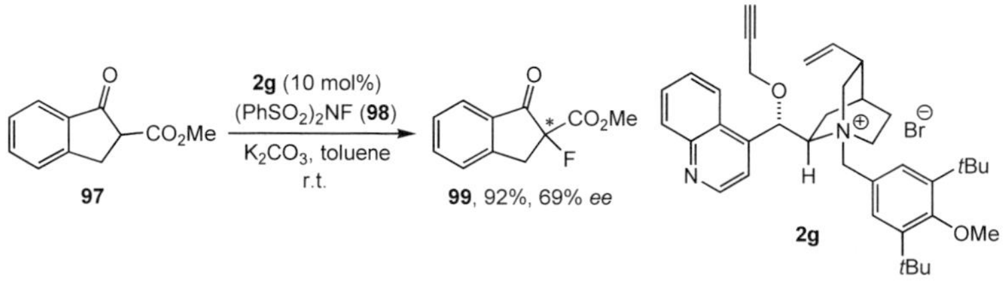

Scheme 10.33 Catalytic asymmetric fluorination of a β-keto ester.

10.4
Conclusion

Asymmetric synthesis of a wide variety of organic molecules containing quaternary stereogenic carbon centers in a truly catalytic manner has been a very challenging issue, and several methods have been elaborated based on different strategies. The stereoselective reactions described in this chapter clearly demonstrate not only the unique ability of chiral phase-transfer catalysis to construct quaternary carbon centers but also the current limitations in terms of reaction modes, usable reagents and substrate generality. However, since the advantages offered by phase-transfer catalysis meet the strong requirement for practical synthetic reactions in the twenty-first century, this methodology will definitely play a more important role even in the general context of organic synthesis, and this should be supported by the rapid development of newly designed chiral catalysts. Its success will solve the current problems and release the full potential of chiral phase-transfer catalysis. This will lead to an ideal process that will establish stereochemically defined quaternary carbons in appropriately functionalized, structurally intriguing compounds, and eventually benefit the progress of various fields of science.

References

1 (a) E. V. Dehmlow, S. S. Dehmlow **1993**, *Phase Transfer Catalysis*, 3rd edn, VCH, Weinheim; (b) C. M. Starks, C. L. Liotta, M. Halpern **1994**, *Phase-transfer Catalysis*; Chapman & Hall, New York; (c) Y. Sasson, R. Neumann (eds.) **1997**, *Handbook of Phase-transfer Catalysis*, Blackie Academic & Professional, London; (d) M. E. Halpern (ed.) **1997**, *Phase-transfer Catalysis*; ACS Symposium Series 659; American Chemical Society, Washington, DC.

2 (a) M. J. O'Donnell **2000**, in *Catalytic Asymmetric Synthesis*, 2nd edn, ed. I. Ojima, Wiley-VCH, New York; Ch. 10; (b) T. Shioiri **1997**, in *Handbook of Phase-transfer Catalysis*, ed. Y. Sasson, R. Neumann, Blackie Academic & Professional, London, Ch. 14; (c) M. J. O'Donnell, *Phases – The Sachem Phase Transfer Catalysis Review* **1998**, Issue 4, p. 5; (d) M. J. O'Donnell, *Phases – The Sachem Phase Transfer Catalysis Review* **1999**, Issue 5, p. 5; (e) T. Shioiri, S. Arai **2000**, in *Stimulating Concepts in Chemistry*, ed. F. Vögtle, J. F. Stoddart,

M. Shibasaki, Wiley-VCH, Weinheim, p. 123; (f) M. J. O'Donnell, *Aldrichimica Acta* **2001**, *34*, 3; (g) K. Maruoka, T. Ooi, *Chem. Rev.* **2003**, *103*, 3013; (h) M. J. O'Donnell, *Acc. Chem. Res.* **2004**, *37*, 506; (i) B. Lygo, B. I. Andrews, *Acc. Chem. Res.* **2004**, *37*, 518.

3 (a) U.-H. Dolling, P. Davis, E. J. J. Grabowski, *J. Am. Chem. Soc.* **1984**, *106*, 446; (b) D. L. Hughes, U.-H. Dolling, K. M. Ryan, E. F. Schoenewaldt, E. J. J. Grabowski, *J. Org. Chem.* **1987**, *52*, 4745.

4 A. Bhattacharya, U.-H. Dolling, E. J. J. Grabowski, S. Karady, K. M. Ryan, L. M. Weinstock, *Angew. Chem.* **1986**, *98*, 442; *Angew. Chem. Int. Ed.* **1986**, *25*, 476.

5 T. B. K. Lee, G. S. K. Wong, *J. Org. Chem.* **1991**, *56*, 872.

6 W. Nerinckx, M. Vandewalle, *Tetrahedron: Asymmetry* **1990**, *1*, 265.

7 (a) P. Bravo, G. Resnati, *Tetrahedron: Asymmetry* **1990**, *1*, 661; (b) G. S. Lal, G. P. Pez, R. G. Syvret, *Chem. Rev.* **1996**, *96*, 1737; (c) V. A. Soloshonok (ed.) **1999**, *Enantiocontrolled Synthesis of Fluoro-organic Compounds*; J. Wiley & Sons, Chichester.

8 S. Arai, M. Oku, T. Ishida, T. Shioiri, *Tetrahedron Lett.* **1999**, *40*, 6785.

9 B. Thierry, T. Perrard, C. Audouard, J.-C. Plaquevent, D. Cahard, *Synthesis* **2001**, 1742.

10 (a) C. Cativiela, M. D. Diaz-de-Villegas, *Tetrahedron: Asymmetry* **1998**, *9*, 3517; (b) U. Schöllkopf, *Top. Curr. Chem.* **1983**, *109*, 65.

11 (a) T. Abellán, C. Nájera, J. M. Sansano, *Tetrahedron: Asymmetry* **1998**, *9*, 2211; (b) R. Chinchilla, N. Galindo, C. Nájera, *Synthesis* **1999**, 704.

12 M. J. O'Donnell, S. Wu, *Tetrahedron: Asymmetry* **1992**, *3*, 591.

13 B. Lygo, J. Crosby, J. A. Peterson, *Tetrahedron Lett.* **1999**, *40*, 8671.

14 S.-s. Jew, B.-S. Jeong, J.-H. Lee, M.-S. Yoo, Y.-J. Lee, B.-s. Park, M. G. Kim, H.-g. Park, *J. Org. Chem.* **2003**, *68*, 4514.

15 M. Nakoji, T. Kanayama, T. Okino, Y. Takemoto, *J. Org. Chem.* **2002**, *67*, 7418.

16 Y. N. Belokon, K. A. Kochetkov, T. D. Churkina, N. S. Ikonnikov, A. A. Chesnokov, O. V. Larionov, V. S. Parmar, R. Kumar, H. B. Kagan, *Tetrahedron: Asymmetry* **1998**, *9*, 851.

17 Y. N. Belokon, K. A. Kochetkov, T. D. Churkina, N. S. Ikonnikov, S. Vyskocil, H. B. Kagan, *Tetrahedron: Asymmetry* **1999**, *10*, 1723.

18 (a) Y. N. Belokon, D. Bhave, D. D'Addario, E. Groaz, M. North, V. Tagliazucca, *Tetrahedron* **2004**, *60*, 1849. See also: (b) Y. N. Belokon, M. North, V. S. Kublitski, N. S. Ikonnikov, P. E. Krasik, V. I. Maleev, *Tetrahedron Lett.* **1999**, *40*, 6105; (c) Y. N. Belokon, R. G. Davies, M. North, *Tetrahedron Lett.* **2000**, *41*, 7245.

19 T. Ooi, M. Takeuchi, M. Kameda, K. Maruoka, *J. Am. Chem. Soc.* **2000**, *122*, 5228.

20 T. Ooi, M. Takeuchi, K. Maruoka, *Synthesis* **2001**, 1716.

21 T. Ooi, E. Tayama, K. Maruoka, *Angew. Chem.* **2003**, *115*, 599; *Angew. Chem., Int. Ed.* **2003**, *42*, 579.

22 (a) K. Manabe, *Tetrahedron Lett.* **1998**, *39*, 5807; (b) K. Manabe, *Tetrahedron* **1998**, *54*, 14465.

23 E. V. Dehmlow, S. Düttmann, B. Neumann, H.-G. Stammler, *Eur. J. Org. Chem.* **2002**, 2087.

24 T. Ooi, T. Miki, M. Taniguchi, M. Shiraishi, M. Takeuchi, K. Maruoka,

Angew. Chem. **2003**, *115*, 3926; *Angew. Chem. Int. Ed.*, **2003**, *42*, 3796.

25 For recent reviews on catalytic asymmetric Michael reactions, see: (a) N. Krause, A. Hoffmann-Röder, *Synthesis* **2001**, 171; (b) M. Sibi, S. Manyem, *Tetrahedron* **2000**, *56*, 8033; (c) M. Kanai, M. Shibasaki **2000**, in *Catalytic Asymmetric Synthesis*, 2nd edn, ed. I. Ojima, Wiley-VCH, New York, p. 569; (d) K. Tomioka, Y. Nagaoka **1999**, in *Comprehensive Asymmetric Catalysis*, Vol. 3, ed. E. N. Jacobsen, A. Pfaltz, H. Yamamoto, Springer, Berlin, Ch. 31.1.

26 D. J. Cram, G. D. Y. Sogah, *J. Chem. Soc. Chem. Commun.* **1981**, 625.

27 R. S. E. Conn, A. V. Lovell, S. Karady, L. M. Weinstock, *J. Org. Chem.* **1986**, *51*, 4710.

28 E. Diez-Barra, A. de la Hoz, S. Merino, A. Rodríguez, P. Sánchez-Verdú, *Tetrahedron* **1998**, *54*, 1835.

29 S. Arai, Y. Shirai, T. Ishida, T. Shioiri, *Chem. Commun.* **1999**, 49.

30 S. Arai, T. Shioiri, *Tetrahedron* **2002**, *58*, 1407.

31 S. Arai, K. Nakayama, T. Ishida, T. Shioiri, *Tetrahedron Lett.* **1999**, *40*, 4215.

32 (a) M. Masui, A. Ando, T. Shioiri, *Tetrahedron Lett.* **1988**, *29*, 2835; (b) E. V. Dehmlow, S. Wagner, A. Müller, *Tetrahedron* **1999**, *55*, 6335.

33 E. F. J. de Vries, L. Ploeg, M. Colao, J. Brussee, A. van der Gen, *Tetrahedron: Asymmetry* **1995**, *6*, 1123.

34 (a) M. J. Porter, J. Skidmore, *Chem. Commun.* **2000**, 1215; (b) T. Nemoto, T. Ohshima, M. Shibasaki, *J. Synth. Org. Chem. Jpn.* **2002**, *60*, 94.

35 (a) B. Lygo, P. G. Wainwright, *Tetrahedron Lett.* **1998**, *39*, 1599; (b) E. J. Corey, F.-Y. Zhang, *Org. Lett.* **1999**, *1*, 1287; (c) B. Lygo, D. C. M. To, *Tetrahedron Lett.* **2001**, *42*, 1343; (d) J. Ye, Y. Wang, R. Liu, G. Zhang, Q. Zhang, J. Chen, X. Liang, *Chem. Commun.* **2003**, 2714; (e) M. T. Allingham, A. Howard-Jones, P. J. Murphy, D. A. Thomas, P. W. R. Caulkett, *Tetrahedron Lett.* **2003**, *44*, 8677; (f) T. Ooi, D. Ohara, M. Tamura, K. Maruoka, *J. Am. Chem. Soc.* **2004**, *126*, 6844.

36 (a) S. Arai, M. Oku, M. Miura,
T. Shioiri, *Synlett* **1998**, 1201;
(b) S. Arai, H. Tsuge, M. Oku, M. Miura,
T. Shioiri, *Tetrahedron* **2002**, *58*, 1623.

37 W. Adam, P. B. Rao, H.-G. Degen,
A. Levai, T. Patonay, C. R. Saha-Möller,
J. Org. Chem. **2002**, *67*, 259.

38 V. K. Aggarwal, C. Lopin, F. Sandrinelli,
J. Am. Chem. Soc. **2003**, *125*, 7596.

39 J.-A. Ma, D. Cahard, *Chem. Rev.* **2004**.

40 D. Y. Kim, E. J. Park, *Org. Lett.* **2002**, *4*,
545.

11
Radical Reactions

Kalyani Patil and Mukund P. Sibi

11.1
Introduction

Radical chemistry, which was once considered to be an underdeveloped field, has seen tremendous progress since the mid-1980s, and has emerged as a powerful branch of synthetic organic chemistry [1]. Construction of quaternary centers stereoselectively [2] is not an easy task even when well-established ionic chemistry is used rather than radical chemistry. Owing to the high reactivity of radicals, it was believed that radical reactions would not proceed with high selectivity. This opinion changed and diastereoselective radical chemistry [3] began around 1985, although enantioselective radical reactions [4] only emerged more recently.

Stereoselectivity in radical reactions can be achieved using several different techniques: substrate control, chiral auxiliary control, and chiral Lewis acid-mediated reactions. In substrate-controlled reactions, the stereogenic element can be present on the radical, attached to the radical trap at the reaction site, or at a site remote from that undergoing reaction. For diastereoselective and chiral auxiliary-mediated reactions, two steps, introduction and removal of the auxiliary is required. In contrast, enantioselective reactions can be potentially carried out using catalytic amounts of a chiral Lewis acid without the need for additional synthetic steps.

Stereoselective formation of quaternary carbon centers using radical chemistry still remains a challenge. This chapter reviews radical methods used for the construction of quaternary carbons, and these include diastereoselective as well as enantioselective bond formation. The organization of the chapter is based on the types of reagents used for the bond construction: (1) tin- and silicon-based reagents and (2) samarium iodide, manganese acetate, titanocene chloride, and cobalt-catalyzed radical reactions. Successful examples of quaternary center formation in the natural-products arena are highlighted [5]. An important class of cyclizations, atom- and group-transfer cyclization, is also discussed and stereospecific formation of vicinal quaternary centers by radical–radical combination is also mentioned. A few general experimental procedures are presented at the end of the chapter.

Quaternary Stereocenters: Challenges and Solutions for Organic Synthesis. Edited by Jens Christoffers, Angelika Baro
Copyright © 2005 WILEY-VCH Verlag GmbH & Co. KGaA, Weinheim
ISBN: 3-527-31107-6

In spite of the high toxicity and difficulty of removal of the tin byproducts, tin hydrides are the most widely used reagents in radical chemistry owing to the low strength of Sn–H bonds and the facility of chain propagation by the tin radical. The Sn–H bond can be cleaved homolytically either photochemically or by chemical initiation. Silanes are good substitutes for stannanes, since Si–H and Sn–H bond strengths are comparable, and silanes are less toxic. More recently, hypophosphorous acid salts have been used as hydrogen-atom donors in radical reactions. 1-Ethylpiperidine hypophosphite (EPHP) is an excellent reagent in this class. It is possible to carry out radical reactions in aqueous media using EPHP [6]. A potential way to synthesize chiral quaternary centers using hydride reagents is the intramolecular addition of a tertiary carbon radical to a carbon-based radical acceptor or carbon radicals to tertiary olefinic acceptors. 5-*exo* Cyclizations are the most prevalent radical cyclizations [7] for kinetic reasons. Thus, most ring constructions are executed by 5-*exo* cyclizations.

Intermolecular radical reactions to form quaternary centers are possible but difficult. The rates for intermolecular radical additions to disubstituted acceptors are low. At present there are no known examples of stereoselective quaternary center formation by this method. However, allyl stannanes or allyl silanes react with tertiary radicals and transfer an allyl group effectively, generating quaternary centers. Some examples are discussed below.

11.2
Radical Cyclization

One of the earliest examples of vinyl radical cyclization to generate a quaternary center was reported by Stork and coworkers [8]. Under standard photolytic conditions, using tributylstannane and azobis(isobutyronitrile) (AIBN), the achiral cyclohexenol substrate **1** was transformed to *cis* product **2** by 5-*exo-trig* cyclization (Scheme 11.1). The *cis* stereochemistry is an outcome of the geometry of approach of the vinyl radical to the ring double bond.

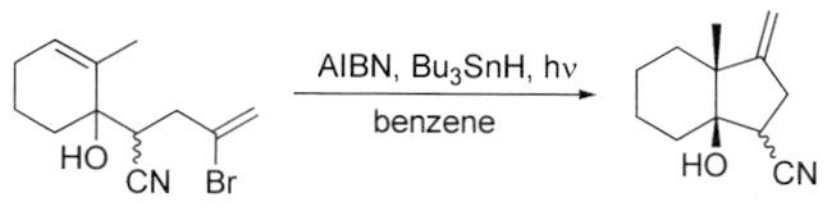

Scheme 11.1 Synthesis of hydrindanes via 5-*exo* cyclization.

Regioselectivity in radical cyclization was controlled in an elegant manner using Lewis acids in constructing cyclopentane and cyclohexane rings (Scheme 11.2). Treatment of optically active α-alkylidenelactone **3** with Bu₃SnH and a catalytic amount of Et₃B favored the 5-*exo* product **4** as the single stereoisomer in the absence of a Lewis acid leading to a quaternary center [9]. However, the regioselectivity was reversed when Et₂AlCl was used, and the cyclization proceeded via a 6-*endo* pathway (Table 11.1). The formation of the 6-*endo* product **5** can be rationalized by the coordination of the monodentate Lewis acid to the carbonyl group, favoring the conjugate addition pathway.

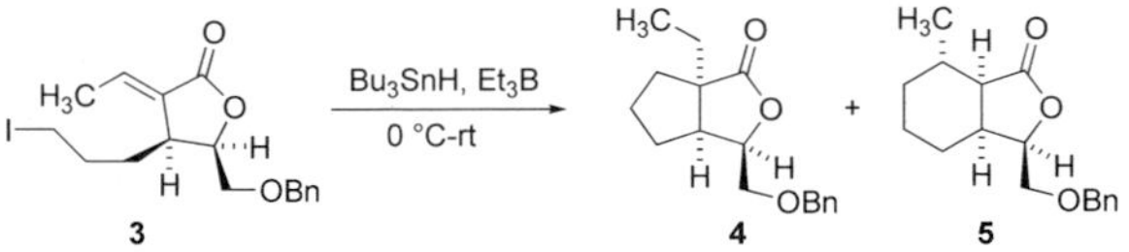

Scheme 11.2 Synthesis of bicyclic lactones: *5-exo* vs *6-endo*.

α-Carbonyl radical cyclization was a key step in an enantioselective synthesis of highly hindered hydrindanone, Schinzer's ketone **9** (Scheme 11.3) [10]. Iodoalkyne **6** synthesized from (*R*)-pulegone was treated with Bu$_3$SnH and AIBN in refluxing benzene to afford the *5-exo dig* cyclization product **7** with *cis* fusion. Desilylation followed by hydrogenation, subsequent conversion to tosylhydrazone, and treatment with *n*-butyllithium provided **8**, which was transformed to Schinzer's ketone by allylic oxidation followed by hydrogenation.

Total synthesis of (±)-merrilactone A **13**, possessing a pentacyclic framework, was achieved efficiently by *5-exo* radical cyclization of the α-bromoacetal **10** (Scheme 11.4). In spite of the steric crowding, the cyclization took place effectively, furnishing the desired isomer **11a** as the major product. The undesired product **11b** was later converted to the desired isomer under acidic conditions [11]. Compound **12** was obtained from (**11a**) by some functional group transformations. Epoxidation of **12** with dimethyl dioxirane followed by epoxide ring opening under acidic conditions furnished the merrilactone A.

The core of a challenging target molecule, (–)-CP-263,114 **16** was synthesized via a radical method, and the efficient installation of the quaternary stereocenter was accomplished by *5-exo* Stork radical cyclization [12]. Treatment of **14** with Bu$_3$SnH and AIBN provided the tetracyclic product **15** in good yield as a mixture of separable diastereomers (Scheme 11.5). The two diastereomers gave keto iodides by irradiation with visible light under Suärez conditions, and were reductively eliminated to produce the core structure.

Clive and coworkers achieved the synthesis of (+)-puraquinonic acid **20** by radical cyclization of bromoacetal **17** (Scheme 11.6) [13]. The *cis*-fused lactol ether **18** was obtained, which was degraded to generate a quaternary center. Compound **18** was converted to **19** in a few simple steps. Barton–McCombie deoxygenation followed by debenzylation and oxidation with CAN yielded (+)-puraquinonic acid **20**.

Lee and coworkers explored the tandem radical cyclizations of (alkoxycarbonyl) methyl radicals and obtained interesting results (Scheme 11.7) [14].

Table 11.1 Regioselectivity in radical cyclization of α-alkylidene lactone **3**.

Entry	Solvent	4:5	Yield (%)
1	THF	70:30	78
2	Toluene[a]	18:82	86

a In the presence of Et$_2$AlCl

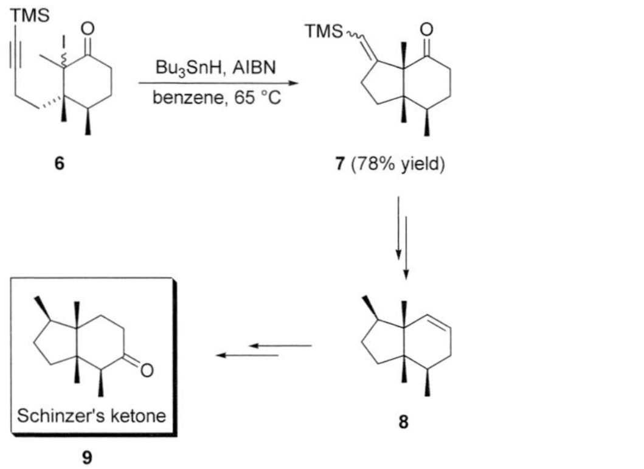

Scheme 11.3 Synthesis of Schinzer's ketone.

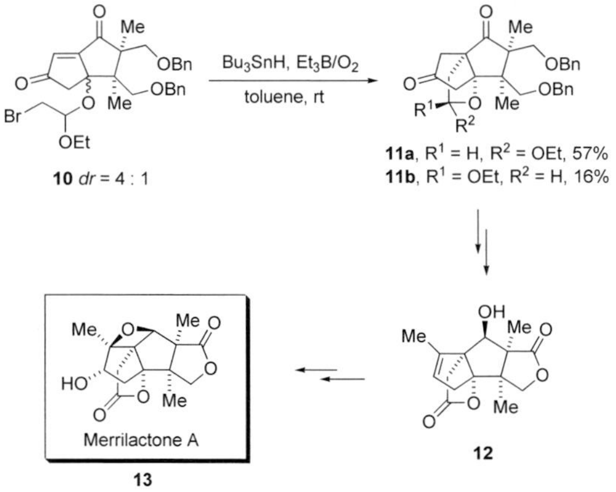

Scheme 11.4 Synthesis of merrilactone A.

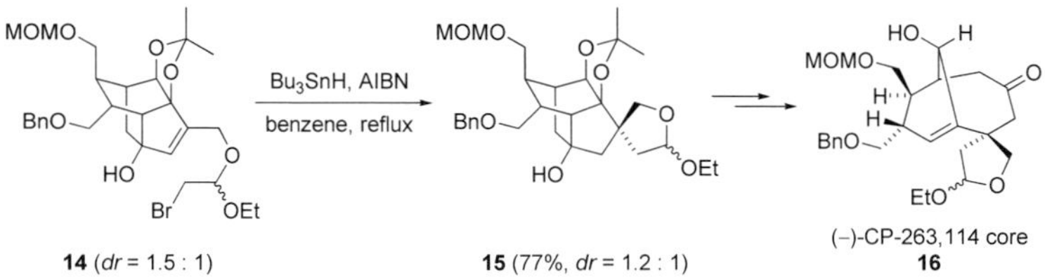

Scheme 11.5 Synthesis of (−)-CP-263,114 core.

Cyclization in an *8-endo trig* conformation was found to be faster than the kinetically preferred *5-exo trig* cyclization, leading to the highly functionalized bicyclic *cis*-fused lactone **22**. This abnormal pathway is most likely a reflection of the conformation of the ester moiety with the ester in an *s-cis* orientation rather than the needed *s-trans*.

Regiochemical control in radical cyclizations is difficult. Ishibashi and coworkers have controlled the regioselectivity of radical cyclization by taking

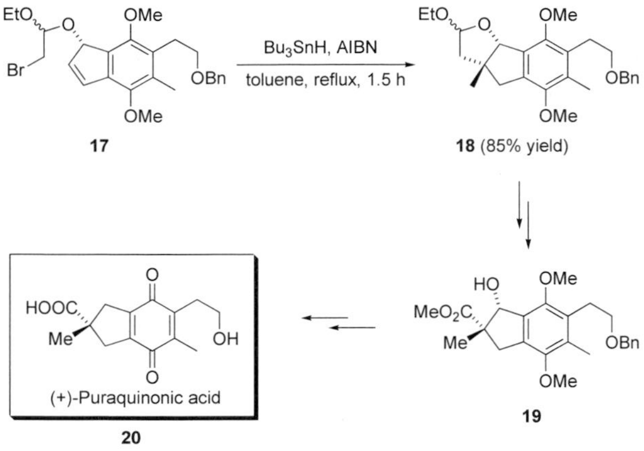

Scheme 11.6 Synthesis of puraquinonic acid.

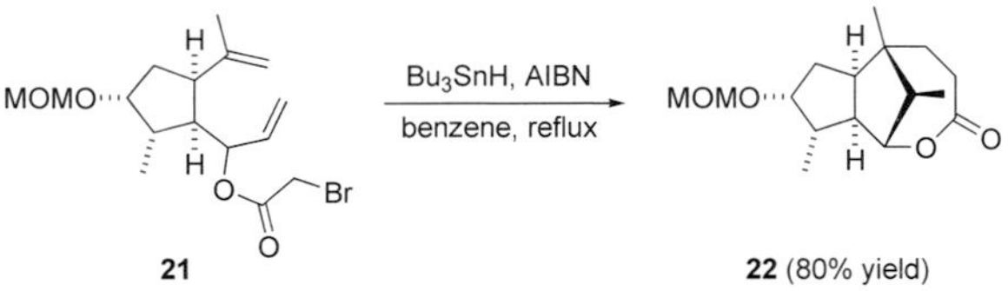

Scheme 11.7 *8-Endo vs. 5-exo* cyclization.

advantage of the stability of a radical adjacent to a sulfur atom (Scheme 11.8) [15]. The aryl radical cyclization of substrates **23a/23b** proceeded smoothly to yield the *exo* products **24a/24b** as the only stereoisomers.

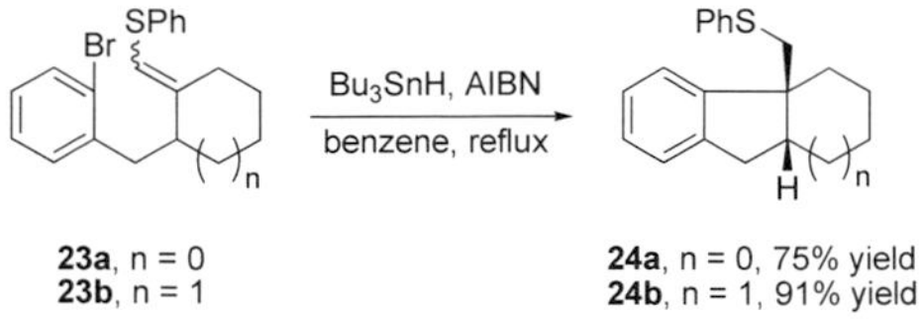

23a, n = 0
23b, n = 1

24a, n = 0, 75% yield
24b, n = 1, 91% yield

Scheme 11.8 Control of regioselectivity in radical cyclizations.

The key transformation for the synthesis of (–)-aphanorphine **28** (Scheme 11.9), which possesses a quaternary center at the benzylic position, involved aryl radical cyclization mediated by Bu$_3$SnH, furnishing the 6-*exo* product **26** [16]. The substrate **25** was derived from *trans*-4-hydroxy-L-proline. Once again, Ishibashi and coworkers took advantage of sulfur-directed *exo*-selective aryl radical cyclization obtaining the *exo* product exclusively to provide the *cis* stereochemistry. Alkaline hydrolysis of the ester group followed by Barton decarboxylation provided **27** which was then treated with Raney nickel in methanol to cause desulfurization, deprotection of the benzyloxycarbonyl group, and reductive methylation simultaneously. The product thus obtained was demethylated to afford the target compound.

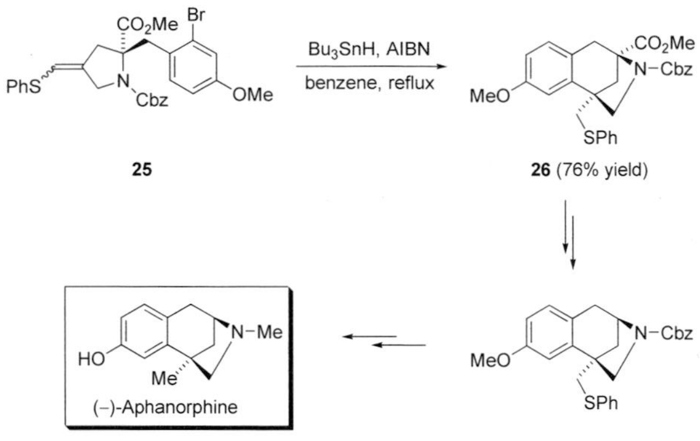

Scheme 11.9 Synthesis of aphanorphine.

An important class of diterpenoids, dolabellanes, consisting of a functionalized cyclopentane ring with a quaternary center, was formed by radical cyclization [17]. The stereoselectivity was dependent on the configuration of the double bond (Scheme 11.10). Using the substrate with an *E* double bond (not shown), the undesired product where the side chains are *trans* to each other is formed. By altering the configuration of the double bond to *Z*-**32**, the boat conformer was more stable than the chair owing to steric interaction between the xanthate and the side chain. The two forms, *exo*-**33** and *endo*-**34**, were similar in energy owing to the repulsion between the hydrogen atom in the 1,3-dioxane and the *exo* ester group or the *endo* vinyl methyl group. Thus, the substrate was changed to the tetrasubstituted olefin **29**, which after Krapcho decarboxylation gave the desired **31a** and undesired product **31b** in 2.7:1 selectivity.

Stereoselective synthesis of (–)-sibirine **39**, a spirocyclic alkaloid, was accomplished by 6-*exo* spirocyclization of an α-aminyl radical (Scheme 11.11) [18]. Subjecting chiral selenide **37** under typical radical conditions provided **38** in 60% yield. The other diastereomer and the uncyclized product were also obtained in 15% combined yield as an inseparable mixture. The regioselectivity was dependent on the bulky phenylsulfonyl group attached at the distal olefin, which assisted in stabilizing the radical adjacent to it. ^{1}H NMR spectroscopy indicated that the OTBS and PhSO$_2$ groups occupied equatorial and axial positions, respectively. Thus, the cyclized radical **41** undergoes ring inversion and the H atom is transferred from the equatorial face. Removal of the phenylsulfonyl group, deprotection of the silyl group, followed by reduction of the ester group completed the synthesis of (–)-sibirine.

Zard and coworkers synthesized a complex lycopodium alkaloid, (±)-13-deoxyserratine **44** by employing tandem radical cyclization as the key step (Scheme 11.12) [19]. Tributyltin hydride promoted the cleavage of the weak N–O bond, and two vicinal quaternary centers were generated by tandem 5-*exo*/6-*endo* cyclization, wherein the ring closure of the intermediate amidyl radical occurred from the least hindered convex face of the molecule. The reaction was carried out

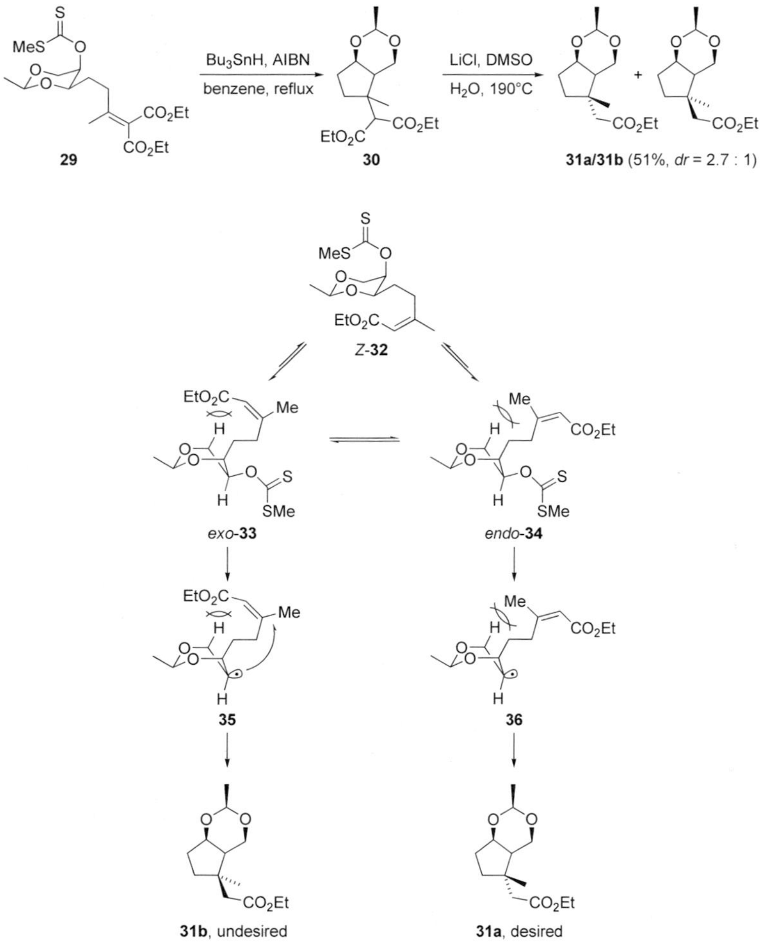

Scheme 11.10 Synthesis of functionalized cyclopentanes.

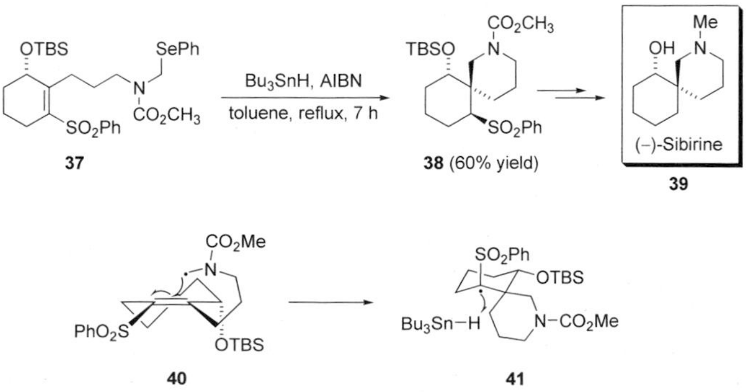

Scheme 11.11 *6-Exo* spiro cyclization. Synthesis of sibirine.

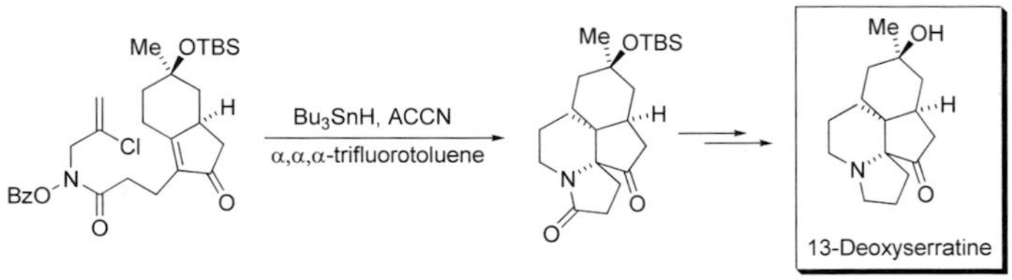

Scheme 11.12 Tandem radical cyclization. Synthesis of
(±)-13-deoxyserratine.

in the presence of 1,1′-azobis(cyclohexanecarbonitrile) as the thermal initiator.
The *6-endo* cyclization was facilitated by a chlorine atom, since in its absence, the
5-exo cyclized product predominated. *In situ* reduction of the chlorine atom took
place with the second equivalent of Bu_3SnH after the tandem reaction. Finally, the
protection of **43** as a silyl enol ether, followed by reduction of the lactam and
deprotection of the silyl group afforded the requisite alkaloid.

The azaspirocyclic core of the biologically important natural products halichlorine and pinnaic acids has been synthesized by 1,5-hydrogen-atom transfer followed
by *5-exo trig* cyclization from a simple piperidine derivative (Scheme 11.13) [20].
Starting with substrate **45**, an α-aminyl radical is generated by [1,5]-radical translo-
cation, which undergoes intramolecular radical cyclization onto the olefinic bond.
The yield of the reaction is increased if an electron-withdrawing group is present.
This sequence generates consecutive quaternary and tertiary centers. The transition
state is depicted where the former **47a** is favored over the latter **47b** owing to the
absence of steric interactions between the R group and an axially oriented proton on
the piperidine ring.

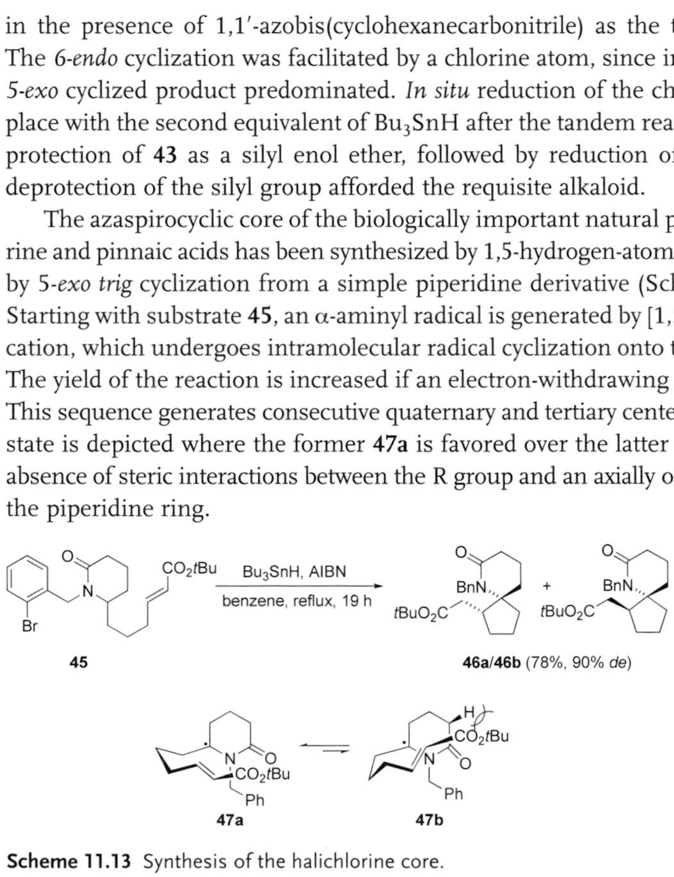

Scheme 11.13 Synthesis of the halichlorine core.

An attractive strategy to construct a diquinane system employed a tandem
radical cyclization–annulation sequence (Scheme 11.14). This one-pot reaction
allowed consecutive formation of four carbon–carbon bonds, with two adjacent
quaternary carbon centers stereoselectively [21]. Exposing this reaction to standard radical conditions followed by Tamao oxidation gave the diquinane in 51%
yield. The tandem radical sequence consisted of *5-exo dig*, vinyl *5-exo trig* cyclization
followed by trapping with acrylonitrile, and then again *5-exo trig* cyclization

followed by hydrogen-atom transfer. The diastereoselectivity was excellent and the *cis–syn–cis* isomer was obtained.

The initial 3,5-*cis* stereoselectivity was explained by a model, wherein a chair-like transition state **55a** is more favored owing to the absence of 1,3-allylic strain. The selectivity in 5-*exo trig* cyclization at the later stage was directed by this 1,3-induction, and intermediate **52** cyclized with the cyano group on the convex face of the tricyclic skeleton.

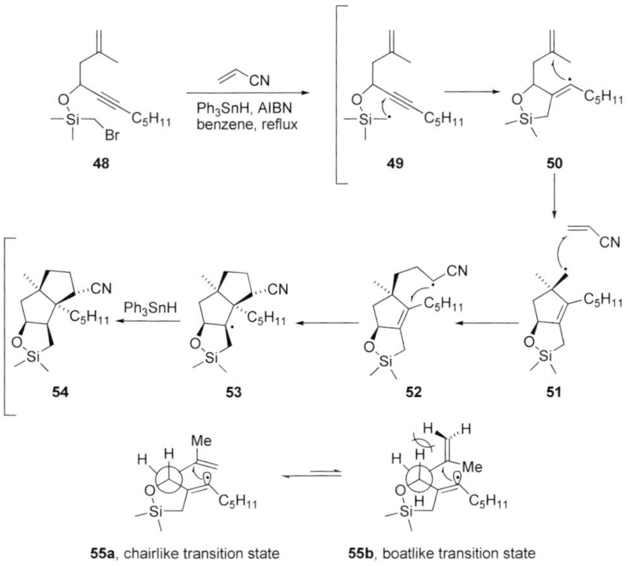

Scheme 11.14 Tandem radical annulation–cyclization.

The use of alternate H-atom donors in the formation of quaternary centers has been explored. Stereoselective 5-*exo-dig* radical cyclizations using chiral tertiary bromides attached to a carbohydrate-based chiral auxiliary have been investigated (Scheme 11.15) [22]. Diastereomerically pure **56**, when treated with AIBN and EPHP in toluene under reflux gave product **57** in high yield. The enantiomeric purity was determined by methanolysis followed by oxidation and was found to be 96%. The observed enantiomer can be predicted by the conformer **58** in which the acetal hydrogen atom is *syn* to the olefinic bond of the delocalized radical. Thus, the chiral auxiliary plays no direct role in the stereoinduction process, but the result is determined by the acetal configuration.

11.3
Atom- and Group-transfer Cyclizations

Atom- and group-transfer radical cyclizations are useful strategies for the construction of cyclic compounds [23]. An added advantage of this methodology over

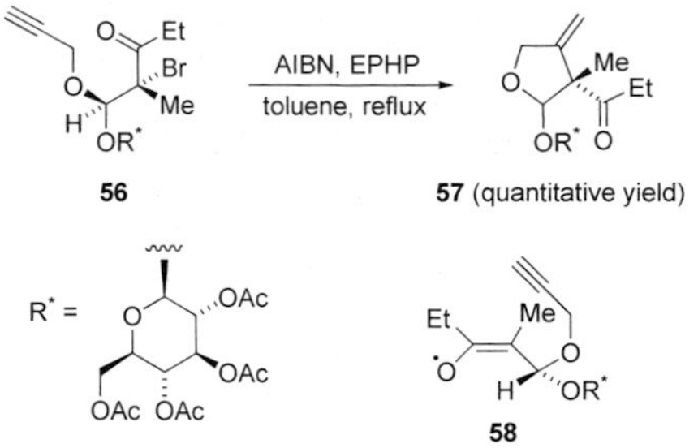

Scheme 11.15 EPHP-mediated stereoselective *5-exo dig* cyclization.

reductive radical cyclizations is further functionalization of the products obtained. Typical reactions in this class involve the transfer of bromine or iodine atoms or the phenylselenyl group.

11.3.1
Diastereoselective Atom- and Group-transfer Cyclizations

Hoffmann has demonstrated the synthesis of complex oxacycles by tandem radical cyclizations of 1,5-enynes tethered to tetrahydropyran (Scheme 11.16). The vicinal quaternary centers were installed via *5-exo-trig* cyclization followed by *6-endo-dig* cyclization [24]. In the second cyclization, ring strain resulted in *6-endo-dig* cyclization dominating over *5-exo-dig* cyclization. Compound **59a** gave *6-endo-dig* product **60** only, whereas its diastereomer **59b** was less efficient and reduced product **61b** as well as *5-exo-dig* product **62** were obtained in low yields, along with the *6-endo* product **61a**.

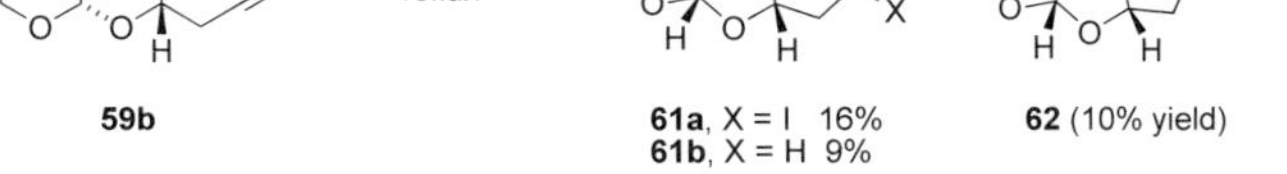

Scheme 11.16 Formation of vicinal chiral centers by *5-exo trig* and *6-endo dig* cyclizations.

Lewis-acid-catalyzed atom-transfer radical cyclization (Scheme 11.17) of unsaturated α-bromo β-ketoamides **63** proceeded in good yields providing excellent *trans* selectivity for the spirolactams [25]. $Yb(OTf)_3$ and $Mg(ClO_4)_2$ provided the best results and in the absence of Lewis acid, no product was obtained. A six-membered chelate, formed by coordination to the Lewis acid, locks the two carbonyl groups into *syn* orientation and the cyclization proceeded via **66** to form the *trans* product **64** exclusively.

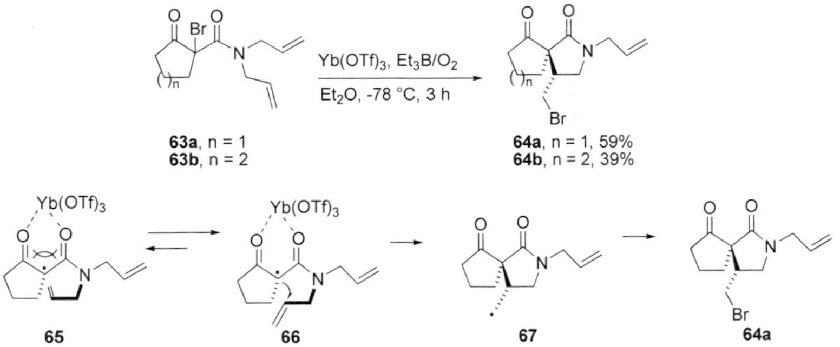

63a, n = 1
63b, n = 2

64a, n = 1, 59%
64b, n = 2, 39%

65

66

67

64a

Scheme 11.17 Lewis-acid-mediated atom-transfer cyclizations.

Yang and coworkers have developed a highly efficient phenylseleno group-transfer tandem radical cyclization in the presence of Lewis acid under photolytic conditions to generate bicyclic products bearing quaternary centers in high regio- and diastereoselectivity (Scheme 11.18) [26]. The tandem cyclization takes advantage of the fact that phenylseleno group transfer is slower than I or Br atom transfer and hence allows enough time for the formation of the second carbon–carbon bond before the radical chain is terminated. The example demonstrates tandem *6-endo* cyclizations starting from **68** to form the *trans*-decalin skeleton **69** with four stereo-centers in a single step selectively. An exocyclic double bond was formed by oxidative elimination of the phenylseleno group, which constitutes the core structure of many biologically active terpenoids. It has been demonstrated that iodine atom transfer cyclization of α-iodocycloalkanones under photolytic conditions affords spirocyclic products in moderate to good yields, albeit with low selectivity [27].

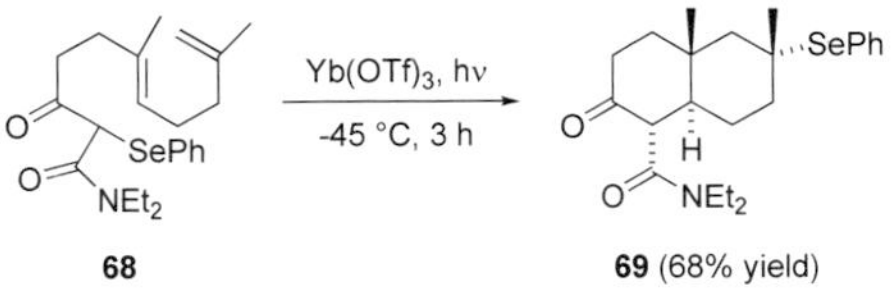

68

69 (68% yield)

Scheme 11.18 Tandem cyclization with group transfer.

11.3.2
Enantioselective Atom-transfer Cyclizations

Highly enantioselective atom-transfer cyclizations mediated by chiral Lewis acids have been reported by Yang and coworkers (Scheme 11.19, Table 11.2) [28]. The unsaturated β-keto ester substrates **70** were treated with Et_3B/O_2 in the presence of $Mg(ClO_4)_2$ and ligand **71** to furnish the products **72** in high diastereo- and enantioselectivity. The two cyclizations, *5-exo* and *6-exo*, proceeded equally well yielding **72** with high enantioselectivity. An interesting observation was that on addition of 1.1 equivalent of water, the enantioselectivity decreased. The selectivity could be restored by the addition of molecular sieves. Another point to note is the use of substoichiometric amounts of the chiral Lewis acid. The high selectivity in the cyclization can be explained by the model **73b** proposed by the authors, in which the magnesium is planar. Owing to the steric bulk of the *t*-butyl groups of the ligand, attack from the *re* face is preferred. Transition state **73b** has the lowest overall steric interactions, and accounts for the product stereochemistry.

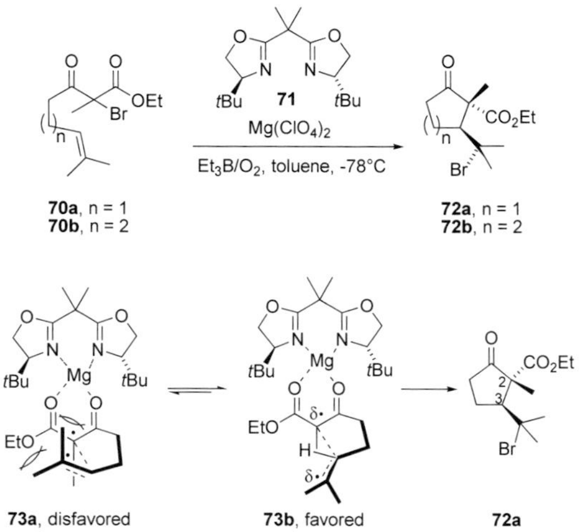

70a, n = 1
70b, n = 2

72a, n = 1
72b, n = 2

73a, disfavored   **73b**, favored   **72a**

Scheme 11.19 Enantioselective atom-transfer cyclizations.

The first examples of enantioselective tandem atom-transfer cyclizations have been reported by Yang and coworkers. Installation of four stereocenters in a single step was the highlight of this work (Scheme 11.20, Table 11.3) [29]. With a combination of Mg(ClO$_4$)$_2$ and ligand **71**, low selectivity was obtained with substrate **74**. In contrast, substrate **75** gave good enantioselectivities in toluene even at higher temperatures, although a low yield of product **77** was obtained.

Atom-transfer cyclizations of substrate **74** were also performed with Yb(OTf)$_3$ in the presence of various ligands (Scheme 11.21, Tab. 11.4). A notable observation was that on addition of molecular sieves, a reversal of selectivity was obtained (entry 3, Tab. 11.4). The authors propose a model to explain the stereochemical outcome in which the Yb complex adopts an octahedral geometry, where *re*-face cyclization is favored for steric reasons. The 6-*endo* cyclization takes place via a chair-like transition state, followed by ring flip and 5-*exo* cyclization.

Table 11.2 Results of enantioselective atom-transfer cyclizations.

Entry	n	Catalyst (equiv.)	Time (h)	Yield (%)	ee (%)
1	1	0.3	7	68	92
2	2	0.5	7.5	53	94

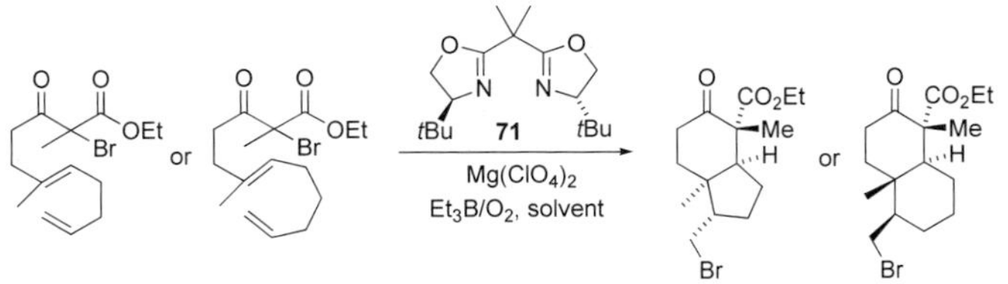

Scheme 11.20 Enantioselective tandem atom-transfer cyclizations.

11.4
Intermolecular Radical Allylations

11.4.1
Diastereoselective Allylation

Few examples have been reported for diastereoselective allylation yielding quaternary centers (Scheme 11.22, Tab. 11.5). Guindon and coworkers have obtained excellent diastereoselectivity for chelation-controlled allylation reactions of α-halo-β-alkoxy esters using $MgBr_2 \cdot OEt_2$ as a Lewis acid and allyltrimethylsilane [30] or allylstannane [31] as the allylating agent. Interestingly, a reversal of selectivity was observed when no Lewis acid was used (entry 2). Allyltrimethylstannane proved to be superior to allylsilane in terms of reactivity, although similar levels of selectivity were achieved. When treated with allylsilane, tertiary acyclic iodides displayed excellent diastereoselectivities in a chelation-controlled reaction; unfortunately, poor yields were obtained owing to competitive formation of the corresponding α-hydroxy esters. The results have been summarized wherein the tertiary iodides display excellent *anti* selectivity through bidentate chelation to the oxygen and carbonyl of the ester, rearranging the R group onto the opposite face of the radical, thus making the attack from the top face more feasible (**90**). However, in the absence of Lewis acid, the *syn* isomer **88b** was obtained as the major product, which took into account the 1,3-allylic strain.

Guindon and coworkers have also examined allylation of a tertiary radical with an adjacent tetrahydropyran ring. These reactions showed excellent selectivity. However, less encouraging results were obtained when the radical was exocyclic to a tetrahydrofuran ring (Scheme 11.23, Tab. 11.6) [31]. The authors suggest that

Table 11.3 Results of enantioselective tandem atom-transfer cyclizations.

Entry	Substrate	T (°C)	Solvent	Product	Yield (%)	*ee* (%)
1	74	−78	CH_2Cl_2	76	41	13
2[a]	74	−78	CH_2Cl_2	76	24	33
3	75	−40	Toluene	77	23	82
4	75	−20	Toluene	77	16	84

a MS 4 Å was added.

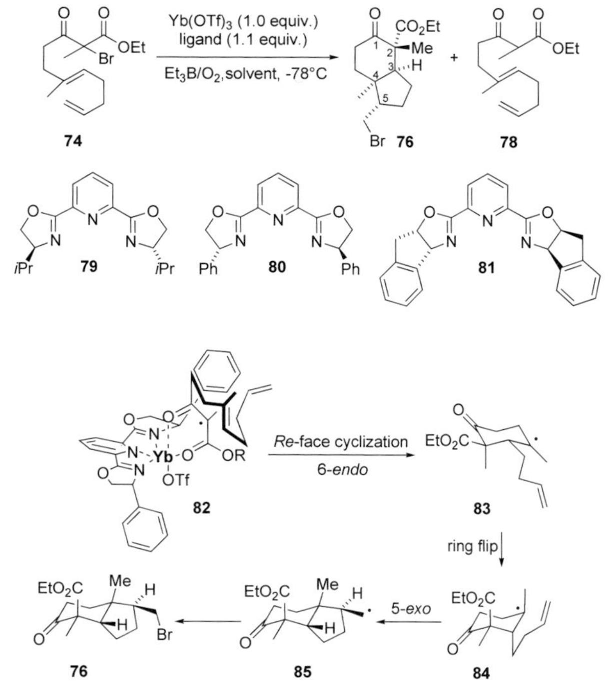

Scheme 11.21 Ytterbium triflate-mediated atom-transfer cyclizations.

the low selectivity in the reaction of **91a** could involve a monodentate transition state **95**. The alternate and potentially more selective bidentate chelate **94** is destabilized by eclipsing interactions between the C–O and the Mg–Br bonds.

11.4.2
Enantioselective Allylation

One of the best examples of quaternary center formation with high enantioselectivity was reported by Hoshino and coworkers, wherein the allylation of α-alkyl-α-iodo-lactone **98** was accomplished by the use of catalytic and stoichiometric amounts of a chiral Lewis acid (Scheme 11.24) [32]. The active catalyst was

Table 11.4 The effect of different ligands on atom-transfer cyclizations.

Entry	Ligand	Solvent	Time (h)	Yield (%) [78]	Yield (%) [76]	ee (%)
1	79	CH$_2$Cl$_2$	13	11	28	−37
2	80	CH$_2$Cl$_2$	15	23	60	66
3^a	80	CH$_2$Cl$_2$	13	68	11	−56
4	80	Toluene	12	30	37	50
5	81	Toluene	10	64	17	−43

a MS 4 Å was used.

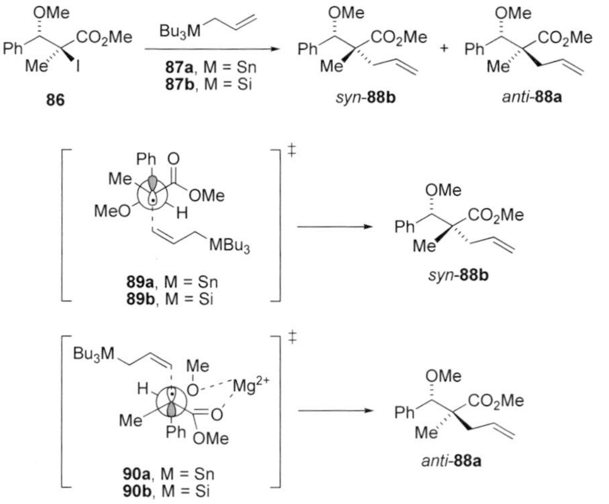

Scheme 11.22 Diastereoselective intermolecular allylations.

generated from a substituted binol-derived ligand **100** and Me$_3$Al in the presence of Et$_2$O. It is speculated that a five-coordinated aluminum complex plays a key role and the substrate is bound in a monodentate manner yielding high selectivity. Executing the reaction using a catalytic amount of the chiral Lewis acid gave similar levels of selectivity. Chiral sulfonamides **102** did not prove to be efficient catalysts for the above substrates and moderate selectivity was achieved with aluminum complexes [33]. The nature of the sulfonyl group had a large impact on the sense of selectivity and the highest enantioselectivity achieved was 54%.

Radical allylation proved to be a valuable method in the synthesis of the unnatural (+)-enantiomer of the marine alkaloid (–)-debromoflustramine B, when various organometallic couplings failed to give any product [34]. A quaternary center was obtained in good yield on addition of allyltributyltin under standard radical conditions. Stereoselective intermolecular addition of radicals followed by trapping with allyl tin would be a valuable method of generating quaternary radicals [35], but there are no reports on this type of transformation.

Table. 11.5 Results of diastereoselective allylations.

Entry	Substrate	M	88a:88b	Yield (%)
1	87a	Sn[a]	> 100:1	76
2	87a	Sn[b]	1:16	75
3	87b	Si[a]	> 100:1	39

a MgBr$_2$·OEt$_2$, Et$_3$B, CH$_2$Cl$_2$, –78°C.
b AIBN, hexanes, reflux.

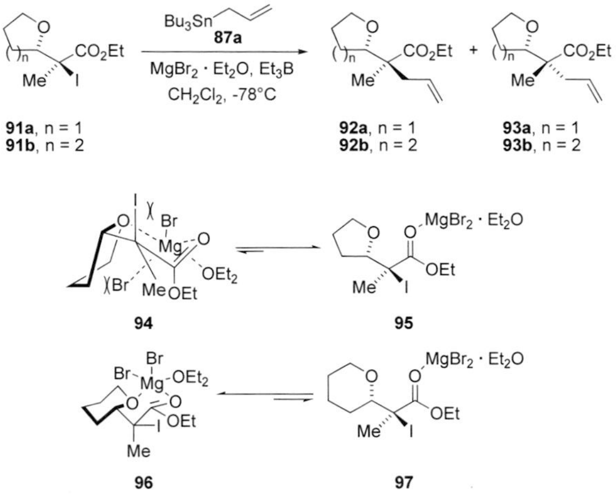

Scheme 11.23 Stereoselectivity in radical allylation with an adjacent chiral ether.

11.5
Other Metallic Reagents

Since the mid-1990s, several research groups have been actively involved indeveloping radical reactions without using tin reagents [36]. Samarium(II) iodide is a versatile reagent that has proved to be useful for inter- as well as intramolecular radical chemistry [37]. Similarly, manganese(III) acetate [38] has been widely exploited in radical cyclizations. Titanocene(III) reagents [39] provide high regioselectivity for epoxide ring opening by forming β-titanoxy radicals, and this transformation has been utilized in natural-product synthesis. Applications of the above reagents will be discussed only briefly, as several reviews have appeared that describe them in detail.

11.5.1
Cobalt-catalyzed Tandem Radical Cyclization/Cross-coupling

An interesting example by Oshima demonstrates the sequential radical cyclization and cross-coupling reaction under unusual cobalt catalysis (Scheme 11.25) [40]. A ketal containing a quaternary center has been constructed stereoselectively by treatment of bromo acetal **103** with phenylmagnesium bromide in the presence

Table 11.6 Results of allylation of a tetrahydrofuran ring.

Entry	Substrate	92:93	Yield (%)
1	91a	1:1	82
2	91b	27:1	54

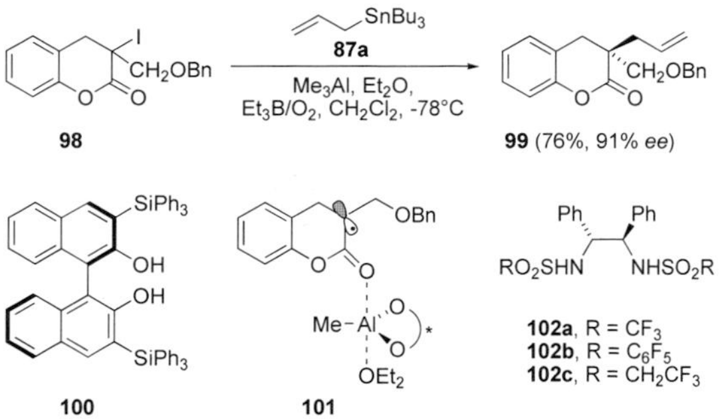

Scheme 11.24 Enantioselective radical allylation.

of CoCl$_2$(dppe). A catalytic cycle using substrate **107** was proposed for this reaction. With PhMgBr (4 equivalent), the 17-electron species **106** is obtained with the release of biphenyl. Single electron transfer generates the anionic radical **108** and Co(I) complex **111**. The loss of bromide from the anionic radical generates the radical **109** which cyclizes instantaneously and combines with **111** to give divalent cobalt species **112**. A reductive elimination furnishes products **113** and **114**. Compound **114** is converted to **106** by reacting with PhMgBr.

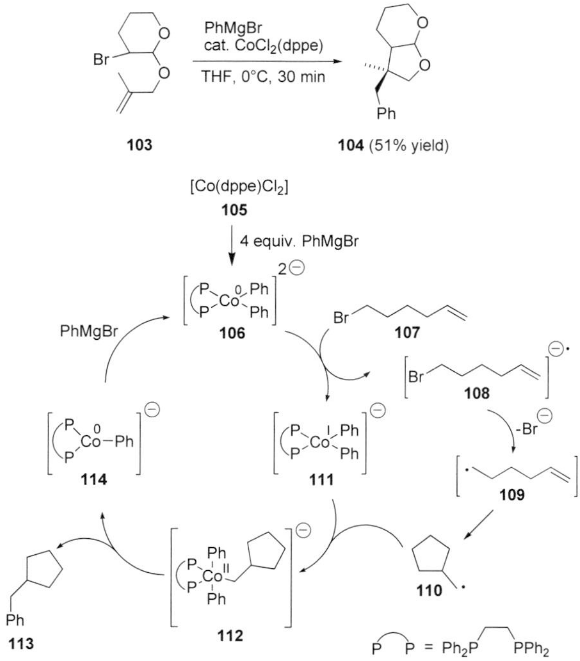

Scheme 11.25 Tandem radical cyclization and cross coupling using cobalt catalysts.

11.5.2
Samarium Diiodide-mediated Radical Reactions

Various bicyclic ethers have been synthesized via ketyl–olefin couplings mediated by SmI$_2$ (Scheme 11.26). Cyclization of **115** provided 2,6-*anti*-5,6-*trans* product **116** as the only diastereomer. The high selectivity can be explained by the non-chelated transition state **117a** which has fewer steric interactions and better dipole orientation than **117b** [41].

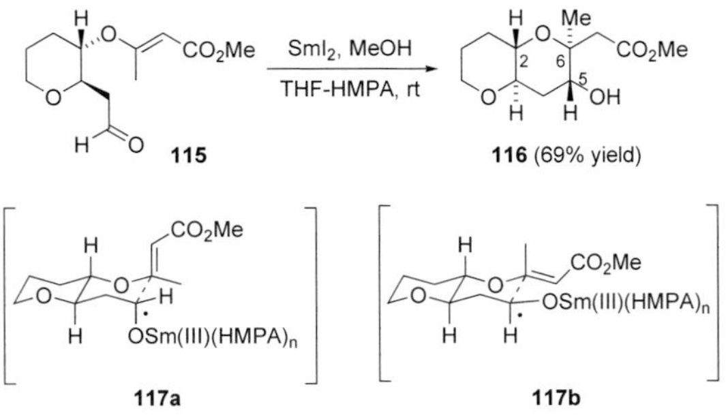

Scheme 11.26 Synthesis of bicyclic ethers using samarium iodide.

Curran and coworkers have developed a novel tandem *5-exo trig/5-exo dig* cyclization sequence *en route* to (±)-hypnophilin **120** with complete diastereocontrol (Scheme 11.27) [42]. The key step involves reaction of aldehyde **118** with SmI$_2$ generating a ketyl radical, which participates in the tandem cyclization. The product vinyl radical either abstracts hydrogen from solvent or undergoes a second electron transfer furnishing an anion, which after quenching provides **119**.

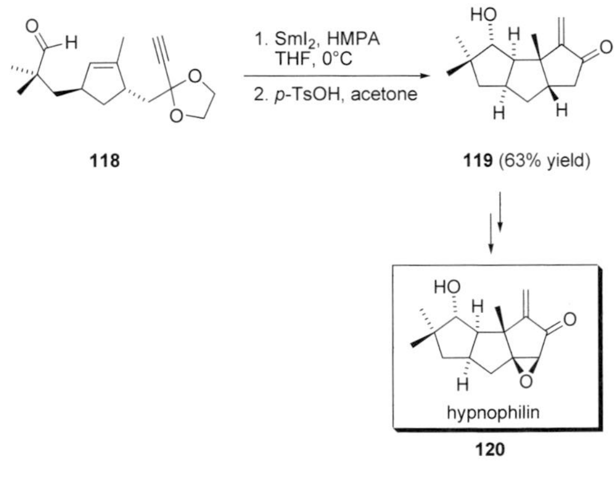

Scheme 11.27 Tandem radical cyclization – the synthesis of hypnophilin.

Synthesis of spirocycles by radical cyclization onto aromatic rings is rather limited owing to the instability of the spirocyclohexadienyl radical intermediate. To achieve the spirocyclization, the radical intermediate should be trapped irreversibly [43]. Substrate **121** undergoes a radical reaction to furnish the spirocyclic product **122** with complete stereocontrol (Scheme 11.28). The ketyl radical intermediate **123** cyclizes onto the carbon *para* to the ester group to generate **124**, followed by a single-electron reduction from a second equivalent of SmI$_2$ to give **125** and **126**. Finally, protonation of **126** followed by 1,4-reduction of the resulting enoate **127** by SmI$_2$, gives the two diastereomeric products **122a** and **122b**.

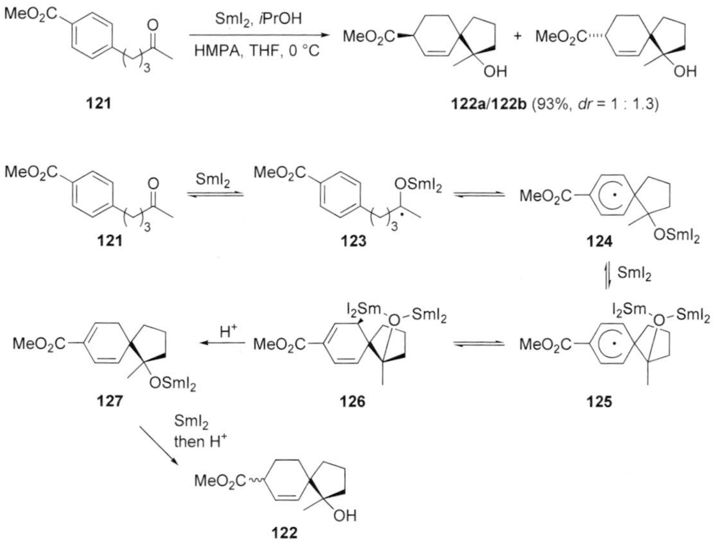

Scheme 11.28 Stereoselective spirocyclization.

Molander and coworkers have investigated intramolecular conjugate additions of alkyl halides onto α,β-unsaturated lactone **128** facilitated by SmI$_2$ in the presence of catalytic amounts of NiI$_2$ (Scheme 11.29). This method provided the complex five-membered lactone **129** as the only stereoisomer [44]. Performing this reaction in the presence of *t*BuOH as the proton source enhanced the selectivity. 6-*exo*-Cyclization was not very effective, although good yields were obtained with minimum steric bulk around the olefin. The function of the NiI$_2$ in this reaction is not entirely clear.

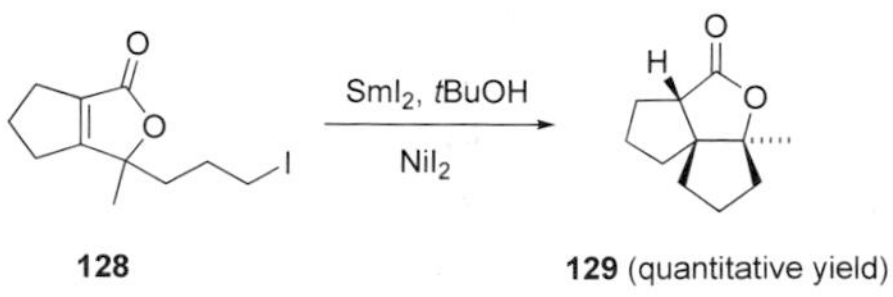

Scheme 11.29 Intramolecular conjugate additions.

11.5.3
Manganese(III)-based Oxidative Radical Cyclizations

Unlike SmI_2-catalyzed cyclizations, which are reductive, $Mn(OAc)_3/Cu(OAc)_2$ cyclizations are classified as oxidative radical cyclizations (Scheme 11.30). $Mn(OAc)_3$ facilitates enol formation, and acts as a single-electron oxidant to generate the electrophilic radical **132** which then adds to the olefinic bond. Under mild conditions, carbon–carbon bonds are formed with excellent stereoselectivity.

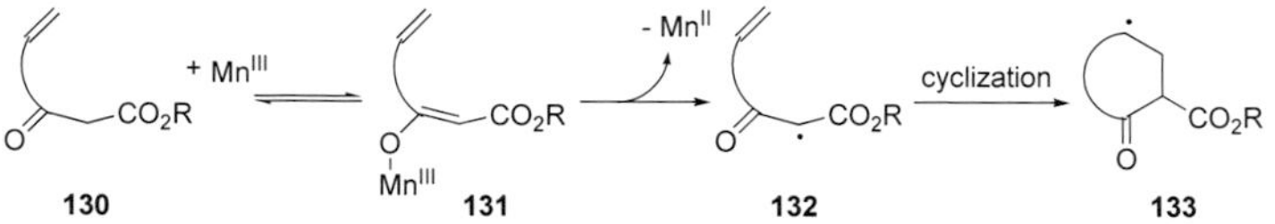

130 131 132 133

Scheme 11.30 Manganese(III)-mediated cyclizations.

Snider and coworkers have shown that β-keto sulfoxides and sulfones can be converted to bicyclo[3.2.1]octanones on treatment with $Mn(OAc)_3/Cu(OAc)_2$ [45]. The sulfoxide chiral center completely controls the stereochemistry of the cyclization. Oxidative cyclization of optically active substrate **134** provides product **135** as a single diastereomer (Scheme 11.31).

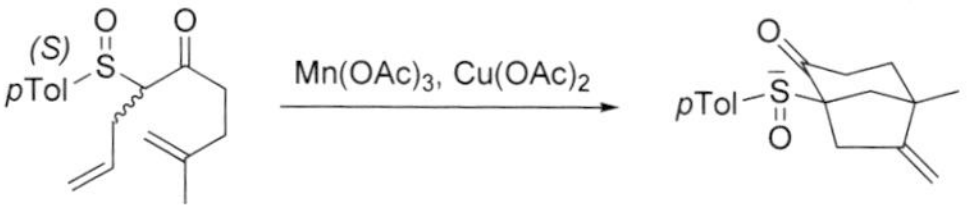

134 **135** (44% yield)

Scheme 11.31 Chiral sulfoxides in manganese(III)-mediated cyclizations.

Efficient alkylation of β-keto esters and β-keto amides with enol ethers in the presence of the $Mn(OAc)_3/Cu(OAc)_2$ system has been investigated by Parsons (Scheme 11.32). Substrate **136** has been converted to substituted pyrrolidinone **138** in good yield and diastereoselectivity in the construction of the quaternary center [46]. A particular advantage of this reaction is the synthesis of tricarbonyl derivatives by alkylation under nonbasic conditions.

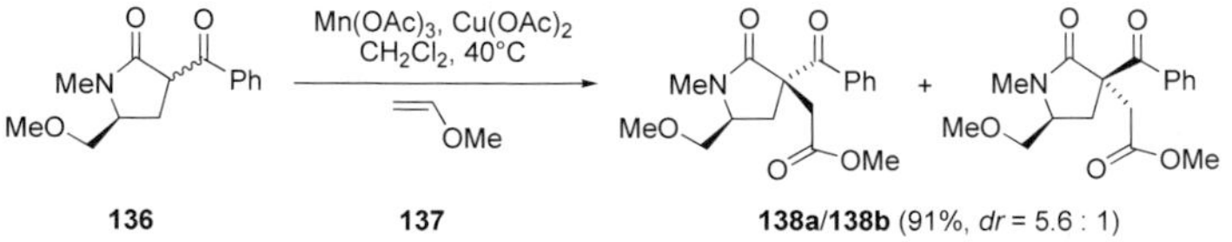

136 **137** **138a/138b** (91%, *dr* = 5.6 : 1)

Scheme 11.32 Intermolecular alkylation using vinyl ethers.

Outstanding research has been done by Yang and coworkers in the field of asymmetric $Mn(OAc)_3$ cyclizations and the methodology has been applied to the total synthesis of a variety of natural products (**139–143**) (Scheme 11.33).

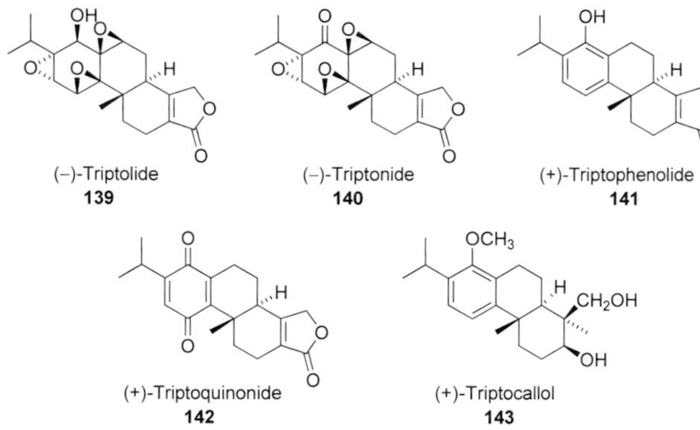

(−)-Triptolide
139

(−)-Triptonide
140

(+)-Triptophenolide
141

(+)-Triptoquinonide
142

(+)-Triptocallol
143

Scheme 11.33 Natural products synthesized by Yang and coworkers.

Both diastereoselective and enantioselective approaches have been implemented. The rate and diastereoselectivity of the radical cyclizations are significantly improved upon addition of lanthanide triflates (Scheme 11.34) [47]. Yb(OTf)$_3$ proved to be the best Lewis acid. For enantioselective radical cyclizations, (−)-8-phenylmenthyl chiral auxiliary was used, and the stereochemical outcome is explained by the model [48]. For the chair-like transition state in the presence of Yb(OTf)$_3$, *syn* orientation **147** predominates owing to the formation of a six-membered chelate by coordination of the two carbonyls, and attack from the *re* face occurs as the *si* face is shielded by the phenyl group of the chiral auxiliary. However, in the absence of Lewis acid, *anti* orientation is preferred and the *si* face **148** is more open for attack. In the synthesis of (+)-triptocallol **143**, the increased steric bulk and electron density of the 8-aryl group has a positive impact on diastereoselectivity [49].

Both enantiomers (>98% *ee*) of a terpenoid, wilforonide, were synthesized by the above method of generating quaternary stereocenters, utilizing the chiral auxiliary derived from *(R)*-pulegone (Scheme 11.35) [50]. (+)-Wilforonide **153a** was synthesized in the absence of Yb(OTf)$_3$ to obtain the correct stereochemistry, whereas the presence of Yb(OTf)$_3$ was necessary in synthesizing (−)-wilforonide **153b**; a methoxy group in the auxiliary was also essential. A drawback of both syntheses was their poor regioselectivity.

As shown in Scheme 11.36 for cyclization of substrate **149**, the phenyl group can shield the *si* face of an Mn(III) enolate thus rendering attack on the *re* face possible, whereas attack on the *si* face is more feasible in the presence of Yb(OTf)$_3$ with substrate **151**.

11.5.4
Titanocene-mediated Radical Cyclizations

RajanBabu and Nugent have devised mild reaction conditions for the opening of epoxides using Cp$_2$TiCl under radical conditions [51]. Substrate **154** generates β-titanoxy radical **155**, which on cyclization followed by proton transfer yields the

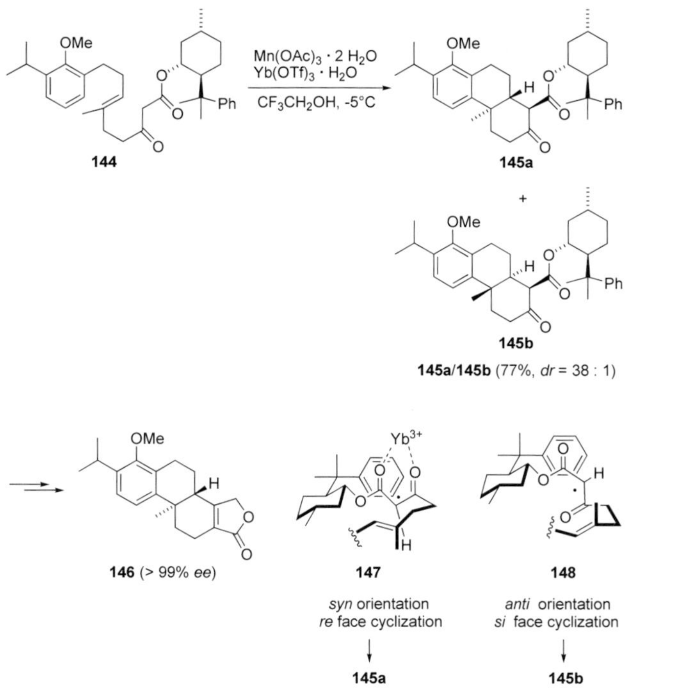

Scheme 11.34 Lewis-acid-mediated radical cyclizations.

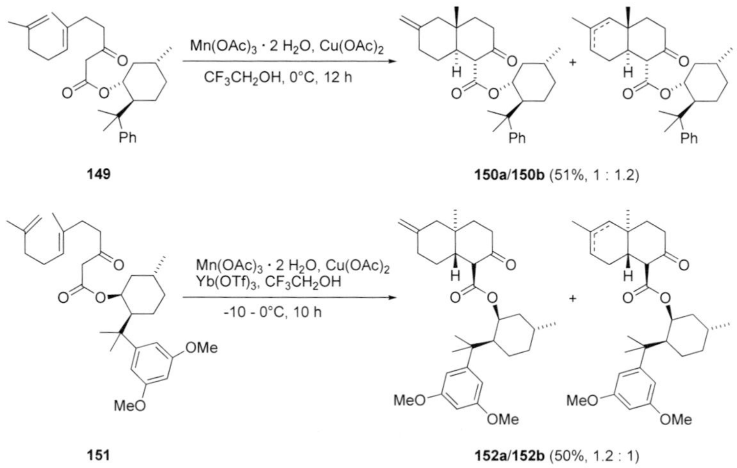

Scheme 11.35 Chiral-auxiliary-mediated radical cyclizations.

cis-fused product **158** (Scheme 11.37). The scheme illustrates the mechanism of titanocene-mediated cyclization to generate a bicyclic ring bearing a quaternary center. Epoxy-olefin cyclization was especially effective for epoxides with a carbohydrate moiety **159** and highly functionalized carbocycles were obtained with good selectivity.

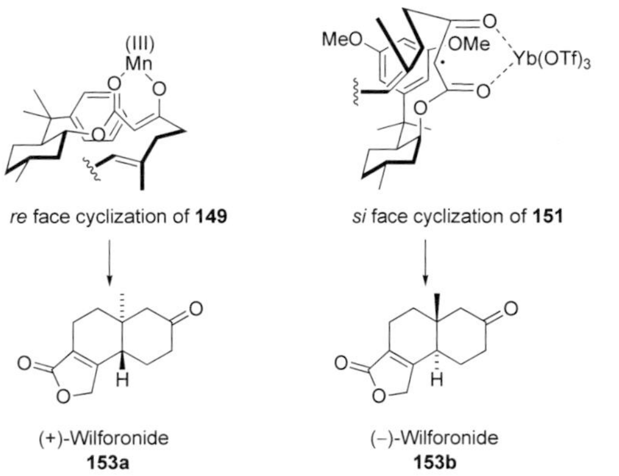

re face cyclization of **149**

si face cyclization of **151**

(+)-Wilforonide
153a

(−)-Wilforonide
153b

Scheme 11.36 Mechanistic rationale for face selectivity.

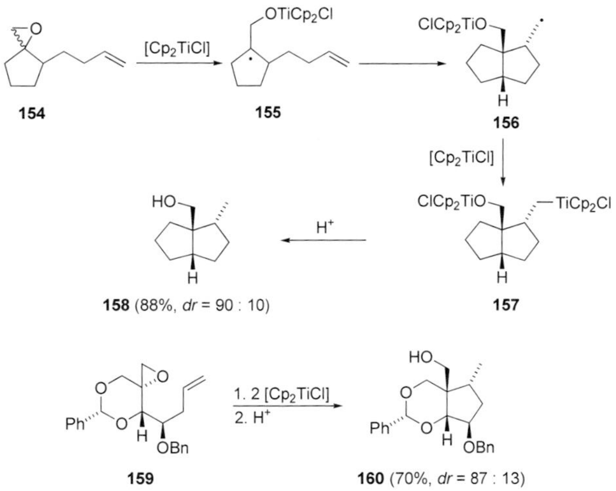

154

155

156

158 (88%, *dr* = 90 : 10)

157

159

160 (70%, *dr* = 87 : 13)

Scheme 11.37 The application of titanocene-mediated radical
cyclization in the synthesis of cyclopentanes.

Tandem reactions have also been employed for the synthesis of complex products using titanocene-catalyzed conditions (Scheme 11.38). On treatment with Cp_2TiCl_2, compound **161** generates a reactive vinyl radical after cyclization, and this then adds to the acrylate to form product **163** with high diastereoselectivity [52]. The steric interactions between the catalyst and the acrylate are important for the selectivity.

11.6
Radical Reactions in the Solid State

An attractive method for synthesizing quaternary centers is by photochemical irradiation of crystalline ketones that have radical stabilizing α-substituents,

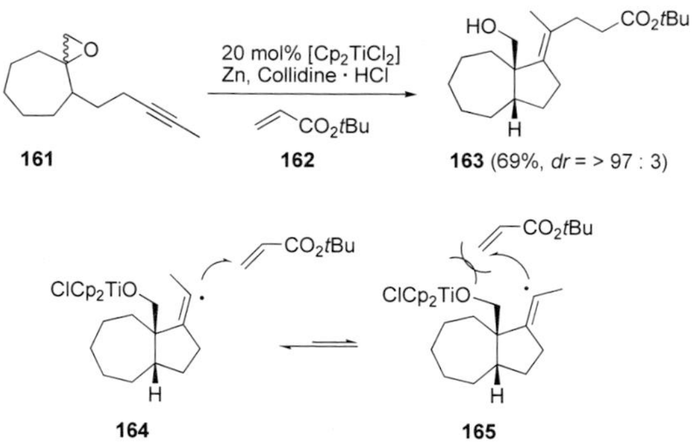

Scheme 11.38 Tandem intra- and intermolecular radical reactions using titanocene reagents.

which induce decarbonylation [53]. This method is highly chemoselective and stereospecific and generates efficient radical pairs that undergo radical–radical combination (Scheme 11.39). Interestingly, these reactions gave various side products when carried out in solution phase. This method illustrates an excellent example of solvent-free and green chemistry, an area of science that has seen rapid growth.

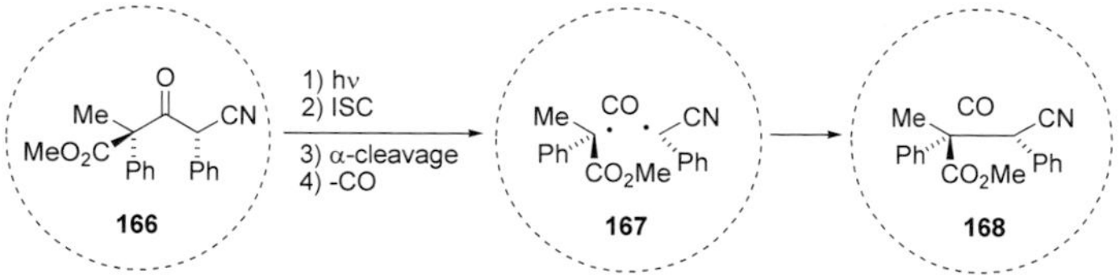

Scheme 11.39 Radical reactions in the solid state.

The same concept has been applied in the diastereoselective synthesis of (±)-herbertenolide **171** and for the first time a solid-state reaction has been employed as a key step in the total synthesis of a natural product (Scheme 11.40) [54]. Two adjacent stereogenic quaternary centers are generated through this method. The synthesis was accomplished in seven steps although the yield of the key step was low owing to subsequent melting of the methyl ester and the generation of small amounts of undesirable solution-phase products.

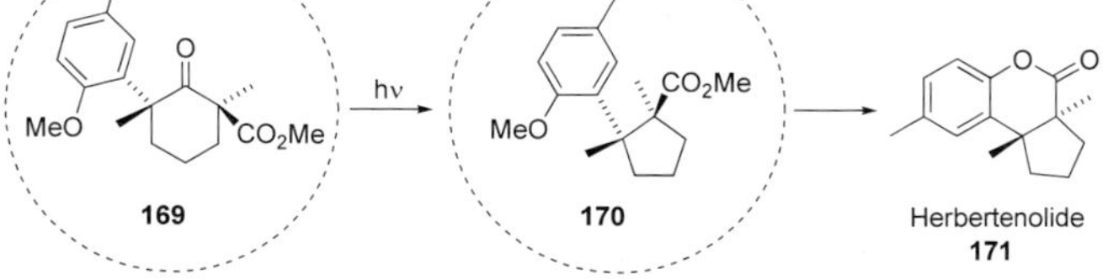

Scheme 11.40 The synthesis of herbertenolide by radical reactions in the solid state.

11.7
Conclusion

In conclusion, radical reactions have been very useful for building quaternary centers and, owing to the mild reaction conditions and the ability to tolerate a number of functional groups, in some instances they even surpass ionic chemistry. Diastereoselective radical chemistry to yield quaternary centers has been well established, although there are only a handful of examples of enantioselective variants. Moreover, not even a single example of enantioselective intermolecular radical addition followed by trapping to form a quaternary center has been reported. Thus, radical chemistry still has plenty of room for improvement.

11.8
Experimental

General procedure for radical cyclization using tributyltin hydride and AIBN as the initiator (excerpted from Ref. [10]): To a solution of (3S,4R)-2-iodo-2,3,4-trimethyl-3-[4-(1,1,1-trimethylsilyl)-3-butynyl]cyclohexan-1-one (1.7 g, 4.4 mmol) in benzene (267 mL) at 65°C was added a solution of Bu$_3$SnH (1.4 mL, 5.3 mmol) and AIBN (90 mg, 0.5 mmol) in benzene (80 mL) with a syringe pump during 6 h. The reaction mixture was heated at reflux further for 2 h and then cooled to room temperature. Benzene was removed with a rotary evaporator. Et$_2$O (25 mL) and saturated KF solution (25 mL) were added. The mixture was stirred for 18 h. Concentration and column chromatography (SiO$_2$, EtOAc/hexane = 1:50) gave the product as yellow liquid in 78% yield.

General procedure for radical-mediated enantioselective allylation using a stoichiometric amount of chiral Lewis acid (excerpted from Ref. [33]): The chiral Lewis acids were generated from chiral sulfonamides (0.2 mmol) when reacted with 1 equivalent of Lewis acid [Me$_3$Al, Et$_2$AlCl or iBu$_3$Al (1 mol dm^{-3} hexane solution)] for 2 h at 80°C in toluene (3 mL), a homogeneous solution. The solution of the chiral Lewis acid was cooled to –78°C, and then stirred for 30 min. A solution of α-alkoxymethyl-α-iododihydrocoumarin (0.2 mmol) in toluene (2 mL) was added, and the mixture was stirred for 1 h at –78°C. Allyltributyltin (0.2 mmol) and Et$_3$B (0.2 mmol, 1 mol dm^{-3} hexane solution) were added successively, and the resulting mixture was stirred at –78°C. When TLC analysis indicated the complete disappearance of iodolactone (1–5 h), saturated NH$_4$Cl was added. The mixture was extracted with benzene (2 × 20 mL). The organic layer was washed with saturated aqueous NaCl and dried over MgSO$_4$. Concentration followed by purification through silica-gel column chromatography (benzene) furnished the desired product as an oil.

General procedure for Mn(III)-based oxidative radical cyclization (Excerpted from Ref. [48b]): To a solution of **144** (163.8 mg, 0.30 mmol) in acetic acid (3 mL) was added Mn(OAc)$_3$·2H$_2$O (182.3 mg, 0.66 mmol) under argon. The mixture was stirred at 50°C. The mixture turned white after 1 h. Saturated NaHSO$_3$ solution was added to the mixture followed by extraction with CH$_2$Cl$_2$. The extract was dried with anhydrous Na$_2$SO$_4$, concentrated, and purified by flash column chromatography (5% to 10% EtOAc in *n*-hexane) to afford *trans*-ring junction products (42% yield) and *cis*-ring junction products (8% yield).

References

1 For general reviews on radical chemistry see: (a) P. Renaud, M. P. Sibi (eds.) **2001**, *Radicals in Organic Synthesis*, Vols. 1 and 2, Wiley-VCH, New York. (b) A. F. Parsons **2000**, *An Introduction to Free Radical Chemistry*, Blackwell Science, Oxford. (c) Z. B. Alfassi **1999**, *General Aspects of the Chemistry of Radicals*, Wiley, New York. (d) J. Fossey, D. Lefort, J. Sorba **1995**, *Free Radical in Organic Chemistry*, Wiley, New York. (e) D. P. Curran **1991**, in *Comprehensive Organic Synthesis*, Vol. 4, ed. B. M. Trost, I. Fleming, M. F. Semmelheck, Pergamon, Oxford, p 715. (f) B. Giese **1986**, *Radicals in Organic Synthesis. Formation of Carbon–Carbon Bonds*, Pergamon, Oxford.

2 For reviews on quaternary centers see: (a) I. Denissova, L. Barriault, *Tetrahedron* **2003**, *59*, 10105–10146. (b) J. Christoffers, A. Baro, *Angew. Chem. Int. Ed.* **2003**, *42*, 1688–1690. (c) J. Christoffers, A. Mann, *Angew. Chem. Int. Ed.* **2001**, *40*, 4591–4597. (d) E. J. Corey, A. Guzman-Perez, *Angew. Chem. Int. Ed.* **1998**, *37*, 388–401. (e) K. Fuji, *Chem. Rev.* **1993**, *93*, 2037–2066. (f) S. F. Martin, *Tetrahedron* **1980**, *36*, 419–460.

3 For reviews on diastereoselective radical chemistry see: (a) P. Renaud, M. Gerster, *Angew. Chem. Int. Ed.* **1998**, *37*, 2562–2579. (b) D. P. Curran, N. A. Porter, B. Giese **1995**, *Stereochemistry of Radical Reactions*, VCH, Weinheim. (c) W. Smadja, *Synlett* **1994**, 1–26. (d) A. L. Beckwith, *Chem. Soc. Rev.* **1993**, *22*, 143–151. (e) N. A. Porter, B. Giese, D. P. Curran, *Acc. Chem. Res.* **1991**, *24*, 296–301. (f) B. Giese, *Angew. Chem. Int. Ed. Engl.* **1989**, *28*, 969–980.

4 For reviews on enantioselective radical chemistry see: (a) M. P. Sibi, S. Manyem, J. Zimmerman, *Chem. Rev.* **2003**, *103*, 3263–3296. (b) A. F. Parsons, S. Bar, *Chem. Soc. Rev.* **2003**, *32*, 251–263. (c) M. P. Sibi, T.

R. Rheault **2001**, in *Radicals in Organic Synthesis*, Vol. 1, ed. P. Renaud, M. P. Sibi, Wiley-VCH, Weinheim, p. 461. (d) M. P. Sibi, N. A. Porter, *Acc. Chem. Res.* **1999**, *32*, 163–171. (e) R. E. Gawley, J. Aube **1996**, *Principles of Asymmetric Synthesis*, Vol. 14, Pergamon, Oxford.

5 For an excellent review on radicals in natural-product synthesis see: C. P. Jasperse, D. P. Curran, T. L. Fevig, *Chem. Rev.* **1991**, *91*, 1237–1286.

6 H. Yorimitsu, H. Shinokubo, K. Oshima, *Synlett* **2002**, 674–686.

7 For a review on radical cyclizations see: L. Yet, *Tetrahedron* **1999**, *55*, 9349–9403.

8 G. Stork, N. H. Baine, *J. Am. Chem. Soc.* **1982**, *104*, 2321–2323.

9 K. Kim, S. Okamoto, F. Sato, *Org. Lett.* **2001**, *3*, 67–69.

10 C-K. Sha, H-W. Liao, P-C. Cheng, S-C. Yen, *J. Org. Chem.* **2003**, *68*, 8704–8707.

11 M. Inoue, T. Sato, M. Hirama, *J. Am. Chem. Soc.* **2003**, *125*, 10772–10773.

12 T. Yoshimitsu, S. Yanagisawa, H. Nagaoka, *Org. Lett.* **2000**, *2*, 3751–3754.

13 (a) D. L. J. Clive, M. Yu, *Chem. Commun.* **2002**, 2380–2381. (b) D. L. J. Clive, M. Yu, M. Sannigrahi, *J. Org. Chem.* **2004**, *69*, 4116–4125.

14 E. Lee, C. H. Yoon, T. H. Lee, S. Y. Kim, T. J. Ha, Y.-s. Sung, S.-H. Park, S. Lee, *J. Am. Chem. Soc.* **1998**, *120*, 7469–7478.

15 H. Ishibashi, T. Kobayashi, D. Takamasu, *Synlett* **1999**, 1286–1288.

16 O. Tamura, T. Yanagimachi, T. Kobayashi, H. Ishibashi, *Org. Lett.* **2001**, *3*, 2427–2429.

17 (a) Q. Zhu, K.-Y. Fan, H.-W. Ma, L.-X. Qiao, Y.-L. Wu, Y. Wu, *Org. Lett.* **1999**,

1, 757–759. (b) Q. Zhu, L.-X. Qiao, Y. Wu, Y.-L. Wu, *J. Org. Chem.* **1999**, *64*, 2428–2432. (c) Q. Zhu, L. Qiao, Y. Wu, Y.-L. Wu, *J. Org. Chem.* **2001**, *66*, 2692–2699.

18 M. Koreeda, Y. Wang, L. Zhang, *Org. Lett.* **2002**, *4*, 3329–3332.

19 J. Cassayre, F. Gagosz, S. Z. Zard, *Angew. Chem. Int. Ed.* **2002**, *41*, 1783–1785.

20 K. Takasu, H. Ohsato, M. Ihara, *Org. Lett.* **2003**, *5*, 3017–3020.

21 M. Journet, W. Smadja, M. Malacria, *Synlett* **1990**, 320–321.

22 R. McCague, R. G. Pritchard, R. J. Stoodley, D. S. Williamson, *Chem. Commun.* **1998**, 2691–2692.

23 (a) A. J. Clark, *Chem. Soc. Rev.* **2002**, *31*, 1–11. (b) M. J. Rudolph, W. C. Agosta, *Chemtracts: Org. Chem.* **1990**, *3*, 79–81. (c) Y. Yamamoto, *Chemtracts: Org. Chem.* **1989**, *2*, 258–260.

24 O. Rhode, H. M. R. Hoffmann, *Tetrahedron* **2000**, *56*, 6479–6488.

25 D. Yang, Y.-L. Yan, K.-L. Law, N.-Y. Zhu, *Tetrahedron* **2003**, *59*, 10465–10475.

26 D. Yang, Q. Gao, O.-Y. Lee, *Org. Lett.* **2002**, *4*, 1239–1241.

27 H.-H. Lin, W.-S. Chang, S.-Y. Luo, C.-K. Sha, *Org. Lett.* **2004**, *6*, 3289–3292.

28 D. Yang, S. Gu, Y.-L. Yan, N.-Y. Zhu, K.-K. Cheung, *J. Am. Chem. Soc.* **2001**, *123*, 8612–8613.

29 D. Yang, S. Gu, Y.-L. Yan, H.-W. Zhao, N.-Y. Zhu, *Angew. Chem. Int. Ed.* **2002**, *41*, 3014–3017.

30 Y. Guindon, B. Guerin, C. Chabot, W. Ogilvie, *J. Am. Chem. Soc.* **1996**, *118*, 12528–12535.

31 B. Guerin, C. Chabot, N. Mackintosh, W. W. Ogilvie, Y. Guindon, *Can. J. Chem.* **2000**, *78*, 852–867.

32 M. Murakata, T. Jono, Y. Mizuno, O. Hoshino, *J. Am. Chem. Soc.* **1997**, *119*, 11713–11714.

33 M. Murakata, T. Jono, O. Hoshino, *Tetrahedron: Asymmetry* **1998**, *9*, 2087–2092.

34 M. Bruncko, D. Crich, R. Samy, *J. Org. Chem.* **1994**, *59*, 5543–5549.

35 Unpublished results from our laboratory.

36 For tin hydride substitutes in radical reactions see: (a) A. Studer, S. Amrein, *Synthesis* **2002**, 835–849. (b) See Ref. [23a]. (c) A. Gansäuer, H. Bluhm, *Chem. Rev.* **2000**, *100*, 2771–2788.

37 (a) D. J. Edmonds, D. Johnston, D. J. Procter, *Chem. Rev.* **2004**, *104*, 3371–3403. (b) G. A. Molander, C. R. Harris, *Chem. Rev.* **1996**, *96*, 307–338.

38 B. B. Snider, *Chem. Rev.* **1996**, *96*, 339–364.

39 A. Gansäuer, T. Lauterbach, S. Narayan, *Angew. Chem. Int. Ed.* **2003**, *42*, 5556–5573.

40 K. Wakabayashi, H. Yorimitsu, K. Oshima, *J. Am. Chem. Soc.* **2001**, *123*, 5374–5375.

41 K. Suzuki, H. Matsukura, G. Matsuo, H. Koshino, T. Nakata, *Tetrahedron Lett.* **2002**, *43*, 8653–8655.

42 T. L. Fevig, R. L. Elliott, D. P. Curran, *J. Am. Chem. Soc.* **1988**, *110*, 5064–5067.

43 H. Ohno, M. Okumura, S-I. Maeda, H. Iwasaki, R. Wakayama, T. Tanaka, *J. Org. Chem.* **2003**, *68*, 7722–7732.

44 G. A. Molander, D. J. St. Jean, Jr., *J. Org. Chem.* **2002**, *67*, 3861–3865.

45 B. B. Snider, B. Y.-F. Wan, B. O. Buckman, B. M. Foxman, *J. Org. Chem.* **1991**, *56*, 328–334.

46 G. Bar, A. F. Parsons, C. B. Thomas, *Org. Biomol. Chem.* **2003**, *1*, 373–380.

47 D. Yang, X.-Y. Ye, M. Xu, K.-W. Pang, K.-K. Cheung, *J. Am. Chem. Soc.* **2000**, *122*, 1658–1663.

48 (a) D. Yang, X.-Y. Ye, S. Gu, M. Xu, *J. Am. Chem. Soc.* **1999**, *121*, 5579–5580. (b) D. Yang, X.-Y. Ye, M. Xu, *J. Org. Chem.* **2000**, *65*, 2208–2217.

49 D. Yang, M. Xu, M.-Y. Bian, *Org. Lett.* **2001**, *3*, 111–114.

50 D. Yang, M. Xu, *Org. Lett.* **2001**, *3*, 1785–1788.

51 T. V. Rajanbabu, W. A. Nugent, *J. Am. Chem. Soc.* **1994**, *116*, 986–997.

52 A. Gansäuer, M. Pierobon, H. Bluhm, *Angew. Chem. Int. Ed.* **2002**, *41*, 3206–3208.

53 M. E. Ellison, D. Ng, H. Dang, M. A. Garcia-Garibay, *Org. Lett.* **2003**, *5*, 2531–2534.

54 D. Ng, Z. Yang, M. A. Garcia-Garibay, *Org. Lett.* **2004**, *6*, 645–647.

12
Enzymatic Methods

Uwe T. Bornscheuer, Erik Henke, and Jürgen Pleiss

12.1
Introduction

An alternative to the stereoselective synthesis of optically pure building blocks with tertiary stereogenic centers is the kinetic resolution of racemic compounds utilizing enzyme-catalyzed processes. In particular, hydrolytic enzymes are versatile biocatalysts and find increasing applications in organic synthesis [1]. A considerable number of industrial processes using these enzymes have been commercialized [2]. Within this class, lipases (E.C. 3.1.1.3) and carboxyl esterases (E.C. 3.1.1.1) are frequently used, as they accept a broad range of non-natural substrates, are usually very stable in organic solvents, and exhibit good to excellent stereoselectivity. However, the vast majority of (chiral) substrates are secondary alcohols, followed by carboxylic acids and primary alcohols. Unfortunately, very few hydrolases have been shown to accept tertiary alcohols **2** or their esters **1** as substrates (Scheme 12.1) [3]. Furthermore, reaction rates or enantioselectivities are quite low and consequently applications are rather unlikely. Finding or developing hydrolases with high activity, high enantioselectivity, and broad substrate specificity toward tertiary alcohols is highly desirable.

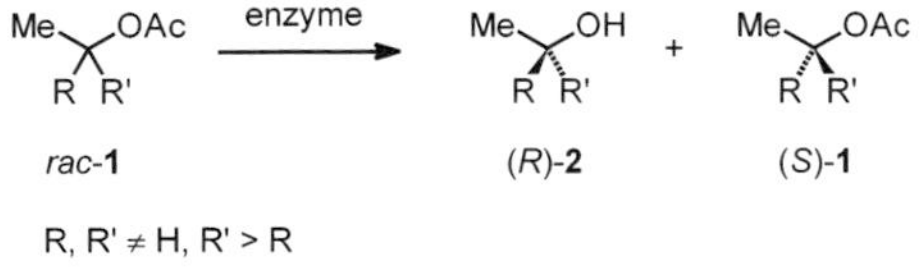

Me、 OAc   enzyme ⟶   Me、 OH + Me、 OAc
  R R'   R R'   R R'

rac-**1**   (*R*)-**2**   (*S*)-**1**

R, R' ≠ H, R' > R

Scheme 12.1 Enzymatic kinetic resolution starting from racemic substrates *rac*-**1** to yield tertiary alcohols (*R*)-**2** as products; priority of R, R' according to Cahn–Ingold–Prelog.

However, only a small fraction of the large natural pool of hydrolases with a known sequence has been biochemically characterized. The largest class of hydrolases, the α/β hydrolases, currently comprises more than 3000 known enzymes [4]. These enzymes differ considerably in sequence and shape of the substrate binding site [5], although they share the same catalytic mechanism and the same general

Quaternary Stereocenters: Challenges and Solutions for Organic Synthesis. Edited by Jens Christoffers, Angelika Baro
Copyright © 2005 WILEY-VCH Verlag GmbH & Co. KGaA, Weinheim
ISBN: 3-527-31107-6

architecture [6]. This rich pool of enzymes with supposedly slightly different properties is by no means completely explored. Thus, further screening for tertiary-alcohol-converting enzymes is a promising approach. The number of potentially useful enzymes can be further increased by creating point mutants or chimeric proteins. It has been shown that even small changes in sequence can lead to large changes in the biochemical properties of the catalyst, and this leads to successful directed-evolution approaches. As a third approach, rational selection and design of the optimal biocatalyst has been shown to be successful. Since the early 1990s, the structures of numerous bacterial and mammalian lipases and esterases have been investigated by X-ray crystallography and by molecular modeling. Analyzing the relationship between sequence, structure, and function has yielded insights into the molecular basis of substrate specificity and enantioselectivity, and explanations for experimental observations such as: why do lipases from *Rhizopus*, *Pseudomonas*, or *Candida antarctica* not convert esters of tertiary alcohols, while lipases from *Candida rugosa* do [7]? Why are secondary alcohols converted with high enantioselectivity, whereas the enantioselectivity toward tertiary alcohols is rather low? Even more it has been shown that single amino acid substitutions can increase selectivity considerably – does this mean that any lipase can be turned into a tertiary-alcohol-converting enzyme with high enantioselectivity just by applying multiple single mutations? Or are activity and selectivity towards tertiary alcohols linked to structural motifs in the enzyme?

In order to develop an enzyme with adequate activity and enantioselectivity towards tertiary alcohols, we tried two different strategies: directed evolution and screening of commercially available and recombinant hydrolases. While the directed-evolution experiments did not yield positive results, screening of a limited number of hydrolases and subsequent sequence/structure analysis proved successful. Starting from our experiences in this elaborate project, we will try to highlight the potential of different methods to improve biocatalytic processes for the resolution of tertiary alcohols.

12.2
Strategies for the Kinetic Resolution of Sterically Demanding Substrates

12.2.1
Kinetic Resolution of Chiral Substrates

To quantify the specificity for one enantiomer in a kinetic resolution, the enantioselectivity E is used. E is defined as the ratio of the catalytic efficiencies (k_{cat}/K_M) towards the two enantiomers:

$$E = (k_{cat}/K_M)_R/(k_{cat}/K_M)_S \tag{1a}$$

or $\quad E = (k_{cat}/K_M)_S/(k_{cat}/K_M)_R \tag{1b}$

if the *(R)*- or *(S)*-enantiomer is preferred, respectively.

For a racemic mixture, E can be simply seen as the ratio of the reaction rates v of both enantiomers:

$$E = v_R/v_S \qquad (2a)$$

or $\qquad E = v_S/v_R \qquad (2b)$

In general, the reaction rate v, the catalytic constant k_{cat}, and the Michaelis–Menten constant K_M of each enantiomer are not accessible. In contrast, the conversion c of the reaction as well as the enantiomeric excess of substrate (ee_s) or product (ee_p) can be determined by chiral analysis. From these data, enantioselectivity is approximately calculated for irreversible reactions [8] as

$$E = \ln\,[(1 - c)(1 - ee_s)]/\ln\,[(1 - c)(1 + ee_s)] \qquad (3a)$$

or $\qquad E = \ln\,[1 - c(1 + ee_p)]/\ln\,[1 - c(1 - ee_p)] \qquad (3b)$

and $\qquad c = ee_s/(ee_s + ee_p) \qquad (4)$

The enantioselectivity depends on the biochemical properties of the enzyme, the structure of the substrate, and other factors including temperature, pH, and cosolvents. If a reaction is not highly enantioselective ($E < 100$), the enantiomeric excess of the product (ee_p) depends on the conversion. In this case, both enantiomers are substrates for the enzyme and the conversion of the non-preferred enantiomer becomes more relevant as the reaction mixture is depleted in the preferred enantiomer. Hence, as the reaction proceeds, the optical purity of the product decreases while the optical purity of the remaining substrate increases.

12.2.2
Directed Evolution

Directed evolution by the creation and subsequent screening of large mutant libraries has been shown to be a successful strategy to optimize an enzyme for a desired activity or selectivity. Several examples of the stepwise improvement of stereoselectivity [9], solvent stability [10], and thermal stability have been published. In these experiments, enhancement of hydrolase stereoselectivity was naturally directed against secondary and primary alcohols and carboxylic acids. However, it has also been demonstrated that activities can be created *de novo* towards formerly non-accepted substrates by directed evolution. Therefore, we employed this powerful method to create a variant of an esterase that catalyzes the direct kinetic resolution of tertiary alcohols. Further rounds of mutagenesis and screening should enhance both activity and stereoselectivity.

For this, enzyme libraries of two esterases from *Pseudomonas fluorescens* [11] and *Bacillus stearothermophilus* [12] were built using error-prone PCR [13] and DNA-shuffling [14] methods. The encoding genes of both enzymes were cloned in a high copy expression vector, and protein expression was controlled by the strong *rhaP* rhamnose inducible promoter [15]. Screening of mutants was based on a pH-indicator assay, which should allow the detection of variants active towards the model substrates by a pH-shift caused by the carboxylic acid released after enzymatic hydrolysis. As model substrates we choose the three acetates **1a–c**

(Scheme 12.2). However, despite the use of various methods for the creation of mutant libraries and the screening of >15000 mutants, no active biocatalyst was found (unpublished).

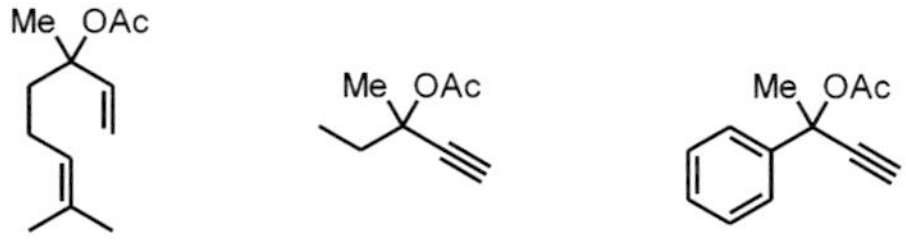

Scheme 12.2 Model substrates for the screening of enzyme-variant libraries derived from random mutagenesis experiments: linalyl acetate **1a**, 1-ethyl-1-methylprop-2-ynyl acetate **1b**, 1-methyl-1-phenylprop-2-ynyl acetate **1c**.

12.2.3
Screening of Biocatalysts

As an alternative route, 22 commercially available and recombinant esterase and lipase preparations were screened for hydrolytic activity towards the three model acetates **1a–c**. Only two enzymes of this set, which broadly covered the spectrum of structural subtypes within the lipase-family, were found to actually exhibit activity towards the tested esters (Tab. 12.1).

The two hydrolases found to be active were lipase A from *Candida antarctica* (CAL-A), an enzyme with unknown structure whose sequence shows low relationship to the α/β hydrolase-family and the well-characterized lipase from *Candida rugosa* (CRL). To obtain a deeper understanding of the structural reasons for the activity of the screened enzymes we concentrated on CRL and the non-active enzymes with known structure (CAL-B, PCL, ROL). CRL is the only member of the GGGX-family [6] of carboxylesterases screened. All non-active enzymes belong to the GX-family, as shown by structure or sequence homology. GGGX- and GX-hydrolases, despite sharing the α/β hydrolase scaffold differ not only remarkably in size (55–65 kDa or 25–40 kDa, respectively) but also in the architecture of the active site.

To verify the hypothesis that in general GGGX-type hydrolases show a higher probability of accepting esters of tertiary alcohols as substrates, the *Lipase Engineering Database* (http://www.led.uni-stuttgart.de) [4] was used as a tool to find more GGGX-hydrolases that could be screened for this activity.

12.2.4
Systematic Analysis of Sequence and Structure

The *Lipase Engineering Database* has been developed as a comprehensive resource on sequence, structure, and function of all α/β hydrolases to allow systematic comparisons for the derivation of sequence–function relationships. In the current release, sequence data on 2313 lipases, esterases, and related proteins are available [4]. They were assigned to 103 homologous families, which are grouped into

Table 12.1 Hydrolases screened for activity towards esters of tertiary alcohols. Abbreviations for the enzymes are given in parentheses.

Preparation	Supplier	Organism/Enzyme
Hydrolases with activity towards **1**		
Chirazyme L-5, lyo	Roche, Mannheim (Germany)	*Candida antarctica* lipase A (CAL-A)
Amano AYS	Amano Nagoya (Japan)	*Candida rugosa* lipase(CRL)
CLEC-CRL	Altus Biotechnics, Cambridge, MA	*Candida rugosa* lipase(CRL)
Hydrolases showing no activity towards **1**		
Chirazyme L-1, lyo	Roche, Mannheim (Germany)	*Pseudomonas cepacia* lipase (PCL)
Chirazyme L-2, lyo	Roche, Mannheim (Germany)	*Candida antarctica* lipase B (CAL-B)
Chirazyme L-6, lyo	Roche, Mannheim (Germany)	*Pseudomonas* sp. lipase
Chirazyme L-7, lyo	Roche, Mannheim (Germany)	Pig pancreas lipase
Chirazyme L-8, lyo	Roche, Mannheim (Germany)	*Thermomyces* sp. lipase
Chirazyme L-9, lyo	Roche, Mannheim (Germany)	*Mucor miehei* lipase
Chirazyme L-10, lyo	Roche, Mannheim (Germany)	*Alcaligenes* sp. lipase
Chirazyme L-12,lyo	Roche, Mannheim (Germany)	
Lipase A Amano 6	Amano, Nagoya (Japan)	*Aspergillus niger* lipase
Lipase AK Amano 20	Amano, Nagoya (Japan)	*Pseudomonas fluorescens* lipase
Lipase D	Amano, Nagoya (Japan)	*Rhizopus oryzae* lipase (ROL)
Lipase G Amano 50	Amano, Nagoya (Japan)	*Penicillium camembertii lipase*
Papain W40	Amano, Nagoya (Japan)	*Carica papaya*
Acylase	Amano, Nagoya (Japan)	*Aspergillus sp.*
Lipase C lipolytica	Fluka, Biel (Switzerland)	*Candida lipolytica* lipase
recombinant PFE		*Pseudomonas fluorescens* esterase
recombinant PFE II		*Pseudomonas fluorescens* esterase
recombinant BsubE		*Bacillus subtilis* esterase
recombinant BsteE		*Bacillus stearothermophilus* esterase

For the enantioselectivities of the active enzymes see Tables 12.2 and 12.3.

37 superfamilies. Although all α/β hydrolases have the same fold and a related catalytic mechanism, their sequences can vary considerably; most superfamilies have less than 10% global sequence identity. As a consequence of their similar enzymatic mechanism, all lipases and esterases have a conserved active-site signature, the GxSxG-motif [16]. They all share the same catalytic machinery consisting of a catalytic triad (serine–histidine–aspartic or glutamic acid) and the oxyanion hole, formed by the backbone amides of two conserved residues. Experimental structure determination of free enzymes and complexes with substrate-analogous inhibitors [17] revealed insights into the structural determinants of enzymatic function, substrate specificity, and selectivity [18]. A systematic analysis of the sequence and structure of lipases has led to a deeper insight into the relation between sequence, structure, and function. From sequence alignment and structure superposition it was observed that the oxyanion hole is highly conserved; only three structural motifs are observed in all α/β hydrolases: the GX motif and the GGGX motif [6], and a third group of hydrolases where the oxyanion hole is formed by a conserved tyrosine side chain rather than a backbone amide. Thus, all lipases and esterases could be assigned to one of three classes, the GX, the GGGX, and the Y class, according to the sequence and structure of the oxyanion hole. The GX class consists of 27 superfamilies and 79 homologous families, which contain 1577 protein entries with 2015 sequences and 258 chains of known structure and 110 homology models. This class comprises mainly bacterial and fungal lipases, eukaryotic lipases (hepatic, lipoprotein, pancreatic, gastric, and lysosomal acid lipases), cutinases, phospholipases and non-heme peroxidases. The GGGX class consists of 6 superfamilies and 20 homologous families including 457 protein entries with 793 sequences and 74 chains of known structure and 93 homology models. It comprises bacterial esterases, α-esterases, eukaryotic carboxylesterases, bile-salt-activated lipases, juvenile-hormone esterases, hormone-sensitive lipases, acetylcholinesterases, and thioesterases, as well as gliotactin, glutactin, neurotactin, neuroligin, and thyroglobulin. The Y class consists of 4 superfamilies and 4 homologous families including 279 protein entries with 340 sequences and 41 chains of known structure. It comprises dipeptidylpeptidases, propylendopeptidases, and cocain esterases.

Structural analysis yielded an interesting difference between the GGGX- and GX-type hydrolases that explained why one class is able to accept esters as substrates, but the other is not. Close to the active-site residues, hydrolases have a binding pocket for the substrate's carboxyl oxygen. The partial anionic character of this oxygen atom is stabilized by hydrogen-bridges towards two close backbone amide groups. In GX-type hydrolases the structural element that builds this "oxyanion hole" consists of a glycine followed by a hydrophilic residue (GX). In GGGX-type hydrolases this binding pocket is formed by a triple glycine (GGG) or a double glycine–alanine (GGA) loop. This leads to a completely different build up close to the position of the C_α-atom of the substrate ester's alcohol moiety. In GX-type hydrolases this area is narrowed by the glycine's carbonyl oxygen, which faces into the binding pocket (Fig. 12.1a). This oxygen occupies the space for the third substituent at the C_α-atom. In GGGX-type hydrolases this oxygen is turned away from the binding pocket (Fig. 12.1b). While this difference might seem marginal, the turned-away carbonyl-oxygen leaves just the spaces for the C_β-group of the third substituent.

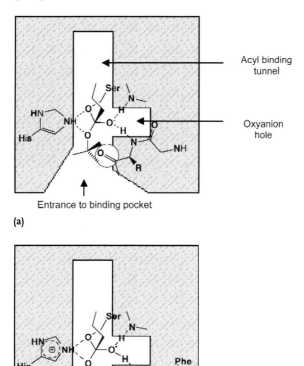

(a)

(b)

Fig. 12.1 Schematic view into the binding pockets of (a) GX-type and (b) GGGX-type hydrolases. In (a) the carbonyl oxygen of the oxyanion hole residue faces inwards into the binding pocket of the alcohol moiety. Hence, interaction of this oxygen with the quaternary C_α of an ester disables appropriate binding of the substrate. In (b) the carbonyl backbone of the flexible triple-glycine motif is arranged in parallel towards the binding-pocket limiting wall. This yields an increased capability to incorporate space-demanding substrates.

Analysis yielded six more GGGX-hydrolases that were either commercially available or in recombinant form. These six enzymes were screened for activity towards esters **1a–c** (Tab. 12.2). With the exception of GCL, all these GGGX-type α/β-hydrolases showed activity towards the model esters.

Table 12.2 Enantioselectivity of GGGX-type hydrolases towards acetates **1a–c** (Scheme 12.2).

	1a E (config.)[a]	**1b** E (config.)[a]	**1c** E (config.)[a]
PLE, E-1	1.5 (*R*)	4 (*S*)	1.3 (*R*)
PLE, E-2	1.7 (*S*)	4 (*S*)	5 (*R*)
bAChE	1.1 (*S*)	7 (*S*)	10 (*R*)
eeAChE	1.5 (*S*)	4 (*S*)	4 (*R*)
hAChE	1.1 (*S*)	3 (*S*)	3 (*R*)
CRL	1.0	2 (*S*)	1.0
GCL	Not active	Not active	Not active
BsubpNBE wild type	1.7 (R)	1.0	3 (*R*)
BsubpNBE A400I	2 (*R*)	1.0	2 (*R*)
BsubpNBE G105A	4 (*S*)	1.3 (*S*)	19 (*R*)

The enzymes used were acetylcholine esterases from three species [electric eel (*Electrophorus electricus*, eeAChE), banded krait (*Bungarus fasciatus*, bAChE) and human (hAChE)], carboxylesterases [*p*-nitrobenzylesterase from *B. subtilis* (BsubpNBE) and pig liver esterase (PLE)], and an additional lipase [from *Geotrichum candidum* (GCL)]. All PLEs and eeAChE were used as commercial-grade lyophilized powders; the other enzymes were produced recombinantly in microbial expression systems. GCL [19], bAChE and hAChE [20] were expressed in the methylotrophic yeast *Pichia pastoris*. The gene encoding BsubpNBE was cloned from genomic DNA from *B. subtilis* DSM 402 and expressed in high yields in *E. coli*. Reactions were carried out at pH 7.5 and 40°C (AChEs: at 25°C). Enantioselectivity *E* was calculated according to Chen et al. [8].

a Configuration of the preferred enantiomer is given.

Screening and sequence/structure analysis have thus shown that this classification has high predictive value for activity towards tertiary alcohols. Interestingly, the GGGX-type hydrolases differ considerably in their overall substrate spectra. CRL is a typical lipase showing a lid mechanism and interfacial activation [21] and prefers long-chain, poorly water-soluble substrates. In contrast, PLE and BsubpNBE are typical carboxylesterases with highest activity towards medium- and short-chain substrates with lower hydrophobicity. AChEs are well known for their very narrow substrate range, i.e. only acetates are converted but not butyrates, as exemplified by their inactivity towards butyrylcholine. This, in turn, means that a knowledge of their usual biocatalytic activity would not suggest that these enzymes had a common activity towards tertiary alcohols. Furthermore, they belong to different classes (carboxylesterases or lipases), and originate from mammalians (acetylcholine esterase, pig liver esterase), bacteria (esterase from *Bacillus subtilis*) or yeast (lipases from *Candida rugosa* and *Geotrichum candidum*), and the overall amino acid sequence similarity is rather low. CRL and PLE share only 24% sequence identity (another 19% amino acids are similar) and only small regions around the GGGX-motif and the active serine are conserved.

12.2.5
Molecular Modeling and Protein Engineering

Protein engineering of the biocatalyst is a third strategy to improve enantioselectivity of the hydrolase-catalyzed racemic resolution of tertiary alcohols. As a starting point, BsubpNBE was selected. This enzyme is available as recombinant enzyme and showed good activity toward **1c**, albeit with low enantioselectivity.

Both enantiomers of **1c** were docked to the hydrophobic-alcohol-binding site of this enzyme and showed a nearly equal geometric conformation, which explains the low enantioselectivity of the esterase towards this substrate (Fig. 12.2a,b). A potential mutation which increases enantioselectivity should prevent binding of the non-preferred (S)-enantiomer and improve binding of the preferred (R)-enantiomer. A promising residue for mutagenesis, G105, was investigated by molecular dynamics (MD) simulations. Replacing one of these amino acids by a more space-demanding residue such as alanine showed a repulsive interaction with (S)-**1c**, but not with (R)-**1c** (Fig. 12.2c). This enantiomer-specific interaction was expected to increase enantioselectivity. Interestingly, the same mutation was predicted to have the opposite effect on substrate **1b**. Here, the enantiomer (R)-**1b**, which is preferred by the wild-type enzyme, is predicted to interact with the methyl group of alanine at position 105, while the mutation has no effect on the binding of the (S)-enantiomer. As a result, a decrease of enantioselectivity or even a switch in enantiopreference from (R) to (S) is predicted. To validate the simulation results, the mutant BsubpNBE G105A was generated and expressed in *E. coli*. Activity and enantioselectivity were determined and compared to the wild-type enzyme. As predicted by molecular modeling, mutation G105A led to a sixfold increase of enantioselectivity towards **1c** ($E = 3$ to $E = 19$ for wild-type and mutant) (Tab. 12.2) while it also displayed the predicted inversion of enantiopreference towards **1b**: BsubpNBE WT preferred hydrolysis of the (R)-enantiomer with low enantioselectivity ($E = 1.7$), the mutant BsubpNBE G105A converted this substrate with $E = 4$, favoring the (S)-enantiomer.

12.2.6
Use of Remote or Alternative Cleavage Sites

All lipases that were found to be active toward esters of bulky tertiary alcohols show only low to moderate enantioselectivities. While improvement of the enzyme by molecular modeling and subsequent introduction of mutations is time consuming, alternative ways of increasing the enantioselectivity are to engineer the reaction conditions or to change the whole reaction pathway. Thus, it was possible to use an alternative cleavage site to achieve the synthesis of optically pure tertiary alcohols. For instance, (–)-kjellmanianone **2d**, which bears a tertiary alcohol group functionality (Scheme 12.3), was kinetically resolved by hydrolysis of the methyl ester at the adjacent carboxylic acid group [22].

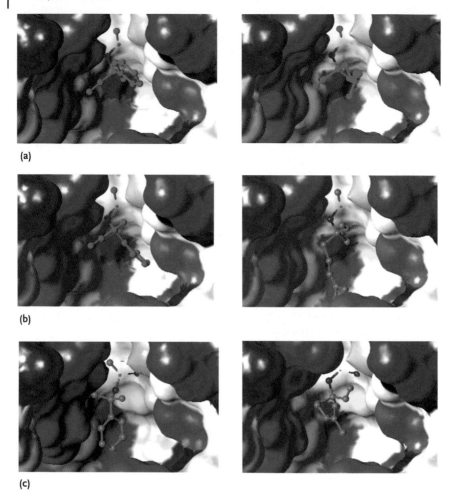

Fig. 12.2 (a) BsubpNBA with preferred substrate enantiomers [(*R*)-**1c**, (*S*)-**1a**]. (b) BsubpNBA WT with non-preferred substrate enantiomers [(*S*)-**1c**, (*R*)-**1a**]. BsubpNBA G105A with non-preferred enantiomers [(*S*)-**1c**, (*R*)-**1a**]. The exchanged residue (G105 in the WT, A105 in the variant) is highlighted in yellow. Other residues are colored according to their hydrophobicity (blue: hydrophilic; red: hydrophobic).

12.2.7
Reaction Engineering

One alternative to hydrolysis of the acetates of the model compounds is the transesterification of the racemic alcohol (Scheme 12.4). This is commonly performed in organic solvents using enol esters as acyl donors, as the reaction becomes practically irreversible owing to the formation of volatile acetaldehyde from the vinyl

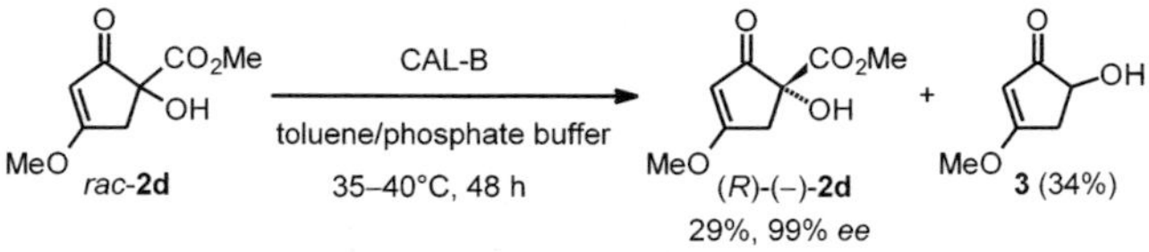

Scheme 12.3 Kinetic resolution of *rac*-kjellmanianone **2d** using lipase B from *Candida antarctica* (CAL-B) to give enantiopure kjellmanianone (R)-(–)-**2d**.

alcohol generated. In addition, reaction conditions can be varied to a greater extent than for a hydrolytic reaction. An additional reason to study this reaction system was the considerable autohydrolysis of substrate **1c**, which makes a hydrolytic reaction less practical. Unfortunately, the recombinant BsubpNBE exhibited only low activity and stability in organic solvents.

Alternatively, the commercial CAL-A was used as model enzyme. With the commercial immobilized enzyme, only low conversion (9%) and an *E*-value of 22 was determined (Tab. 12.3). However, careful optimization of the reaction system with respect to organic solvent, temperature, and carrier, allowed a substantial increase in enantioselectivity. Under optimum conditions, the resolution proceeded with an *E*-value of 87, yielding the alcohol almost optically pure. However, the conversion is not satisfactory as at best 25 % could be achieved. Nevertheless, these results demonstrate that "classical" methods for reaction engineering provide an useful alternative to rational protein design or directed evolution.

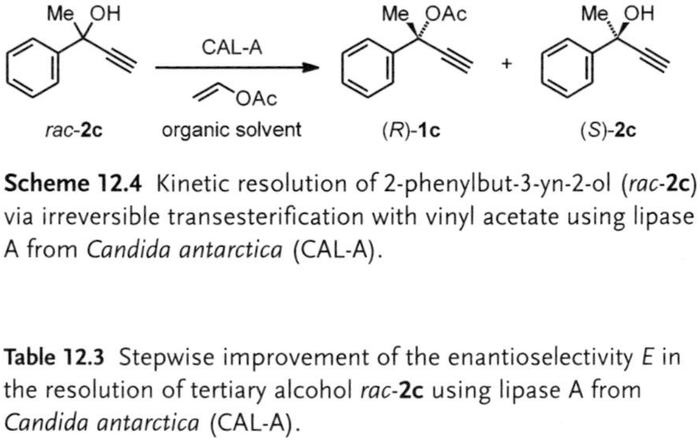

Scheme 12.4 Kinetic resolution of 2-phenylbut-3-yn-2-ol (*rac*-**2c**) via irreversible transesterification with vinyl acetate using lipase A from *Candida antarctica* (CAL-A).

Table 12.3 Stepwise improvement of the enantioselectivity *E* in the resolution of tertiary alcohol *rac*-**2c** using lipase A from *Candida antarctica* (CAL-A).

Parameter	Conversion (%)	Enantiomeric excess *ee* (%)		*E*
		(S)-**2c**	(R)-**1c**	
CAL-A-C1[a]	9	9	91	22
Isooctane	21	26	95	49
20°C/EP100[b]	25	32	96	65
20°C/EP100[b]	18	21	97	87

a Standard reaction conditions: CAL-A immobilized on carrier C1, vinyl acetate, 20°C, *n*-hexane. b EP100: polyproplyene carrier.

12.3
Conclusion

Combining extensive screening of commercially available enzymes, subsequent engineering of the recombinant enzymes, and optimization of reaction conditions has been shown to provide an innovative route to enantiopure tertiary alcohols. In addition, the rich natural pool of enzymes can be further exploited by screening strain collections or by metagenome screening. However, as for directed-evolution experiments, high-throughput screening needs a cheap, sensitive, and robust assay. Therefore, it is highly desirable to enrich a library prior to screening. A comprehensive analysis of sequence and function of all lipases and esterases has shown that a single sequence motif (GGGX) is sufficient to identify active enzymes, which hydrolyze esters of tertiary alcohols. Starting with an enzyme that is active but not yet enantioselective, mutating single key residues leads to a considerable increase of enantioselectivity. Those key residues can be predicted by molecular modeling, or can be found by directed-evolution experiments. A further improvement can be achieved by reaction engineering, which allows numerous parameters to be modified to increase activity and (enantio-)selectivity. Besides the discussed possibility of switching from hydrolytic reactions in aqueous solutions to ester synthesis in organic solvents, it has been shown that temperature, cosolvents, and the acid moiety of the substrate ester influence the enantioselectivity of hydrolase-catalyzed reactions. The examples described above demonstrate the progress that has been made in utilizing hydrolytic enzymes in the synthesis of optically pure tertiary alcohols. They further illustrate the high potential of integrating classical methods such as screening and process engineering with newer, technology-driven approaches such as directed evolution and *in silico* prediction to develop enzymatic processes that are efficient enough for industrial applications.

References

1 (a) U. T. Bornscheuer, R. J. Kazlauskas **1999**, *Hydrolases in Organic Synthesis - regio- and stereoselective biotransformations*, Wiley-VCH, Weinheim; (b) U. T. Bornscheuer, C. Bessler, R. Srinivas, S. H. Krishna, *Trends Biotechnol.* **2002**, *20*, 433–437.

2 A. Liese, K. Seelbach, C. Wandrey **2000**, *Industrial Biotransformations*, Wiley-VCH, Weinheim.

3 (a) Y. Hotta, S. Ezaki, H. Atomi, T. Imanaka, *Appl. Environ. Microbiol.* **2002**, *68*, 3925–3931; (b) M. Pogorevc, U. T. Strauss, M. Hayn, K. Faber, *Chem. Monthly* **2000**, *131*, 639–644; (c) A. Schlacher, T. Stanzer, I. Osprian, M. Mischitz, E. Klingsbichel, K. Faber, H. Schwab, *J. Biotechnol.* **1998**, *62*, 47–54.

4 M. Fischer, J. Pleiss, *Nucleic Acids Res.* **2003**, *31*, 319–321.

5 J. Pleiss, M. Fischer, R. D. Schmid, *Chem. Phys. Lipids* **1998**, *93*, 67–80.

6 J. Pleiss, M. Fischer, M. Peiker, C. Thiele, R. D. Schmid, *J. Mol. Catal. B* **2000**, *10*, 491–508.

7 E. Henke, J. Pleiss, U. T. Bornscheuer, *Angew. Chem. Int. Ed.* **2002**, *41*, 3211–3213.

8 C. S. Chen, Y. Fujimoto, G. Girdaukas, C. J. Sih, *J. Am. Chem. Soc.* **1982**, *104*, 7294–7299.

9 K. E. Jaeger, T. Eggert, A. Eipper, M. T. Reetz, *Appl. Microbiol. Biotechnol.* **2001**, *55*, 519–530.

10 F. H. Arnold, *Acc. Chem. Res.* **1998**, *31*,
125–131.

11 (a) N. Krebsfänger, K. Schierholz, U. T.
Bornscheuer, *J. Biotechnol.* **1998**, *60*,
105–111; (b) N. Krebsfänger, F. Zocher,
J. Altenbuchner, U. T. Bornscheuer, *Enzyme
Microb. Technol.* **1998**, *22*, 641–646;
(c) I. Pelletier, J. Altenbuchner, *Microbiology*
1995, *141*, 459–468.

12 Y. Amaki, E. E. Tulin, S. Ueda, K. Ohmiya,
T. Yamane, *Biosci. Biotech. Biochem.* **1992**,
56, 238–241.

13 J.-P. Vartanian, M. Henry, S. Wain-Hobson,
Nucleic Acids Res. **1996**, *24*, 2627–2631.

14 (a) W. P. C. Stemmer, *Nature* **1994**, *370*,
389–391; (b) W. P. C. Stemmer, *Proc. Natl.
Acad. Sci. USA* **1994**, *91*, 10747–10751;
(c) H. Zhao, F. H. Arnold, *Nucleic Acids Res.*
1997, *25*, 1307–1308.

15 T. Stumpp, B. Wilms, J. Altenbuchner,
BioSpektrum **2000**, *6*, 33–36.

16 C. Chapus, M. Rovery, L. Sarda, R. Verger,
Biochimie **1988**, *70*, 1223–1234.

17 (a) U. Derewenda, A. M. Brzozowski, D. M.
Lawson, Z. S. Derewenda, *Biochem.* **1992**,
31, 1532–1541; (b) J. Uppenberg, S.
Patkar, T. Bergfors, T. A. Jones, *J. Mol.
Biol.* **1994**, *235*, 790–792; (c) D. A.
Lang, M. L. Mannesse, G. H. de Haas,
H. M. Verheij, B. W. Dijkstra, *Eur. J.
Biochem.* **1998**, *254*, 333–340.

18 (a) H. Scheib, J. Pleiss, P. Stadler,
A. Kovac, A. P. Potthoff, L. Haalck,
F. Spener, F. Paltauf, R. D. Schmid,
Protein Eng. **1998**, *11*, 675–682;
(b) T. Schulz, J. Pleiss, R. D. Schmid,
Protein Sci. **2000**, *9*, 1053–1062.

19 E. Catoni, C. Schmidt-Dannert,
S. Brocca, R. D. Schmid, *Biotechnol.
Technol.* **1997**, *11*, 689–695.

20 C.O. Weill, S. Vorlova, N. Berna,
A. Ayon, J. Massoulie, *Biotechnol.
Bioeng.* **2002**, *80*, 490–497.

21 (a) R. D. Schmid, R. Verger, *Angew.
Chem. Int. Ed.* **1998**, *37*, 1608–1633;
(b) R. Verger, *Trends Biotechnol.* **1997**,
15, 32–38.

22 J. Christoffers, T. Werner, W. Frey,
A. Baro, *Chem. Eur. J.* **2004**, *10*,
1042–1045.

Index

Quaternary Stereocenters: Challenges and Solutions for Organic Synthesis. Edited by Jens Christoffers, Angelika Baro
Copyright © 2005 WILEY-VCH Verlag GmbH & Co. KGaA, Weinheim
ISBN: 3-527-31107-6